A SHAU

A SHAU

CRUCIBLE *of the* VIETNAM WAR

JAY PHILLIPS

IZZARD INK
PUBLISHING

IZZARD INK PUBLISHING
PO Box 522251
Salt Lake City, Utah 84152
www.izzardink.com

Library of Congress Cataloging-in-Publication Data

Names: Phillips, Jay, 1949- author.
Title: A Shau : crucible of the Vietnam War / Jay Phillips.
Description: First edition | Salt Lake City : Izzard Ink Publishing, [2021] |
Includes bibliographical references and index.
Identifiers: LCCN 2020054291 (print) | LCCN 2020054292 (ebook) |
ISBN 9781642280449 (hardcover) | ISBN 9781642280432 (paperback) |
ISBN 9781642280425 (epub)
Subjects: LCSH: Vietnam War, 1961–1975—Campaigns—Vietnam—A Shau Valley. |
A Shau Valley (Vietnam)—History, Military—20th century. |
Civilian Irregular Defense Group Program (Vietnam). |
Vietnam War, 1961–1975—Commando operations—United States. |
Counterinsurgency—Vietnam.
Classification: LCC DS557.8.A2 P45 2020 (print) | LCC DS557.8.A2 (ebook) |
DDC 959.704/342—dc23

Designed by Alissa Theodor
Cover Design by Andrea Ho
Cover Image: Bettmann/Bettmann Collection/Getty Images
Smoke flares mark the landing spot for an evacuation helicopter coming
in to take out U.S. cavalrymen wounded in the battle for control
of the vital A Shau Valley

First Edition

Contact the author at info@izzardink.com

LC record available at https://lccn.loc.gov/2020054291
LC ebook record available at https://lccn.loc.gov/2020054292

Hardback ISBN: 978-1-64228-044-9
Paperback ISBN: 978-1-64228-043-2
eBook ISBN: 978-1-64228-042-5

This book is dedicated to my wife Marilyn

and to 1LT Angelo M. "Mike" Patacca and PFC Glenn B. Dale, Jr., who risked their lives to pull me to safety from a clearing in the jungle in War Zone C on Easter Sunday 1969.

CONTENTS

ACRONYMS AND ABBREVIATIONS

AAA	anti–aircraft artillery
AAR	after action report
ACD	air cavalry division
ACL	aircraft load
AHB	assault helicopter battalion (UH-1 "Hueys")
AHC	assault helicopter company
AO	area of operations
ARA	aerial rocket artillery (Huey or Cobra helicopter gunships)
ARC LIGHT	B-52 heavy bomber air strikes
ARP	aero-rifle platoon, also known as "Blues"
ARVN	Army of the Republic of Vietnam (South Vietnamese Army)
ASHB	assault support helicopter battalion (CH-47 Chinooks)
AW	automatic weapons (i.e., machine guns or automatic rifles)
AWOL	absent without leave
BDA	bomb damage assessment
BS	border surveillance
C&C	command and control
CBU	cluster bomb unit
CCN	Command and Control, North (the northern HQs of MACVSOG)
CCP	Combined Campaign Plan
CG	commanding general
CIA	Central Intelligence Agency

CIDG	Civilian Irregular Defense Group
CINCPAC	Commander in Chief, Pacific
COMINT	communications intelligence
COMUSMACV	Commanding General, MACV
DASC	direct air support center
DMZ	Demilitarized Zone (Not!) (the border between North and South Vietnam)
DoD	Department of Defense
DSC	Distinguished Service Cross (2nd highest medal for Valor)
FAC	forward air controller
FLIR	forward-looking infrared radar
FO	forward observer
FOB	forward operating base
FSB	fire support base (also FB, firebase)
GWAO	guerrilla warfare area of operations
HMM	Marine Medium Helicopter Squadron
KBA	killed by air
KIA	killed in action
LLDB	Lac Luong Dac Biet (RVN Special Forces)
LOC	line(s) of communication
LOH	light observation helicopter ("Loach")
LRP	long-range patrol
LRRP	long-range reconnaissance patrol
LZ	landing zone
MACV	Military Assistance Command, Vietnam
MACVSOG	MACV Studies and Observations Group
MAF	Marine Amphibious Force (III MAF was the top USMC HQs in RVN)
MGF	mobile guerrilla force
MIA	missing in action
MOH	Medal of Honor
MRTTH	Military Region Tri-Thien-Hue (the PAVN Headquarters for Thua Thien)

MSF	mobile strike force (aka "Mike" force)
NCO	noncommissioned officer
NDP	night defensive position
NSA	National Security Agency
NVA	North Vietnamese Army (see PAVN)
OPCON	operational control (re: a unit temporarily under command of another)
OPN	operation
OPORD	operational order
OR-LL	Operational Report – Lessons Learned
PAVN	People's Army of Vietnam (North Vietnamese Army)
PCV	Provisional Corps, Vietnam (later XXIV Corps)
POW	prisoner of war
RIF	reconnaissance in force
RT	recon team
RTO	radio telephone operator (radioman)
RVN	Republic of Vietnam (South Vietnam)
RVNAF	Republic of Vietnam Armed Forces
RZ	reconnaissance zone
SAR	search and rescue
SF	Special Forces
SIGINT	signals intelligence
SLAR	side-looking airborne radar
SOG	Studies and Observations Group
SPOS	Strong Point Obstacle System (a component of the "McNamara Line")
TF	task force
USSF	United States Special Forces
VC	Viet Cong (derogatory term for South Vietnamese Communists)
WIA	wounded in action
WO	warrant officer

PREFACE

"We were winning when I left." How often one heard this mantra when Vietnam Veterans began to "come out of the closet" and reunite with their wartime buddies in the early 1980s. And we all meant it. We had accomplished our missions. We had never lost a major battle. We killed 10 "gooks" for every 1 of our own that fell. We could have taken Hanoi, if only "they" had let us. These were not only the thoughts of the former grunts (Army and Marine infantrymen), but also thoughts that were and are echoed in much of the literature of the war, especially the official histories and many of the narratives by those who were there. If you pick up works of that era, and even many from more recent years, regardless of whether the stories end in 1963 or 1973, most will leave you with the impression as you turn the final page that all was well and would end well. But alas, we lost the war.

As someone who was there and who knows how it ended, revisiting the war in search of a different outcome is not my endeavor. Rather, I seek to analyze how, despite the commitment of enormous resources over many years, the United States and its South Vietnamese ally were never able to wrest control of one strategically critical location on the enemy line of communications, the A Shau Valley, away from the Communists after they overran the Special Forces camp in March 1966, and how that failure contributed, in a major way, to the final outcome of the war. For those who participated in the campaigns in the A Shau – the fallen and the survivors – I hope to shed light on *what* happened in the A Shau, and equally important, *why* the possession of that hostile, forbidding valley deep in the jungle near the Laotian border had to be contested, without giving a false impression that all of the sacrifices would be rewarded with victory. I also hope to trace the threads of defeat even as in real time, events appeared to the participants, especially U.S. military leaders, to augur victory. In so doing, we learn why our strategies and tactics failed and what signs portending our ultimate defeat we overlooked or misinterpreted. For this reason, the relevant developments on the Communist side are an important part of this study.

A Shau was not a microcosm of the Vietnam War. There was no "pacification," as the sparse local population was already co-opted by the Viet Cong (South Vietnamese Communists) at the time the battle there began. The valley was, for all intents and purposes, a "free fire" zone for both sides, as everyone there was a combatant. On the other hand, what transpired in the A Shau may well be described as metaphor for much of the failed strategy of General Westmoreland, who ran the Military Assistance Command, Vietnam (MACV) from 1964 to 1968. While Westmoreland was counting bodies, looking for the "crossover point" when his forces could kill the enemy faster than the Communists could replace their losses, the North Vietnamese Army (NVA) logistical system never ceased its slow but ominous march to victory. While American generals "won" their battles at places like the Ia Drang Valley, Khe Sanh, and Hamburger Hlll, and then abandoned the battlefields, proclaiming the irrelevance of occupying terrain, the NVA remained in the A Shau Valley and the surrounding hills, slowly constructing the largest enemy base area inside South Vietnam. In the course of the battle for the A Shau, every "scientific" trick known was used to try to alter the enemy's ability to supply and reinforce his army in the south; none had more than marginal success. In 1969, when Westy's successor, "Abe" Abrams, sent in the airmobile infantry at Hamburger Hill, the price of victory was finally exposed as greater than the American public was willing to pay to preserve the existence of the Republic of Vietnam.

So much for the big issues; for those who were there, the war always had a very personal meaning. I cannot recall exactly what was said or when, but I do have a distinct memory that while serving in Binh Dinh Province with the First Cavalry Division in the summer of 1967, the last place anyone ever wanted to be sent to was the A Shau Valley. The rumors about the valley most certainly dated to the overrunning of the Special Forces Camp in March 1966. After all, the Green Beanies were the best of the best, and nobody could whip them. Except the NVA in the A Shau. Somehow linked to that, too, were stories of entire recon teams disappearing without a trace in that terrifying place nestled up against Laos. Even later on, when the Marines were getting their butts kicked at Khe Sanh, it was the A Shau that scared the heck out of you. At Khe Sanh, either your number was up or it wasn't. But the A Shau was a terrifying place, where the enemy was everywhere, and help was nowhere to be found. And death was more certain than life. All of this mystique was around before the Cav got shot out of the sky on the first day of Operation DELAWARE and long before Hamburger Hill or RIPCORD. In Nam, if an officer or NCO told you to do something you didn't agree to, the most common retort (best left unspoken, of course) was "What are you to going to do? Send me to Vietnam?" Nobody ever said or thought "Send me to the A Shau?"

What was behind this mystique? What really happened in the A Shau? Although a few published works have told parts of the story, none have fully documented it or recognized the significance of its events to the war's final outcome. As such, filling in this gap in the historiography of the Vietnam War is a major goal of this study. As I hope this book will show, the A Shau Valley was one of the most consequential pieces of real estate in all of South Vietnam because of its role in the battle for Hue in Tet 1968, the turning point of the war, in the acceleration of the American "redeployment" a year later in the wake of Hamburger Hill, and above all, as the epicenter of the Communist supply network for the entire northern part of South Vietnam.

It is widely accepted that the 1968 Tet Offensive marked the turning point in the war. The fighting in Saigon and Hue dominated the concerns of both the generals in Vietnam and the public in the United States. As graphic as the photos from the American Embassy in Saigon and of the execution of the Viet Cong terrorist by National Police Chief Nguyen Ngoc Loan were, those events occurred within the first 48 hours of the offensive. The Battle for Hue went on for 26 days and demonstrated, even more graphically to the American television audience, the strength and resilience of an enemy that the American people had been led to believe was all but defeated. The soldiers of the Viet Cong and the People's Army of Vietnam (PAVN) in Hue infiltrated and were supplied through the A Shau Valley. Without the North Vietnamese logistical infrastructure that had been built up in and around the A Shau Valley between the fall of the Special Forces Camp there in March 1966 and the Tet Offensive in February 1968, the enemy could neither have taken nor held Hue. The PAVN decision to occupy the A Shau, made as early as 1959, was one of the first steps leading to the defeat of the United States and the demise of the Republic of Vietnam in 1975.

While the allied cast changed over time, the PAVN, also known as the North Vietnamese Army (NVA), was the constant foe. Except for the very earliest period, there were no Viet Cong units present. While the allied presence in the valley waxed and waned in large degrees, the enemy was always there. Always ambushing, always building trails and stocking caches, always watching and waiting for the opportunity to strike, or to be more or less overt depending on the allied threat level, and above all, always waiting for the time or the season to build more roads and stockpile more supplies until the moment of final victory.

As the ensuing pages will reveal, every U.S. unit that spent time in the A Shau Valley was elite, and almost every type of elite unit, sooner or later, would be there.* The Green Berets, in all of their guises—A-Teams, Mobile Guerrilla Force, Project Delta, MACVSOG—were there. Rangers/"Lurps" (Long-Range Patrol units) were there. Marine Recon and Force Recon were there, along with the Ninth Marines. The two finest divisions of the United States Army, the 1st Cavalry Division (Airmobile) (the "fire brigade" of the Military Assistance Command, Vietnam [MACV]) and the 101st Airborne Division (Airmobile) (the last complete American division in Vietnam), were there. Only the best could be sent into the A Shau. This is their story. It is also, as far as I have been able to document, the story of the Soldiers of the Army of the Republic of Vietnam who were in the valley. The ARVN 1st Infantry Division, the only

---

* The Navy SEALS being the sole exception.

ARVN unit that sent troops into the A Shau, was the best division in our ally's army.

In the long run, the best wasn't good enough, but that did not and could not become evident until the withdrawal of the American forces beginning in late 1969 cut short plans to retain a permanent presence along Route 548. For six long years, as each successive group of contestants for hegemony in the valley came and went, some thought we were winning, some just thought "we haven't lost yet," and many died. For the last two years that Americans operated in the A Shau (1969–1971), we could only try to slow the enemy down. The period from April 1968 till September 1969 was the high-water mark of allied "boots on the ground" in the valley, but men fought there, in the air and on the ground, for many years both before and after those operations. This is the story of all of those brave men.

For the record, I was never in or near the A Shau Valley during my duty in Vietnam. For that, I am grateful.

1

EARLY DAYS: 1961–1965

1959–1961 – In the Beginning

In May 1959, the People's Army of North Vietnam (PAVN) formed the 559th Transportation Group for "the infiltration of men and material from North Vietnam into South Vietnam." This action marked the renewal of the North Vietnamese Communist leadership's prioritization of the unification of Vietnam, which had taken a back seat to Hanoi's internal affairs since the defeat of the French colonial government and the partition of the country in 1954.

Over the next sixteen years, the initial trail blazed from the North through Laos and into South Vietnam would expand into a honeycomb of thousands of kilometers of roads and paths, becoming famous as the Ho Chi Minh Trail. Route 9, the northernmost east-west road in South Vietnam, connected the Vietnamese lowlands on the South China Sea with the Laotian village of Tchepone, passing through a village called Khe Sanh. In Laos, "below Route 9, the trail extended along the present alignment of Routes 92 and 922 into Thua Thien Province, South Vietnam."[1] The area where Route 922 crossed into Thua Thien, the first terminus of the Ho Chi Minh Trail in the South, was the A Shau Valley (see Map 1).

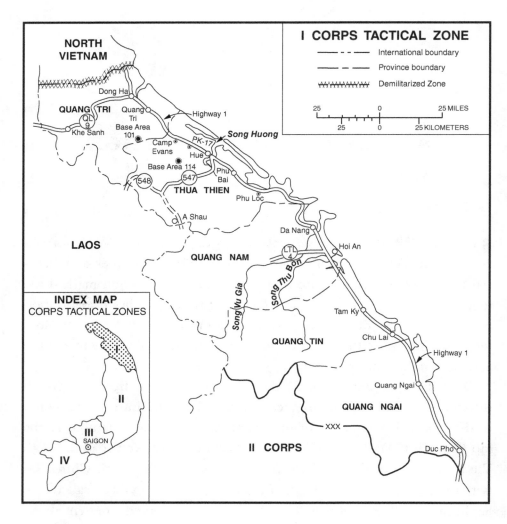

MAP 1

Source: Villard, Eric *The 1968 Tet Offensive Battles of Quang Tri City and Hue,* (Fort McNair, DC: U.S. Army Center of Military History, 2008), p. 3.

By February 1961, when the PAVN force developing the trail had grown to regimental strength (2,000 men), the die was cast for war in Thua Thien. For more than a decade, the valley, the hills, and the mountains around it would be the most important area in South Vietnam on the Trail.

1963 – The American Advisor

Long before the United States committed its own ground forces to the war in Vietnam, American advisors "humped the boonies" with infantry units of the Army of the Republic of Vietnam (ARVN). The ARVN 2nd Battalion, 3rd Infantry Regiment, 1st Infantry Division was stationed near A Shau village at the south end of the river valley of the same name in late 1962, searching for the elusive Viet Cong (VC—a derogatory term for South Vietnamese Communists) guerrillas in the heavily jungled western reaches of Thua Thien, the second-most northern province in South Vietnam. In January 1963, after a lengthy delay in obtaining transport to this remote outpost, a new U.S. Army advisor joined that unit in the field. "I had arrived in Vietnam in the rainy season, and getting to A Shau was not easy. You could either fly there in thirty hair-raising minutes or take weeks to walk in. The bad weather grounded flights for days. . . . " The advisor was 25-year-old Captain Colin Powell, future chairman of the Joint Chiefs of Staff and secretary of state.[2]

Captain Powell's initial conversation with Captain Hieu, commander of the Vietnamese battalion, was circuitous. When Powell asked why the battalion base was established in such a vulnerable location—a valley with higher ground all around—this exchange ensued:

> "Very important outpost," Hieu assured me.
> "But why is it here?"
> "Outpost is here to protect airfield," he said. . . .
> "What's the airfield here for?" I asked.
> "Airfield here to supply outpost."[3]

Captain Powell was not the first to hear this aphorism. Former Special Forces advisor Ken Haines, who served with Bill Patience, the first American to die in the A Shau Valley (see below), told author Ray Bows: "The reason we were there was absurd and we all knew it. 'The outpost at A Shau protecting the airstrip there to supply the outpost . . . ' was repeated many times and became a standard gag. General Colin Powell mentions

[it], . . . but the first time I heard it was from Bill Patience. I have always considered him the originator of that wisecrack."[4]

As with much black humor, this seemingly nonsensical exchange reveals a deeper truth. Friendly units operating in the A Shau could only be supplied by air, and with enemy sanctuaries close by and jungled ridges all around, an inordinate proportion of their forces were required for base camp perimeter defense and local patrolling, leaving few if any troops for offensive operations, even though, as Powell notes, "we were supposed to engage the Viet Cong to keep them from moving through the A Shau Valley and fomenting their insurgency in the populated coastal provinces."[5] Comparisons to the French dilemma at Dien Bien Phu are not out of place. Indeed, if there was any outpost in South Vietnam that resembled Dien Bien Phu (an isolated camp in a valley far from allied bases) geographically, it was A Shau, rather than the infamous Khe Sanh.

Captain Powell was soon patrolling the valley and nearby hills with the South Vietnamese regulars. It was a most inhospitable environment, with all the heat, dense undergrowth, steep mountains, rough ground, bugs, spiders, snakes, leeches, enemy booby traps, and snipers one could possibly imagine. On the sixth day out, the battalion column was ambushed, losing one man killed in action. Two days later, they were hit again and two more soldiers fell. Thereafter, for several weeks, the VC ambushed the column almost every day. The enemy had the initiative; the ARVN had all the casualties. When the ARVN came across abandoned Montagnard villages (the tribes living in the remote western regions of the Vietnamese highlands), they burned them and destroyed any crops. There was no serious attempt to gain the support of or "pacify" the indigenous population in the A Shau, just a "scorched earth" fate for anything that might benefit the guerrillas. There was no struggling here for the "hearts and minds" of the sparse inhabitants, either—just a slowly escalating battle for life or death between the South Vietnamese troops and the Communist Viet Cong.

The 2nd Battalion eventually moved about eight kilometers from A Shau, establishing a new camp at Be Luong near the Rao Nai, the next river valley to the northeast. Contact with the enemy continued, almost invariably initiated by the VC. Finally, on 18 May, the battalion recorded its first confirmed VC killed in action (KIA). On 23 July 1963, as his unit was at last departing the area on foot for a well-deserved rest, Captain Powell stepped on a punji trap, a poisoned bamboo spike. The spike pierced his foot, and he was medevaced to Hue, the province capital, his duty as a field advisor at an end. But for the United States, this was just the end of the beginning in the A Shau Valley.[6]

The Green Berets Come to the A Shau Valley

The 14th Special Forces Detachment, which conducted commando training for an elite group of South Vietnamese warriors at Nha Trang in the summer and fall of 1957, was the first American Special Forces team in Vietnam.[7] The leader of this team, Captain Harry G. "Hairbreadth" Cramer (United States Military Academy, class of 1946), was the first Special Forces soldier to die in the Vietnam War and is the first soldier killed in action listed on the Vietnam Memorial in Washington, D.C. His unit was ambushed while on a training exercise on 21 October 1957. That was almost two years before the oft-cited "first" American KIAs of the war, Major Dale Buis and Master Sergeant Chester Ovnand, were killed in a Viet Cong attack at Bien Hoa in July 1959.*[8] Thereafter, Green Berets continued to rotate from Okinawa and stateside duty to temporary duty (TDY) assignments training South Vietnamese troops through the beginning of the 1960s, when the one-year tour of duty became standard for the U.S. Army in Vietnam.

In 1961, as the insurgency in South Vietnam worsened, the Central Intelligence Agency (CIA) conceived the idea of creating a Village Defense Program among the Montagnard tribesmen, whose settlements were scattered over much of the area along the Cambodian and Laotian borders in the highlands of Vietnam.[9] The "Yards" resided in remote mountainous regions of the country and were generally despised and discriminated against by the Diem regime, but many tribes were receptive to service defending their home villages as de facto mercenaries for the Americans. This force would be designated the Civilian Irregular Defense Group (CIDG) and

---

* Captain Cramer is still listed on the Virtual Wall as a "non-hostile" casualty, but Chalmers Archer, Jr., who was present at the time, has set the record straight.

would be led by U.S. Special Forces troops and by ARVN Special Forces soldiers, also trained by the Americans and referred to by their Vietnamese abbreviation, LLDB (Lac Luong Dac Biet). The CIDG force included both minimally prepared local hamlet militia and better-trained "strike force" troops, for mobile reaction to enemy attacks. By the end of 1962, as the program accelerated, there were 21 CIDG camps scattered throughout the four corps areas of the country. During 1962 and 1963, under Operation SWITCHBACK, control of the Special Forces teams and the CIDG passed from the CIA to the Military Assistance Command, Vietnam (MACV), the new American headquarters established in February 1962.[10]

Several of those camps, including Khe Sanh in I Corps (the five provinces in the most northern of the four ARVN corps zones), also had border surveillance (BS) duties. The BS units evolved from an earlier trail-watcher program, where SF trainers prepared indigenous recruits to serve in long-range, intelligence-gathering missions along the Laotian and Cambodian borders.[11] As plans for the employment of the "strikers" (as members of the CIDG mobile strike forces were called) developed, it was the hope of General Paul Harkins, the first commander of MACV (COMUSMACV), "that with the Montagnards plus recently opened Special Forces camps at Aluoi and Ashau in the Annamite foothills, North Vietnamese access across the border would *certainly* be blocked [emphasis added]."[12] Unfortunately, what was *certain* was that if the camps were not strong enough, the North Vietnamese and Viet Cong would simply force their evacuation or overrun them in order to preserve and expand the lines of supply from their Cambodian and Laotian sanctuaries into South Vietnam. That was to be the fate of A Luoi and A Shau.

While Captain Powell and the ARVN infantry were patrolling just to the south, the first American unit to be stationed in the A Shau Valley arrived in March 1963 to build and man a CIDG camp at Ta Bat (YC425945), near the center of the valley.† The twelve members of the First Special Forces Group's Team A-433 were led by Captain (later Lieutenant Colonel) Jerome

---

† Grid references are used throughout this book and refer to military map sheets 6441-I, 6441-II, 6441-III, and 6441-IV. The standard six-digit coordinates identify locations to the nearest one hundred meters.

"Jerry" Bruschette. A Special Forces A-Team consisted of two officers and ten enlisted men, specializing in weapons training, medicine, communications, intelligence, and demolitions. At first, the Green Berets, who were accompanied by a 12-man Vietnamese LLDB team and a CIDG strike force company of about 100, thought they would have a relatively easy time filling the roster of their new Montagnard unit with local tribesmen. The summary report for mid-May to mid-June 1963 opined: "It is apparent from reception received on patrols that the potential for a CIDG program in this area is *quite good* [emphasis added]."[13] It wasn't long, however, before there were omens of the difficulties to come. The following month, while reporting "a large number of patrols" but no contact with enemy forces, it was also noted that the province chief "has not allowed any weapons to be issued to trainees, [which] is a contributory factor to AWOLs and causing ill-feeling with CIDG personnel." The province chief was likely not dissuaded from this position, if not outright encouraged, by President Diem, as "the Saigon regime held all minority groups in contempt as either primitive savages or potential dangers to the central government."[14] These early patrols—part on-the-job training and part "live-fire" with a very real enemy, as would be the case for the entirety of the CIDG history in the valley—remained fairly close to the camp and reported that there were no other hamlets nearby, the nearest being Kon Tom, 4 km to the southeast, although aerial reconnaissance spotted "many villages" to the west.

By the end of August, a second, incompletely trained strike force company had arrived from outside the valley. While construction on the camp finally was progressing, with the completion of the team house and dispensary and with advanced weapons training and refresher training on patrol tactics underway, the potential for recruiting among the *local* population had become "doubtful." Most ominously, the A-Team's planning for future operations now for the first time indicated not only the presence of the enemy nearby, but also his dominance. In the Rao Lao Valley to the northwest of the camp, where the A Shau drained to the west into Laos: "Two operations . . . have shown the need for helicopter support for the purpose of gaining surprise and reducing the number of casualties during operations in the Viet Cong dominated area. The enemy have a very clever and intense early warning system against troops approaching from the East." Hence, the proposed plan of maneuver called for helicopters to land the strike force "behind the enemy and walk East towards Camp Ta Bat." With the marginal flying weather at this time of year in the valley and the minimal helicopter support available to Special Forces prior to the commitment of regular American ground units, this must have seemed an improbable scheme. Indeed, the following month, while training and construction continued apace, the plan for an operation to the west "along the Vietnam-Laos border" was deferred.

On 3 October 1963, it was time for Team A-433 to be replaced. Its successor was Team A-434, led by Captain Harry Hilling. The new camp dispensary began to treat local villagers as well as the camp's troops, reporting 422 patients the first month, and averaging over 800 patients monthly for the next six months. Local inhabitants turned out for medical care but didn't want to serve the Republic of Vietnam. Although none could be described as fond of the Saigon regime, the Montagnard tribes throughout South Vietnam presented varying degrees of willingness to serve the *Americans*. The dominant tribe in the A Shau was the Pacoh. Unfortunately for the CIDG recruitment effort, the Pacoh were exceedingly ill disposed to join the fight against the Viet Cong. "In the mid-1960's . . . it was reported that the Viet Cong had extensive control of the Katu, as they had over groups [e.g., the Pacoh] in the same general area."[15] The Katu/Pacoh alliance with the NVA/VC dated to the French conflict in Indochina and was well-established and unbreakable.[16] Thus, it is not surprising that Team A-434's first monthly report (20 Oct 1963 to 20 Nov 1963) also indicated "potential doubtful" for signing up local CIDG "strikers."

One month after arriving at Ta Bat, Team A-434 was tasked with splitting like an amoeba, with half the team going south 13 km to set up a forward operating base (FOB) at A Shau (YC525812). The two bases were to work together, patrolling to cover the entire southern half of the A Shau Valley. On 1 November 1963, the Border Surveillance Program in I and II Corps was turned over to MACV by the CIA as part of Operation SWITCHBACK, phasing out CIA control of Special Forces operations in South Vietnam.[17] The area of every one of the six camps involved—Khe

Sanh, A Shau, Kham Duc, Ta Bat, Dak Pek, and Dak To—would later be the site of intense battles.

Meanwhile, to the north, trouble was brewing. In mid-November, an ARVN outpost 7 km north of Ta Bat, in the vicinity of A Luoi, was attacked by an estimated VC battalion. The enemy's campaign to purge the Americans and ARVN from the valley had begun, and it was just a matter of time before combat between the SF/CIDG and VC would occur. As 1963 came to a close, the pace of war was quickening in the A Shau Valley.

1964 – Special Forces at A Shau and ARVN at Ta Bat

Not long after the bifurcation of Team A-434 and the CIDG contingent in the valley, the local Viet Cong commenced testing the defenses at both Ta Bat and A Shau. By 20 January 1964, they had already launched 11 probes, as well as four ambushes of CIDG patrols. The first SF casualty in the valley came on 15 January, when SFC Thierault of A-434 was wounded. As with the ARVN patrols accompanied by Captain Powell the previous year, there were no confirmed enemy losses.

Plans were underway to reshuffle the ARVN and CIDG units in the valley. The Special Forces and CIDG would consolidate at the A Shau camp in the south, and the ARVN 1st Division would take over Ta Bat in the valley center. At A Shau, the SF FOB and CIDG Strike Force had assumed responsibility on 16 February for five outposts on higher ground around the camp and were conducting joint patrols with the ARVN near the southern end of the valley. While the monthly team report cited "good results" in local patrolling, practice alerts, weapons training, and coordinating intelligence with the ARVN, it also noted: "No extended patrols or operations were conducted against the VC primarily because of the [Tet] holiday season and the number of troops on leave." Responding to the transfer of the outposts to the CIDG, the Viet Cong did not take "leave" or a holiday season vacation and wasted no time in launching six probes of the outposts at A Shau, testing the new defenders.

On 19 March 1964, another A-Team rotation took place, with team A-421 from Okinawa under Captain Harry Ching replacing Captain Hilling's men. The following day, Captain Ching's entire team was brought together at A Shau, as the 1st Battalion, 1st Infantry Regiment of the ARVN 1st Division took over the Ta Bat camp and the reorganized valley responsibilities took full effect. The omens for the two camps were mixed. At Ta Bat, in the days before the transfer, the VC for the first time employed ground fire against aircraft, hitting CV-2 Caribous (small army-piloted transports capable of landing and taking off from short, unpaved airstrips) on their approach to landing three times. But at A Shau, the Special Forces medics were able to conduct daily sick-call for villagers, and plans to use nightly patrols and ambushes to replace the camp's outposts were being prepared.

In March and April, the VC were "fairly active in the valley area between ASHAU and ALOUEI," although they refrained from attacking the camp or outposts directly. The troops were busy building and repairing barracks and fortifications; local security patrols had no contact. It was so quiet that a school was even constructed for the few village children still in the A Shau area, and rice cultivation was expanded. The interlude of relative peace was soon to end. With the completion of the camp transfers, the time had come to step up operations against the VC in the valley. First, however, for the Green Berets, there was the matter of placating a slightly irate deputy commanding general.

In late April, as MACV Deputy Commanding General William C. Westmoreland's Caribou was preparing to leave from a visit to Camp A Shau,‡ it was taken under fire by enemies concealed near the airstrip. The general was not hit, but four Americans (the pilot, copilot, crew chief, and Special Forces Major George Maloney) as well as two CIDG personnel were wounded. In his book *A Soldier Reports*, Westmoreland describes this as his "closest call" with death during his entire tour of duty:

> The first rounds tore through the nose of the plane, ... slightly wounding pilot and copilot. The two nevertheless managed somehow to swing the plane about and gun the engines for takeoff while bullets ripped through the thin walls of the craft and the other passengers and I sat helpless. As

‡ Westmoreland would assume command of MACV in June 1964.

we lifted into the air, bullets were still striking the plane, wounding the crew chief and several Vietnamese soldiers and missing [Barry] Zorthian, [Alfred] Hurt, and me only by inches.[18]

As a result, all other activities at the base were moved to the back burner, so that brush could be removed from around the airfield. It was not a promising introduction to the valley for "Westy."

There is a story that when Westmoreland asked the camp commander why the foliage near the camp from which the VC fired upon his plane had not already been cleared, the response was that the area was occupied by an old, unmapped French minefield.§ The general replied that he would "send something" that would take care of that problem. A few days later, several large drums of Agent Orange were delivered for spraying on the old minefield area. Almost two years later, when the camp was overrun, several of those drums remained near the camp center. Perforated by bullets and shrapnel during the battle, they were emptied, creating one of the most toxic dioxin hot spots in all of Vietnam.[19]

As discussed previously, CIDG camps might have both a Village Defense (local security and denying the enemy access to the local population and crops) and a Border Surveillance (armed reconnaissance) mission. Unlike the camp at Ta Bat, which had been expected to serve in a Border Surveillance role but then was transferred to the ARVN, the A Shau location evolved rapidly as a planned staging point for longer and more aggressive patrols. Training of a reconnaissance platoon began, and once elements of two ranger battalions (11th and 32nd) of ARVN regulars arrived, a multi-company operation in the valley commenced. The 32nd Ranger Battalion included on its advisory team WO2 J.W. Roughley of the Australian Army Training Team, and this was probably the first time allied forces personnel were engaged in the A Shau.[20] Operation Lam Son 119 took place 25 May–9 June 1964. "This first attempt at a joint [ARVN/CIDG] operation by the Strike Force of this camp revealed problems arising from a lack of clear-cut command channels and inadequate joint prior planning and coordination."

---

§ When ARVN engineers subsequently cleared minefields around A Shau, they removed 1,000 American M16 mines planted earlier by the South Vietnamese, but "did not have instructions or diagrams with which they could clear remaining [French-era] minefields."

Back at the camp, VC sniping and harassment continued. There were several ambushes and frequent pot shots directed at the outposts, resulting in three CIDG KIA, nine CIDG wounded in action (WIA), and one USSF WIA. At the end of May, C-123 planes sprayed Agent Orange for the first time at the area right around the camp. "The spray appears to have been effective on trees but only partially effective on the elephant grass under the trees."[21] Later, the camp's troops burned the defoliated trees. By this time, there were four companies and over 550 CIDG assigned to A Shau. None were natives of the area.

Thus, the pattern for the next several months was set. At A Shau, the U.S. continued, with little or no success, to woo the few local inhabitants to "our side" (A Shau village had just 21 remaining residents in July). Sick call was held, psywar (psychological warfare) leaflets were distributed, and houses were built. But the Pacoh weren't biting, as their alliance with the Viet Cong's predecessors, the Viet Minh, endured. Operationally, training, patrolling, and fortification construction continued for the CIDG, while ambushing and sniping continued for the VC. The casualties were usually friendly. Only three of A Shau's outlying outposts remained, with one of those (Charlie) due to be abandoned by 10 July.

In August, it was time to rotate the A-Teams once again. Captain Ching's SF team left the country, but not before suffering the first U.S. soldier killed in the A Shau. On 21 August, just a week before his A-Team was scheduled to depart, SFC William R. Patience, Jr., the Intelligence Sergeant for A-421, was killed by a sniper while on patrol. Ken Haines, who was also on that mission, tells the story: "We were in the field for fourteen or fifteen days and were moving out after a night of bivouac in the A Shau Valley. Our job was to take up rear security for the ARVN battalion with our CIDG troops. We weren't out very long that morning when we were ambushed and Bill was killed by small arms fire."[22] Sergeant Patience had served an earlier tour of duty with the Special Forces in Laos on *Operation White Star*. On the 29th of the month, Captain Arley Harper and Team A-113 came into town to continue the struggle for control of the A Shau Valley.

For the last several months of 1964, weather became the dominant factor in operations. Typhoon Tilda made landfall in Vietnam on 23 September, damaging about 80% of the structures at Camp A Shau and diverting much of the camp's energies into rebuilding. A policy of rotating Strike Force companies between camps also created significant problems in conducting operations. The three companies at A Shau regularly traded places with those from other bases, including Ta Ko. These rotations were often delayed by weather, and the units that arrived were sometimes at only one-third strength. In November, only local patrols were sent out, and in December there were just three 4-day company-size longer-range patrols. On one of those patrols, on 23 December, the unit ran into a VC platoon. In the ensuing firefight, Sergeant Emmett H. Horn of Team A-113 was killed in action, becoming the second U.S. soldier to lose his life in the A Shau Valley.

As the U.S. and its CIDG strikers chased an elusive enemy at the southern end of the A Shau, the ARVN soldiers operating in the Ta Bat vicinity were equally challenged. After the ARVN 1st Division took over the Ta Bat Camp, the patrolling activities of that unit (one of the best in the ARVN) typically mirrored those of their CIDG counterparts to the south. For most of 1964, they coped with the ambushes and booby traps of an enemy who seldom made contact, and then usually only briefly and on his own terms. One difference between the ARVN and USSF operations was that in the area around Ta Bat, there were still a number of Montagnard villages where pacification efforts continued in theory, although over the long run razing was usually the final result.[23]

1965 – The Enemy Turns Up the Heat

The 1982 *Study of Strategic Lessons Learned in Vietnam* prepared for the Department of Defense identified the "overall military objectives" of the NVA as of January 1965 as: "Saigon (the political heart and mind), the Delta (food and people), Hue (historical and psychological) and Da Nang (port and airfield)."[24] The enemy's "long-term strategy for Phase III [conventional warfare] consisted of creating three main centers of gravity. . . ," one of which, in the north, would include "flanking movements through the A-Shau and Elephant Valleys [that] would threaten Hue and Da Nang."[25] This enemy strategy, linking the critical role of the

A Shau Valley with the key objectives, Hue and Da Nang, would still be in effect a decade later when the final offensive of the war was launched. It was Hanoi's insight about the future decisive role of the valley that guided the buildup of NVA forces there in 1965. This strategy was unknown and the importance of the valley unappreciated by the allied high command at that time. It would continue to be so until 1968. And so in 1965, a small force of ARVN regulars and a handful of Green Berets and their often reluctant indigenous troops were camped in the A Shau Valley, unaware that the entire 325th Division of the People's Army of Vietnam (PAVN) would soon sweep them aside.

During the opening months of the year, Team A-113 under Captain Harper continued to train CIDG companies and the reconnaissance platoon, all recruited from outside the valley. The recon platoon was manned by Nung mercenaries, a tribal group of Chinese extraction who readily enlisted as "guns for hire," filling the CIDG Mobile Strike Force companies and becoming known for their fierce dedication to their American advisors. Local and occasional long-range patrols were conducted. Attrition (desertion) from the CIDG contingent was an ongoing problem. In March, due to troop shortages related to attrition, it was decided that the last two outposts, on mountains to the east and west, would be abandoned.[26] This was done on 21 March.

In May, major changes were in store for Special Forces operations in the A Shau. Team A-113's tour of duty came to an end. A-102, commanded by Captain Edward Short, replaced them. It is at this point that the existing records of Special Forces units and operations in the valley become opaque, perhaps because elements of two Special Forces Groups (1st and 5th) were operating at the same time but under different reporting systems. According to the Summary Reports for Camp A Shau, "An advance party from this location arrived at the New FOB located at A Luoi on May 2 and two CIDG companies from Ashau closed on the FOB 4 May." There is also evidence that a second SF FOB was established at Ta Bat (although in a different site than the camp in 1963–64) on May 2, and this may have been run by staff from the SF higher headquarters, Team B-11 at Quang Ngai.[27]

In any case, Captain Short continued to command at A Shau and A Luoi and to engage in heavy local patrolling and company-sized longer-range patrols through the end of September. During June, two VC "communications complexes" were found. One had previously been destroyed by artillery fire from A Shau, while the other was flattened by an airstrike. Around both camps, C-123s regularly sprayed Agent Orange to defoliate fields of fire, especially near the airfields. Keeping the high grasses and undergrowth down was a never-ending struggle, as is reflected repeatedly in the monthly reports, and in spite of the Ranch Hand (Agent Orange) sorties. Local recruiting was nonexistent, and new CIDG troops were regularly flown in to replace those lost through enemy fire, disease, or desertion. The camps' combined strength was five understrength companies (with average strength about 85–90 per company, compared to TO&E [Table of Organization and Equipment] of 150). Construction of the new camp at A Luoi proceeded.

In August, in response to an intelligence report that two VC battalions were preparing to attack one of the camps, a Strike Force company of Nungs was flown in. It was already becoming clear that the camps lacked power not only to defend themselves against a determined enemy but also to impede in any material way the growing power of the VC in the valley.

October was a tough month for the new commander of Team A-102, Captain Ivan Jennings. The Vietnamese LLDB honcho at A Luoi was seriously wounded when a grenade was thrown into his barracks. An investigation revealed the presence of nine VC among the CIDG camp defenders. October also brought the replacement of Camp A Shau's two 105mm howitzers from the ARVN 12th Field Artillery Battalion (with a range of 11,500 meters, or far enough to support patrols almost to Ta Bat) with 4.2" mortars (with a range of less than 6,000 meters). Although no typhoon hit the camps this year, "rains . . . caused heavy damage at both A Shau and A Luoi. The protective wall around both camps [was] washing away."[28] If the enemy outside or camp traitors inside didn't get you, mother nature would.

After seven months of trying to keep the FOBs open at A Luoi and Ta Bat, where a great deal of time was certainly spent in just building and maintaining camp structures, fortifications, and fields of fire, it was time to pack up and leave the center of the valley to the VC. On 8 December, both sites were abandoned, and the forces from A Luoi marched back to A Shau. On Christmas Day 1965, the ARVN, too, packed their bags and left the A Shau, knowing that their positions there were indefensible against the growing presence of the North Vietnamese Army (NVA).[29]

After almost three years in the A Shau Valley, the Americans and their South Vietnamese ally had next to nothing to show for their efforts. SF/CIDG destruction of enemy camps amounted to two "communications complexes," and by far the largest part of the troops' energies had been devoted to camp construction and maintenance and to local (defensive/training) patrols. The enemy dictated the when and where of combat.

December 1965 – The View from Headquarters

According to the U.S. Army's volume on Special Forces in Vietnam, "the [A Shau] camp's mission was border surveillance and the interdiction of infiltration routes."[30]

The 5th Special Forces Group Quarterly Command Report for October–December 1965 cites seven specific missions identified by COMUSMACV for the detachments and CIDG units under its command.[31] Enclosure 6 designates the role of Camp A Shau under the command of Captain Jennings as border surveillance (Mission 2). Enclosure 12, which defines each of the seven designated missions in detail, says this about Mission 2:

PROVIDE ASSISTANCE IN THE ESTABLISHMENT OF BASES FOR THE CONDUCT OF BORDER SURVEILLANCE OPERATIONS:

In carrying out this mission, USASF detachments are deployed along with VNSF detachments and CIDG troops in a series of camps along the Cambodian and Laotian Border areas. *Past experience has demonstrated the need for establishing border surveillance camps near population resources* [emphasis added]. These resources can then be exploited for intelligence as well as serving as a recruiting source for CIDSG companies.

The second counterinfiltration mission would have fallen under the aegis of Mission 4: "Provide assistance in the establishment of bases for conducting operations to interdict VC internal movement corridors." For Mission 4, it was noted that "such camps must be located in populated areas for the same reasons discussed under Mission 2."

The A Shau Valley, with its small, scattered Pacoh population already under the domination of Hanoi for many years, did not offer any realistic prospects for the development of local recruits or intelligence sources; this was well known by the A-Teams in the valley by the end of 1963. The Operational Briefing Narrative accompanying the Quarterly Report offers no explanation as to how the square A Shau peg could fit into the round border surveillance/interdiction hole. In fact, it gives the impression that all of the BS camps in I Corps had already been closed (Para 7):

> Border surveillance was developed in 1964 in IV Corps. The original plan was to extend the border program along the entire Vietnamese border; however, because of a lack of population in I & II Corps and the dense vegetation in the border area, it was not feasible to conduct border surveillance in these regions.

I Corps border camps at Ta Ko and A Ro in Quang Nam province to the south were abandoned in February and April 1965, neither having existed for a full year. But A Shau (A-102), Kham Duc (A-105), and Khe Sanh (later moved to Lang Vei, A-101) remained. All three were engaged in border surveillance, all three were devoid of a requisite local indigenous population, and all three would be overrun by the NVA. Lang Vei fell in February 1968, and Kham Duc in May of that year. The camp at A Shau would not survive nearly that long. These were the only Special Forces camps in Vietnam to be overrun after mid-1965, and the only ones in I Corps to fall to the enemy.

The briefing narrative continued:

> We have 3 special missons also [i.e., in addition to those designated by COMUSMACV]. The first of these is the Montagnard program. This is a carryover from the days when Special Forces worked for the CIA back in 1961.

At that time we went into areas that were isolated from GVN control either by VC activity or terrain. The areas were primarily in the western portions of I and II Corps. This did not work too well because when the Americans phased out, the Vietnamese also left and the Montagnards went back to their tribal ways of life. The emphasis [now] is on the Piedmont region of the highlands, closer to the east coast.

An early work (1983) on the Green Berets in Vietnam had this to say about the difficulty of operating CIDG camps in the Pacoh country: "Without people, the effort became a conventional military operation, with no real stake in the area, no human intelligence nets, no source of recruits, no hearts and minds to win from the VC."[32] So was it perhaps just bureaucratic inertia that left these camps so exposed? Were they being kept around just to prevent the ARVN from pulling out and the locals from reverting to "tribal" life? Were they to serve as expendable "tripwires" to alert COMUSMACV when the enemy was initiating a major incursion into the border regions? Whatever the reason, the I Corps camps became victims of being assigned to what higher headquarters, by its own parameters, already knew was "mission impossible."**

Perhaps Bill Patience and Captain Hieu had it right all along. The allied camps in the A Shau Valley were there because they were there. They would remain there only so long as the enemy suffered their presence.

** According to Simpson, *Inside the Green Berets*, p. 108: "A Shau was kept open [after the closing of Ta Ko and A Ro] over the objection of SF." There is no citation to indicate the source of this information. Neither is there any reason to doubt it.

2

THE A SHAU VALLEY

The Streams

The usual river valley is a canyon with an upstream end, which might or might not include the river's origin, or headwaters, and a downstream end. Correspondingly, the terms *upper valley* and *lower valley* apply. That does not, however, describe the A Shau Valley. The A Shau is in fact a T-shaped geographic feature, with two sets of headwaters at the northwest and southeast ends, and two streams meeting roughly in the middle to form a river that exits the valley to the west-southwest (see Map 2). The headwaters of the larger stream in the southern end of the valley lie as far south as YC502770 and originate, on the western and southern sides, in the mountains along the Laotian border. War-era maps call this stream Rao Lao (A Sap [A Shau]). The northern, smaller of the two streams rises in the vicinity of YD330066. Era military maps provided no name for this branch, but the major village in the area is A Luoi, the modern name for the entire valley, so it may be presumed that this is what it was called. The junction of the streams is on the west side of the valley at YC403951, from which point the Rao Lao flows to the west-southwest, crossing the border into Laos at YC302912. The southern tributary of the Rao Lao is about 22 kilometers (14 miles) long; the northern valley about 14 kilometers (8.5 miles).

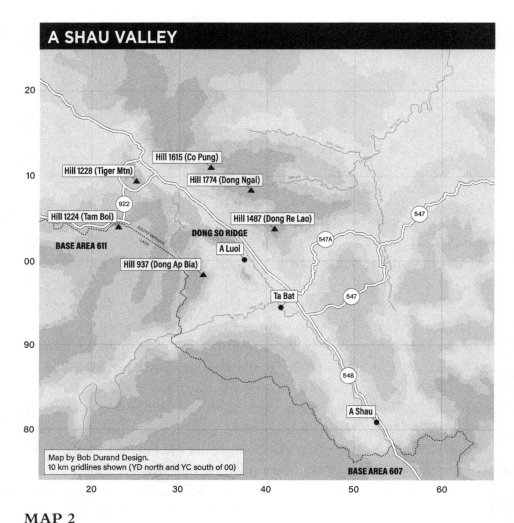

MAP 2

Source: Bob Durand Design

After a rise of roughly 10 meters in altitude (YD323065), the valley that might be expected to terminate at the headwaters of the A Luoi stream merges immediately into a yet another multiheaded river valley, the Xe Sap, which for all geographic intents and purposes extends the valley with several meandering tributaries another 9 kilometers (5.5 miles) to the border of Quang Tri Province (YD255123). The Xe Sap also drains to the south and southwest, serving as the Vietnam/Laos border for several kilometers before being joined by the Rao Lao where it enters Laos. The total length of the A Shau Valley, as the term was understood and used, is about 45 kilometers (28 miles), equally distributed to the north and south of the confluence of the A Luoi and A Shau streams. The valley extends the entire length of the western Thua Thien border with Laos, from Quang Tri Province to the north to the Laotian mountains to the south. The width of the valley floor ranges from less than 200 meters north of A Luoi to almost 3 kilometers near the abandoned villages of A Luoi, Ta Bat, and A Shau.[1]

Speaking of "up" and "down" the valley in traditional (upstream and downstream) usage was, for obvious reasons, a nonstarter for military planners, and they may be forgiven if they chose to equate up with north and down with south. It made things much simpler.

One additional peculiarity characterized the A Shau/Xe Sap/Rao Nai river system. Whereas the other rivers in western Quang Tri and Thua Thien flowed to the east into the South China Sea, the A Shau drained into the mighty Mekong River system in Laos and then through Cambodia until the waters returned to South Vietnam in the vast Mekong Delta south of Saigon.

Thua Thien Province

Thua Thien Province was the second northernmost province in the Republic of Vietnam (South Vietnam), bordered to the north by Quang Tri Province, to the east by the South China Sea, to the south by Quang Nam Province, and to the west by the Kingdom of Laos. This simple fact of geography, combined with the spiritual and symbolic importance of the former Vietnamese imperial capital of Hue as the province capital, imbued the province with a strategic importance to both sides second only to Saigon. Strategically, a Communist force in Thua Thien posed a threat not only to Hue City, but also to South Vietnam's second largest City, Da Nang, in Quang Nam just to the south. Enemy troops gathered here were on the flank and rear of the entire northern defense line along the DMZ (including, of course, Khe Sanh), in Quang Tri to the north.

Roads and Trails

In 1961, the only "road" in the valley was provincial Route 548, a cart track that ran the length of the valley floor paralleling the streams. The trail connecting the center of the valley to Hue, Route 547, was overgrown and in disuse for much of its length on the A Shau end. The Montagnard valley inhabitants had little use for commerce with the Vietnamese lowlanders, and that was even truer in the obverse.

So it was, then.

The potential, however, was almost unimaginable. While seemingly lost on the allied military leaders, the North Vietnamese planners understood early how the valley's proximity to both the Laotian sanctuaries to the west and the populous lowlands to the east could be utilized to great advantage.

A 1969 MACV J-2 (Intelligence) study, "Avenues of Approach: A Shau Valley" would later identify multiple routes between the valley and Laos:

1) **Avenue "Alpha": Route 922, joining Route 548 near the northern end of the valley. This would always be the best maintained and most heavily traveled route.**

2) **A network of trails due west of A Luoi where the border was less than 10 km from Route 548 and much of the area was clear of jungle. Just north of Dong Ap Bia (see below).**

3) **Avenue "Bravo": Another network of trails paralleled the Rao Lao from the border to the vicinity of the rivers' confluence. Just south of Dong Ap Bia.**

4) **In the central part of the southern valley, there were more trails through the mountains, although the going was generally rougher here than farther north.**

5) **A well-used trail ran from the border 5 km west of A Shau into the valley.**

6) Avenue "Delta": South of the valley, Route 548 connects to Laotian Route 923 and then to South Vietnam Route 614A, passing across a finger of Laotian territory and entering Quang Nam Province.[2]

Most of these routes were not usable during the monsoon, and most could not accommodate vehicular traffic, but all served at one time or another to move troops and supplies into the A Shau.

By 1971, the enemy had opened up additional connections between the A Shau and the next valley to the north, the Da Krong, where more new routes were being opened from Laos.

The multiplicity of paths between Laos and the valley made it virtually impossible to completely close the Communist lines of communication with the enemy base areas 611 to the north and 607 to the south. In addition, the myriad hideaways concealed in and around the valley rendered impractical any attempt to trap enemy transportation and engineer units working in the area.

If there were many entrances from the west, there were even more exits to the east. The old Route 547 ("Approach Charlie") was overgrown in the early 1960s, at which time the Communists were using trails connecting the southern end of the A Shau Valley with the next valley to the northeast, the Rao Nai, sometimes called the Be Luong Valley after the largest village there. This route was developed by NVA engineers in 1970 and was known to the allies as "Gorman's Road." Another track ran from the north of old Route 547 along a tributary of the Rao Nho. The Communists developed this prior to the Tet offensive of 1968, but U.S. engineers later did much work here and established several fire bases along Route 547A. The entire southern part of the A Shau runs diagonally away from the north-south Rao Nai Valley to the east, but there are multiple trails via streambeds connecting the two regions, which abut at their southern terminuses. At the southernmost end of the valley, a network of trails connects Vietnam with a salient of Laos that protrudes between Thua Thien and Quang Nam. Here, Route 548 connects with Laotian route 923 and then with Vietnamese Route 614 ("Approach Echo"), later to be known as the "Yellow Brick Road," extending the Communist line of communication into the southern provinces of I Corps.

Key Terrain Features: Villages (abandoned), Mountains, and Hills

As discussed in chapter 1, almost all of the inhabitants of the A Shau Valley had departed during the early 1960s. The sites of the three largest villages were the locations for early allied bases, and each had an airstrip prior to abandonment. A Luoi (YD375001) had a north–south airstrip one kilometer to the east. Ta Bat (YC415948) had a short NE/SW airstrip one kilometer to the east. A Sap (A Shau) (YC497830) had an NW/SE airstrip to the north. There were other abandoned villages, most notably Con Tom [Kon Tom] in the southern valley and La Dut north of A Luoi, as well as several villages along the Rao Lao west of the stream junction.

Military maps (1:50,000) denote prominent hills by their height in meters. The average elevation of the valley floor is 575 meters, so hilltop heights should be considered relative to that base. On the western (Laotian) side of the valley, the peaks that played prominent roles in the campaigns in the valley included (from north to south):

1) **Co A Nong (Hill 1228) (YD253090). Known as Tiger Mountain and prominently situated between the original Route 548 and the "Route 548 Bypass," along the most important route into the valley from Laos.**

2) **Dong So Ridge (Hills 1126 to 975) (YD325032 & YD353024). Also known as Razorback Ridge. To the east, Dong So towers over the narrowest part of the valley, where allied air tried to create a choke point for Communist vehicles; to the south, it overlooks the "Punchbowl," the early NVA headquarters and major base in the valley.**

3) **Dong Ap Bia (Hill 937) (YC327982). The infamous "Hamburger Hill." Located between two major trail networks connecting the valley with Laos.**

4) **Hill 996 (YC314939). Southwest of Dong Ap Bia, rising above the valley where the Rao Lao and Xe Sap merge, an important trail corridor.**

On the east side of Route 548, virtually every prominent hill became, at one time or another, an allied fire support base. Those are covered in the relevant chapters. The most dominant peaks are the following:

1) **Co Pung (Hill 1615) (YD336107). A towering mountain not occupied by the allies until 1970, but a key NVA base proximate to the "Warehouse" and FSB RIPCORD.**

2) **Dong Ngai (Hill 1774) (YD382083). The highest point in Thua Thien Province.**

3) **Dong Re Lao (Hill 1487) (YD406035). First occupied by the 1st Cavalry Division as Signal Hill and later repeatedly used as Eagle's Nest by the 101st Airborne Division. It served as an important radio relay site, as tactical radios could not communicate from the valley floor or lower hills to the allied headquarters in bases to the east.**

All of the peaks named were located along the northern or central valley. The hills along the eastern side of the southern valley, between the A Shau and Rao Nai Valleys, were lower, topping out at 800–900 meters.

Weather

The A Shau Valley owned the distinction of suffering from the worst of both the northeast and southwest monsoons.

Either low overcast and fog or thunderstorms and heavy rain hid the valley most of the year. Air access to the valley was curtailed nearly the whole year because the A Shau Valley lay exposed to the influence of the northeast and southwest monsoons. By all approaches, the winds had to climb to reach the valley and this upslope wind-flow contributed to thunderstorm activity and the formation of low clouds and fog. From May to November, heavy rains made the poorly drained valley very marshy. When the rains decreased in November, fog and low overcast returned with the northeast monsoon off the South China Sea. Just such poor weather had prevented tac air from successfully defending the A Shau Special Forces Camp in March 1966.[3]

Ground conditions in A Shau Valley are affected by both monsoon seasons. April through December is the wet season with September, October, and November receiving the heaviest rainfall (14.5 to 21.5 inches). January, February, and March are dry season months and receive 2.0 or less inches of rain. The dry season months experience the most frequent and persistent low ceilings, below 1000 feet.[4]

The low ceilings would present a greater challenge to air operations than the rainfall, something that would not be appreciated by allied planners until after the first operation in the valley, DELAWARE, in April 1968. For the Communist logisticians, the low ceilings would be welcome; the rainfall would not. While I do not dwell in the chapters that follow on the impact of weather on allied operations, it was a constant theme, playing in the background throughout the war. For fixed-wing aircraft, extensive cloud cover often precluded ground support (tactical air strikes, aka tac air). For the helicopter and air cavalry units, whose pilots who were over the valley virtually every day from April 1968 through late 1971, "weather birds" went out first in the mornings to determine which areas could or could not support air operations. It was common for operations in the mornings to be delayed until some of the clouds and fog burned off. Insertions and extractions of long-range patrol teams were also impacted by the A Shau climate.

Communist Base Areas

"When viewed as a unit, Base Area 611, A Shau Valley and Base Area 607 provide a formidable complex for logistical and operational activities."[5] A bit of an understatement, that.

Base Area 611 was, next to Base Area 604 southeast of Tchepone, the most important Communist base area in Laos. As defined by allied planners, it extended westward from the Vietnam border, where Route 922 intersects with Route 548, to the junction of Routes 92 and 922, and then north along Route 92. Route 92 was the major NVA north-south line of communications in the Laotian Panhandle. The area where 92 and 922 meet was the location of the forward headquarters of NVA operations on the Ho Chi Minh Trial, the 559th Transportation Group. Known to the secret warriors of MACVSOG as "Target Oscar Eight," it was the most fearful place in all of Southeast Asia for Green Beret missions.

Base Area 607 was located at the opposite end of the A Shau Valley, in a salient of Laos that abuts Thua Thien to the north and Quang Nam to the east. This base was connected with Route 92 via Route 165 at Chavane, and hence could be used either as a transshipment point for troops and supplies coming south down Route 548—its usual purpose—or, in a pinch, as an alternate connection between Route 92 and the NVA logistical network in South Vietnam.

Both base areas were connected with the A Shau Valley by mazes of well-camouflaged roads and trails and served as havens whenever allied troops made the valley uncomfortable for Communist forces. Both bases were also well protected by anti-aircraft weapons and counter-reconnaissance teams.

Not the A Shau Valley

This book is exclusively about the A Shau Valley. Included are actions that took place between the A Shau and the Laotian border, including along the Rao Lao Valley, but not those on the east side that were outside the immediate confines of the valley, e.g., battles around fire support bases such as RIPCORD, VEGHEL, and BASTOGNE. Many other authors have taken liberal views of what constitutes "near" the valley or "on the east wall" of the valley and have included in their "A Shau" stories events that took place in what we might broadly define as the A Shau campaign area. In other words, events that were directly related to what was happening in the valley, but not necessarily within the valley or the immediately adjacent hills and watershed. With very few well-identified exceptions, such events are excluded from this history.

3

THE FALL OF CAMP A SHAU

Camp A Shau in Early 1966

In early 1966, "no military mission was more difficult or potentially more dangerous than border duty. . . . The northwestern outposts, in close proximity to North Vietnam and the dreaded A Shau Valley, were the most feared assignments."[1] Captain James Davis Blair had assumed leadership of A-102 from Captain Jennings on 2 February 1966, not knowing that this would be just in time to command the camp during its final days. There were only 10 soldiers assigned to A-102, 2 short of a full roster.

Steadily increasing pressure had already forced the abandonment of the hilltop outposts that helped to protect the ill-situated camp at A Shau, and intelligence suggesting an NVA buildup had triggered the closure of the bases in the central part of the valley, at A Luoi and Ta Bat, in December 1965.[2] As a staging point for interdiction and border surveillance, the A Luoi/Ta Bat area (those villages are about 6–7 km apart) made a much more logical location for a camp, which is undoubtedly why the NVA prioritized the elimination of the allied presence there. Located near the valley's center, where the streams that flow north and south merge and drain westward into Laos and, most critically, near the junction of Routes 547 and 548, A Luoi and Ta Bat occupied key terrain. Route 547 would be the main approach and lifeline for the Communists when they attacked Hue in 1968 and for conventional North Vietnamese Army offensives in 1972 and later. A Shau, on the other hand, was in the southeast corner of the valley, and though the route southward from there would eventually be known as the "Yellow Brick Road" and would connect the Ho Chi Minh trail with Quang Nam province (where the major allied population center and base at Da Nang was located), in early 1966, that development was still years in the future. Once the howitzers had been removed from A Shau and replaced with heavy mortars, patrols could only receive indirect fire support for about six kilometers around the camp, with no prospect for supporting fires if they ventured as far as the abandoned posts to the north.

Not only was the location of the camp less than ideal from a perspective of offensive reconnaissance or interdiction, but also, the site selected was tactically very difficult to defend. Bennie G. Adkins, who would receive the Medal of Honor from President Obama almost a half century later for his actions at A Shau in March 1966, recounts a story from another SF Soldier, George Pointon, who had been part of the original site selection team: "We were looking at five different potential locations for camps in the A Shau valley and we had ordered them from the best locations to the worst. The irony is that some moron picked the one location [where] we had said, 'Don't, under any circumstances, put the camp here.'"[3]

The position had been chosen due to its proximity to the airstrip built for the ARVN units serving earlier in the area, as it could only be supplied by air. Adkins remarks, "I can tell you that none of us were happy to be in that camp; it was about 30 miles from another friendly camp, was bordered by high mountains on the east and west, and was surrounded by a triple-canopy jungle. We were like fish in a barrel."[4] In Captain Blair's 1968 monograph,[5] he noted several aspects of the campsite's geography that played into the hands of the NVA. "Observation into the camp from cleared spots and towers on the ridgelines is very good. . . . Conversely, observation of the ridges from the camp is extremely poor."[6] There were concealed approaches to within 75 meters of the perimeter, and while the old minefields had proven dangerous and unclearable to the CIDG force, "the enemy, on the other hand, seemed to move with impunity through the minefields."[7]

If the location of the camp was dismal, its readiness for the impending battle was abysmal.

Elephant grass reaching 8 to 12 feet high covers most of the valley floor around the camp, therefore, observation from the ground and air is very difficult and detecting movement of even large units is practically impossible unless they are detected while moving on trails in the valley. On the east side of the airstrip and the south side of camp were old mine fields that were overgrown with dense, high grass, which could not be cut because of the danger to friendly forces.[8]

Repeated application of defoliants, including forced curtailment of patrolling for over two weeks in September 1965 to allow aircraft to douse the valley with Agent Orange, had done little or nothing to preserve clear fields of fire around the perimeter, nor had month after month of hard labor by the camp contingent. Captain Blair observed: "The elephant grass around the camp was constantly cut, burned, and defoliated but still remained thick and tall."[9] Inside Camp A Shau, the buildings and fortifications were also in noticeable disrepair.

Intelligence collection through patrolling and air reconnaissance was weak, although there were several harbingers of the coming battle in the month prior. Had A Shau been located in an area where there were local villagers sympathetic to the Americans or the RVN (as was prescribed in the SF mission guidelines), there would likely have been rumbles of a strengthening enemy presence well ahead of time. But "the only local population that existed prior to the attack were an unknown number of secretive and hostile Katu [Pacoh] tribesmen who were either Viet Cong or Viet Cong sympathizers."[10] The camp was compelled to rely on aerial observers and its own local patrols to find out what the enemy was up to. These sources would eventually alert team A-102, but with little time to prepare and no specific knowledge about the enemy's strength or plans.

The Intelligence and Camp Reinforcement

Patrols provided the first inkling that the People's Army of Vietnam (PAVN) was about to make its initial appearance in force in the A Shau. On 18–19 February and again a week later, documents were captured revealing that the enemy had commenced reconnaissance of Camp A Shau in preparation for an assault. More alarming, but absent in the After

Action Report due probably to disbelief by headquarters, Bennie Adkins recorded: "At night . . . I could hear large convoys of trucks going. I reported their coordinates to higher headquarters, but they didn't believe it because there were no roads in the A Shau Valley. What they didn't realize was the North Vietnamese were building a series of roads under the A Shau's triple canopy. Even though we in the camp knew it, we didn't have a large enough force to interdict those convoys. . . . "[11]

On 5 March, less than four days before the battle at Camp A Shau began, hard information about the enemy strength and intentions finally arrived in the form of two NVA defectors. They reported that the 95B Regiment of the 325B NVA Division was going to attack with four battalions on 11–12 March. Forewarned with this information, air and ground recon around the camp was stepped up, finding numerous excavated but unoccupied heavy weapons positions, including anti–aircraft emplacements, and signs of the enemy "preparing the battlefield" by clearing and marking approaches to the camp perimeter. The enemy fortifications, including the alleged headquarters of the 325th NVA Division, were hit with air strikes, but with undetermined results.[12]

The Special Forces camp commander, Captain Davis Blair, requested reinforcement by at least two companies and at least two or preferably six howitzers. The ARVN I Corps Headquarters declined to provide any units, but the Special Forces command for the corps area, Detachment C-1, was able to obtain a company from the CIDG Mobile Strike Force (MIKE Force) in II Corps, which arrived with seven SF advisors (Team A-503 under Captain Tennis [Sam] Carter) and 141 troops on the afternoon of 7 March. The Nung MIKE Force was an outstanding unit but would be too little, too late.

The Battle of 9–10 March 1966

It is not my intention to recount here, blow-by-blow, the fall of Camp A Shau and the evacuation of its survivors. That story has already been told, and told well, in several other works, notably James Blair's unpublished monologue for the United States Army Infantry School, by Medal of Honor recipient Bennie Adkins, and in Thomas Yarborough's *A Shau Valor*. Additional details may be found online at the 220th Aviation Company's site.[13] The defense of the camp was remarkable for the seemingly bottomless well of courage that was displayed by the Special Forces troops, as well as by Air Force, Army, and Marine Corps air crews.

A brief recap of the battle is presented, with an emphasis on issues that would continue to affect allied operations in the valley throughout the war and/or that would be seen in similar actions at other A-Team camps at later dates.

By the first week of March 1966, when the attack on A Shau was recognized as highly probable, the size of the assigned CIDG force had declined from a peak of five companies and almost 500 men to just three companies (each of about 60 men compared to TO&E of 132) and a 30-man recon platoon. Operating with poor morale and at less than 50% strength, it is not surprising that every activity at the camp (i.e., patrolling, local defense, and camp maintenance) was imperiled. The two heavy 4.2 inch mortars at the camp had been reduced to one before the battle commenced. There were also two 81mm mortars – not much indirect fire with which to face a division. Ammunition for the mortars was only about 300 rounds total, but pleas for additional ammunition were denied. The difficulty was that a large store of artillery or mortar ammunition, while a necessity when under ground attack or supporting patrolling units, could also lead to devastating losses when ammo dumps were hit by enemy fire (U.S. ammunition storage was often done without overhead cover) or could be lost in the event that a base was overrun.

The NVA assault began just before 0400 on 9 March with a heavy and accurate 82mm mortar barrage that continued until 0630, killing or wounding over 50 of the defenders. An NVA ground assault by two companies at around 0430 hours was repulsed without additional friendly casualties. During the night, however, the cloud ceiling over the camp was down to 300–500 feet, so no air support was available.

With the coming of dawn, the enemy attack slacked off, but throughout the day on 9 March only 29 air sorties were directed against the besiegers, and bombs had to be dropped from above the clouds and adjusted by sound because of heavy ground fog.[14] Meanwhile, the defenders attempted to repair their fortifications and steel themselves for the next round. Several resupply drops by CV-2 Caribou light transports were attempted, and medical evacuation by helicopters had some success, although NVA .51-cal anti-aircraft fire was intense. During the afternoon, members of CIDG Company 141 within the camp were observed firing at the transports, but when captured by one of the Green Berets and taken before the Vietnamese LLDB Camp Commander for requested execution, that officer merely dismissed them and stated his belief that the company was loyal.

At about 0400[15] on 10 March, the enemy mortar barrage resumed with even greater accuracy, and the fire from NVA 57mm recoilless rifles knocked out over 50% of the camp's crew-served weapons. The ground assault began again at 0500. As the enemy attack developed, Company 141 openly joined the attackers, changing sides en masse and opening an irreparable gap in the camp defenses. This unreliability of the CIDG irregulars was a problem that would plague Special Forces through its operations in Vietnam, notwithstanding the intense loyalty that many SF and Montagnard soldiers still express for each other to this day.* Hand-to-hand combat followed, until at about 0800, the south wall was completely overrun.

Weather on 10 March continued to frustrate American efforts to provide accurate air strikes. While 210 sorties were flown that day, including several directed at enemy concentrations inside the partially overrun camp, low heavy clouds persisted, forcing the fighter-bombers to either fly at extremely low altitudes or to drop their loads from above the clouds. Heavy fighting within the camp continued throughout the day, with the Special Forces troops and their remaining loyal forces squeezed into a smaller and smaller area, and with food, water, and ammunition running low. At 1500 hours, III Marine Amphibious Force (III MAF) Headquarters, the U.S. command corresponding to the ARVN I Corps, decided to evacuate the remaining camp survivors.

---

* The official After Action Report of the 5th SF Group (OR-LL, 3) makes no mention of the treachery of Company 141, saying only that "CIDG Company 141 ceased all effective resistance."

At 1700 hours, while SF and Mike Force troops provided cover from along the north wall, the remaining survivors moved to a location about 300 meters to the north, awaiting pickup by the Marine choppers. NVA fire was heavy and two Marine CH–34s were destroyed, including that of the squadron commander, LTC House (see below). As panic and chaos spread among the CIDG troops, only 69 men were evacuated before the operation was called off and the camp was pronounced officially "closed" at 1745 hours.

At 1800, approximately 100 survivors of Camp A Shau, including 7 Green Berets and the crews of the two Marine helicopters, commenced escape and evasion to the north. Most of them were rescued on 11 or 12 March. Of a total camp population of 434 troops, only 186 escaped. There were 17 Americans at the Camp on 9 March; 12 made it out alive and all were wounded.

The overall experience at A Shau exposed manifold shortcomings in the soldierly aptitude of most CIDG recruits. Although training was a more-or-less continuous process at the camp, the ineptitude of the CIDG in fire discipline and marksmanship was notable. While the morale of many of the friendly troops in the battle remained surprisingly strong in the beginning, it waned rapidly as the enemy advanced and no reinforcements or artillery support and little air support were evident. It did not help that the enemy in general was much better armed than the camp defenders, who relied primarily on World War II era .30 caliber carbines. The camp itself had lost a very high percentage of its force in the preceding months to attrition (mainly "sick" troops who somehow never returned to duty) or to outright desertion when the opportunity arose. It was also noted that the commander of the ARVN Special Forces Team and designated camp commander was guilty of outright cowardice during the fight.[16]

The weather was not neutral in this battle. With no available support from U.S. or ARVN artillery, only the ground attacks of Marine, Air Force, Navy, and VNAF planes were available to break up the enemy troop concentrations and destroy his heavy weapons positions. Ground fog, low ceilings, and heavy clouds forced pilots to release their loads in attacks of doubtful accuracy, or to approach the target area at altitudes that exposed them to intense .51-cal anti-aircraft fire from the hills around the valley.

By the time the battle ended, nine Americans had lost their lives. Five Green Berets were killed in the ground battle, including SFC Raymond Allen, SSG Billie A. Hall, SGT Owen F. McCann, SP5 Phillip T. Stahl, and SGT James L. Taylor. Hall and Stahl were posthumously awarded the Distinguished Service Cross, our nation's second highest award for valor. An AC-47 gunship from the USAF 4th Air Commando Squadron was shot down on 9 March, resulting in the deaths of CPT Willard M. Collins, 1LT Delbert R. Peterson, and SSG Robert E. Foster. LT Peterson, who survived the crash and then gave his life to save three other crew members who were successfully rescued, was awarded the Air Force Cross. 1LT Augusto M. Xavier, a marine pilot flying an A-4 Skyhawk, was shot down and killed while supporting the camp in extremely marginal weather conditions and subsequently received a Silver Star.

In what was a unique event for the Vietnam War, two Medals of Honor were awarded to participants in the battle, both of whom survived. Major Bernard F. Fisher landed his A-1E Skyraider on the shell-torn remnants of the A Shau airstrip to pick up a fellow aviator, MAJ Dafford Myers, shot down moments earlier, in what must forever stand as one of the most audacious rescues in Air Force history. As previously noted, Bennie Adkins would also be presented with the Medal (upgraded from an earlier award of the DSC) for his selfless actions during the battle for the camp and subsequent escape and rescue of many camp personnel. Finally, mention must be made of USMC LTC Charles A. House of Marine Medium Helicopter Squadron 163, who after his helicopter was shot down at the camp, took command of the surviving U.S. and indigenous troops, including many wounded, and led them in a successful escape and evasion action until they were rescued two miles from the camp on 11 March. For his courage and leadership, LTC House received the Navy Cross, was relieved of command, and received a letter of reprimand for disclosing that American Soldiers had fired on their CIDG troops when the latter panicked and fought with each other as the rescue pickup was underway. According to his commanding officer, LTC House made "some rather emotionally charged statements to authority about the wisdom and futility of the mission."[17] This was a fitting epitaph, perhaps, to the final action of the Special Forces A-Teams and the CIDG in the A Shau Valley.

Allied Command in Denial

Although Captain Blair later held ARVN I Corps (which was operationally responsible for the defense of the camp), III Marine Amphibious Force (the senior US command in I Corps), and 5th Special Forces Group Detachment C-1 responsible for failing to provide adequate reserves for the defense of A Shau, there is some truth in the observation that in the short run the camp lacked adequate space to hold any additional infantry or artillery.[18] Only with much longer notice of the enemy's plans might additional troops have been fruitfully deployed to occupy key terrain features and conduct spoiling operations against the NVA. Yet even those actions, if implemented, might just have played into the NVA concept for the battle. Blair himself had these comments on the NVA's intentions:

It is thought that the attack on A Shau was a 325B Division controlled operation that was intended to be a major battle to destroy A Shau, *to draw friendly reinforcements into isolated terrain organized and fortified by the enemy and on which the friendly forces would be destroyed* [emphasis added], to remove the threat to the 325B Division's line of communications, and to create a psychological victory for the enemy and a corresponding psychological defeat for the friendly forces. The enemy obviously intended to take advantage of an adverse weather period to minimize the effectiveness of friendly air support.[19]

The NVA committed only four battalions of the 325B, although it is suspected that the entire division was available if needed. Captain Davis thought it probable that the third regiment of the 325B was deployed along likely ground reinforcement routes or around potential landing zones to ambush would-be rescuers.[20] Thus, the commitment of a few more U.S., CIDG, or ARVN reserves might have been more likely to increase the magnitude of the defeat than to prevent that outcome.

The fall of Camp A Shau in March 1966 was entirely predictable. Any outcome other than complete VC/NVA supremacy in the valley was unthinkable once the enemy matched its resources to the goal. The allies signaled their proclivity to abandon rather than reinforce camps in the valley with the ARVN withdrawal in December 1965. The cards were stacked so strongly against the Special Forces that one must stand in awe of their dedication in trying to prevail against odds that were just too high. A small force, heavily outnumbered, reliant on aerial supply and reinforcement in an environment where those could be delayed for days by weather, beyond range of supporting artillery or reach of ground reinforcement, with a force made up largely of reluctant, poorly trained and motivated irregular mercenaries, undermined by dozens of turncoats, performing a mission that lacked a key element for success (a friendly local population), Camp A Shau fell.

Shelby Stanton has observed: "The A Shau Valley dominated allied strategy for the duration of the Second Indochina War, but it was swept out of Special Forces existence in less than fifty tragic hours on 9 and 10 March 1966."[21] It would probably be more precise to say that the A Shau played a dominant role in National Liberation Front strategy. As future events would show, adequate allied attention to the threat it represented was absent until the wake-up call in early 1968, by which time that barn door could never be completely closed, nor the horse returned to the stable.

If the fall of A Shau was a harbinger of the North Vietnamese offensives to come in I Corps, resulting from the enemy possession of that critical valley, the lack of planning of *any* Allied higher headquarters for the reinforcement and/or relief of that outpost, or for alternative ground force deployment(s) to counter the loss of the final base there, was emblematic of the absence of unity of command pervading the American and ARVN forces in I Corps. There were three Allied commands with the potential to respond operationally: Military Assistance Command Vietnam (MACV), ARVN I Corps, and the Third Marine Amphibious Force (IIIMAF). None did. The 5th Special Forces Group (5SFG), the only element that took any action at all prior to 9–10 March 1966 (to provide a single company of additional CIDG strikers) lacked authority or responsibility to take any further steps. But it, too, was part of the command friction, as there was no love lost between the Green Berets and the Marines.

It is difficult to see how the American commanding general, who had visited the camp two years earlier and was aware of the waning friendly presence and waxing enemy strength in the valley, could have failed to foresee the battle there in March 1966. But according to one biographer, "Westmoreland was . . . stunned by the fall of the camp at A Shau. Since the camp had represented the only Allied presence in the strategic valley, he saw the attack as a prelude to a larger, more concerted attack on all of northern I Corps, with Hue being the ultimate prize."[22] It would not be the last time that Westmoreland sorely underestimated the enemy's capabilities and intentions. In spite of his "high" level of concern, the A Shau was an American "economy of force" zone for the next two years, and there was not a single allied ground operation (just reconnaissance patrols, air strikes, and Agent Orange) until *after* the feared attack on Hue took place two years later.

4

AFTER THE FALL: PROJECT DELTA AND THE MOBILE GUERRILLA FORCE

Reaction to the Fall of Camp A Shau – Project DELTA Arrives

With the fall of Camp A Shau, the anticommunist forces had been swept from the valley, not to return in force until after the watershed events of Tet 1968. For more than two full years during the height of the conflict, there was no allied unit larger than a company to hinder the development of the A Shau as a major NVA base area and line of communications hub. Throughout this period, a number of smaller American and ARVN units, almost exclusively elite, long-range reconnaissance teams, would be sent into the valley to try to measure enemy infiltration and base development and to identify bombing targets.

During the battle for Camp A Shau, several U.S. units had been alerted to provide possible reinforcement for the camp. One such unit was Team B-52, a special operations component of the 5th Special Forces. In May 1964, the U.S. mission had formed a unit composed of U.S. Special Forces, ARVN Special Forces (LLDB), and CIDG strikers to conduct long-range reconnaissance operations. Under the code name LEEPING LENA, this top-secret group was initially tasked with reconnaissance operations in Laos. LEEPING LENA was an unmitigated disaster. In June 1964, recon teams consisting solely of Vietnamese Special Forces troops were inserted into the Laotian panhandle. They did provide some valuable intelligence but suffered heavy losses while remaining in Laos for much shorter periods of time than planned.[1] After a short appearance, LEEPING LENA was history.

By October 1964, Team B-52, Project DELTA, had changed roles. Unlike (but sometimes confused with) the Studies and Observation Group (MACSOG), which was to conduct its business "over the fence" in Laos and Cambodia, Team B-52 was henceforth to operate only *inside* South Vietnam. Also unlike SOG, which reported directly to MACV and operated from forward operating bases (FOBs) that generally remained fixed, Project DELTA was part of the 5th Special Forces Group and would move its single FOB from place to place as operational assignments warranted.

On 17 March 1965, one year before the battle at Camp A Shau, in one of its earliest operations, three Project DELTA recon teams had entered the A Shau Valley area "with the mission to confirm suspected enemy units, capture enemy personnel, equipment and documents, and to be prepared to direct air strikes on lucrative targets within operational area."[2] This would be a common template for recon team missions for the next six years. One team was inserted near the Rao Nai Valley and operated within a mile of the valley floor. A second team, Team 4, landed on a hill midway between A Shau and Ta Bat and about 800 meters from the A Shau River. It then moved to the northeast, finding numerous old booby traps all along the way until it was exfiltrated in the Rao Nai valley four days later. Team 6 flew into an LZ about two miles southeast of the Ta Bat Airstrip and within 200 meters of Route 548. It spent the next four days searching several kilometers to the northeast before returning to the A Shau Valley and crossing to the west side of Route 548 for pickup on 19 March. Team 6 saw a few enemy observation posts and signs of habitation but encountered no live enemy.

Almost exactly one year later, on 12 March 1966, the elite warriors of B-52, answering the call to support the besieged Camp A Shau, completed their move from II Corps to a new FOB at Phu Bai. Camp A Shau, however, had fallen before the FOB was up and running at its I Corps location. With COMUSMACV concerned about the NVA's future intentions, especially with respect to Hue, DELTA's mission was changed to "assess[ing] the ability and likelihood of the enemy [i.e., the 325th NVA Division in the A Shau Valley] to continue northeast toward Hue and capture the city."[3]

Beginning on 15 March, nine recon teams, five Roadrunner teams (CIDG only), and four company-size search and destroy missions by the ARVN 91st Ranger Battalion were inserted to the east and northeast of the A Shau Valley.[4] Most teams operated some distance from the A Shau, with a concentration in the Song Huu Trach Valley, running north to south about 20 km to the east of the A Shau SF camp. The Intelligence Conclusions must have been reassuring to the MACV commander: "a. That NVA occupy A Shau Valley in multi-battalion strength with anti-aircraft weapons. [this was confirmation, not news!] . . . f. There is no evidence to support the probability that the Hue-Phu Bai complex is endangered by a major VC offensive campaign at this time."[5] What is notable is that from the moment the A Shau camp was lost, there was an awareness of an implied menace to Hue and Phu Bai.

An additional line item in the Intelligence Analysis alerted HQ: "c. A major East-West trail which leads from a highly cultivated area, vicinity coord YD5202, to the road net running to HUE is not being interdicted or surveilled." That location is about 1 km from Route 547 in the Rao Nai Valley. The Rao Nai flows south to north from just east of the southern end of the A Shau Valley until intersecting Route 547, the path to Hue, about 20 km to the north. Fifteen months later, when the next Project DELTA mission was targeted at this area, an enemy concentration was reported in the same "mountain and valley area immediately southwest of the confluence of the Rao Nai and Rao Nho Rivers at YD5203."[6] Despite the concern about a threat to Hue and the identification of a probable key point on the NVA A Shau–Hue line of communication, over a year later, that threat had still not be addressed. Operation 4-66 ended on 28 March 1966.

Operations BLUE STAR and TURNER: The Sequels to the Fall of Camp A Shau

Two months after the fall of A Shau, a special mission was planned to recover the bodies of the Americans, bury the CIDG KIAs, and search any remaining NVA bodies for intelligence information. Four of the SF troops who had been present during the battle were part of Operation BLUE

STAR: CPT Sam Carter, LT Lewis Mari, Wayne Murray, and SFC Vic Underwood. Captain Blair and SFC Adkins requested to be included but were turned down. On 14 May 1966, after aerial reconnaissance indicated that the enemy appeared to have departed the area, three teams, one for the recovery of the Americans, one for the burial of the indigenous troops, and one for security, were helicoptered into the camp. In about 20 minutes on the ground, they recovered the bodies of all Americans except SGT Taylor. No contact was made with the enemy.[7]

Although the NVA had removed much of the equipment abandoned when the camp was evacuated on 10 March, the Special Forces mission to recover the MIAs also revealed that substantial amounts of weapons and ammunition remained at the site. On 1 June, LTG Walt, III MAF CG, "decided to send a Marine demolition team later in the month into the camp to destroy the ammunition." It was suggested that this could be accomplished with air strikes, but General Walt preferred a ground mission. An attack force, including a platoon of Company I, 3/4th Marines, engineers, demolition experts, and a forward air control (FAC) team was assembled.

> After extended rehearsals and one or two postponements, Operation Turner went off smoothly. On 23 June, under cover of Marine fixed-wing aircraft and armed Hueys from VMO-2, six CH-46s from HMM-164 lifted the attack force into A Shau. Arriving at their destination prior to 0700, both groups [a separate team was sent to destroy two pallets of ammo that had landed some distance from the camp] completed their missions and were once more airborne within two hours ... having seen no sign of any enemy force.[8]

The Mobile Guerrilla Force Is Formed: Operation LONGSTREET

According to the U.S. Army history of combat operations in Vietnam, by the fall of 1966 General Westmoreland had begun to consider the "strong possibility" that "the enemy appears to be seeking a Dien Bien Phu in Quang Tri [the province to the north of Thua Thien and adjacent to the DMZ]." At that time, he authorized the Mobile Guerrilla Force operations into the A Shau Valley,[9] as that emerging enemy base was unpropitiously situated deep in the rear of the allied DMZ strong points. Eight months after Project DELTA's mission to gather intelligence on what direction the enemy might take after capturing Camp A Shau, the members of SF Detachment A-100 and their Mobile Guerrilla Force were the next to be tasked with entering the valley for reconnaissance.

In July 1964, the enemy attacked and almost overran the Special Forces/ CIDG camp at Nam Dong, some 20 miles to the east of the A Shau Valley. The commander of SF Team A-726 at the camp, Captain Hugh H.C. Donlon, was awarded the first Medal of Honor of the Vietnam War for his fearless leadership during that action. The Nam Dong battle was similar to the March 1966 Camp A Shau action in two ways: (1) a large number of CIDG irregulars switched sides to the enemy midbattle, and (2) the carbines of the defenders were no match for the AK-47 automatic rifles of their enemies. There were also differences: (1) at Nam Dong, reinforcements did reach the camp during the battle, and (2) the camp was held.

One of the "lessons learned" in the after action report (AAR) for Nam Dong was that the Special Forces higher headquarters (C-Teams: one in each of the four corps areas) needed to have direct control of their own reserves to enable timely employment in support of threatened A-Team camps. It was also important to avoid the confusion that inevitably occurs when multiple headquarters are responsible for units fighting a single battle. From those realizations, by mid-1965, came the establishment of a mobile strike force (MSF or MIKE Force) under the 5th Special Forces C-Team within each corps, and a fifth unit to serve as a general reserve based at Nha Trang. In the summer of 1966, the 5th Special Forces Group further expanded its stable of directly controlled CIDG strikers to include a new mobile guerilla force (MGF). Operationally, the MGF differed from the MIKE Force in two respects. While the MSF companies were under the "joint" command of the South Vietnamese Special Forces (LLDB) and the Americans, there would be no ARVN SF (LLDB) detachments in the guerilla force, with Green Berets and specially trained Australian warrant officers commanding rather than "advising." In addition, while the MIKE Force companies were assigned missions where they would operate in conjunction with local A-Teams and/or in multicompany deployment, the MGF companies were expected to operate independently, with no recourse to reinforcements or ground support.[10]

In September 1966, Detachment A-113, the controlling element for the I Corps MSF, was split and the new Team A-100 was formed, responsible for training and leading the nascent I Corps MGF. The team commander was Captain Alvin Morris.[11] There were problems right from the start.

One would expect a small long-range mission unit destined to operate independently deep in enemy territory for weeks at a time to be commanded by a very experienced leader, to consist of highly trained troops with a high percentage of combat veterans, and to be thoroughly trained *as a unit* in patrolling and ambushing. Not so for the I Corps Mobile Guerrillas. Captain Morris arrived in the country on 15 September and took command of A-100 on 27 September, lacking both in-country and combat experience. The rest of the American team consisted of roughly 50% volunteers and 50% cadre hand chosen by the team executive officer (XO). Only three members, including the XO, had seen combat. Three Australians, all recently assigned to the MIKE Force, joined the 14 Americans on Team A-100. The indigenous personnel included three Rhade rifle platoons, one Koho weapons platoon, and one Koho reconnaissance platoon, a total of 148 men. The Montagnards had been trained only in defensive tactics (to serve as part of the MIKE Force reserve for A-Team camps under attack), lacked patrolling and ambushing skills, and "not more than a handful . . . had been under fire." There were noticeable challenges during the short training period (29 September–24 October 1966):

1) **The Americans, Australians, and Montagnards never trained "as a complete unit."**

2) **Training at Kham Duc was so fraught, with friction between the CIDG and a co-located ARVN training camp, that the final exercise had to be moved to another location after the MGF troops finally moved to the camp airstrip en masse and against orders, refusing to continue training until flown out.**

3) A number (30?) of the Montagnards quit before the first mission was launched.

4) The recon platoon had no training in scouting operations and did not even complete patrol training.

5) There was friction between CPT Morris and his commanding officer, LTC Stein.

6) Overall, it was estimated that about half the necessary training time was lost due to (1) through (5).

7) A final 4–5 day patrol exercise at Gia Duc did appear to be well executed.

And so on 27 October 1966, against the pleas of Captain Morris and his entire team for additional training time, Special Forces Team A-100 and the I Corps Mobile Guerrilla Force set out on Operation BLACKJACK 11/LONGSTREET. In turning down the request for postponement, LTC Stein expressed the opinion that the MGF "were as adequately prepared then to go on the mission as they ever would be."

Insertion into the GWAO (guerrilla warfare area of operations) began smoothly with the recon platoon on 27 October, with the rest of the unit following on the 29th. On 30 October, they set off through the jungle, receiving resupply by helicopter on 2, 6, and 11 November. It became apparent to Captain Morris that the assigned force was so strung out on the trail that he could not effectively manage the column. The 165 men were spread out over 800–1,000 meters, at a standard distance of 5–6 meters between troops. While noise discipline was good, the Montagnards were not focused on the mission and left behind a trail of trash that would cause an urban litterbug to blush.

In the end, though, the weather and starvation rather than the under-training, poor security, or combat inexperience brought BLACK-JACK 11 to a premature halt. In the operational order (OPORD) for

LONGSTREET, there was no discussion of weather or terrain. It was the rainy season in this part of Southeast Asia, and after the third resupply on 11 November, the weather closed down and prevented any further replenishment. Perhaps a patrol commander knowledgeable in the weather patterns of western Thua Thien Province would have been aware of this possibility and had a contingency plan. Captain Morris was not and did not. The troops were expected to carry food for five days, but the Montagnards often went through their rations in three to four days. By 15–16 November they were out of food.

On the morning of 18 November, when they had been without food for two or three days and were in dire need of receiving resupply, the unit had formed a perimeter. Suddenly, two or three VC opened fire, killing two and wounding two of the strikers. SGT Donald P. Smith, a Special Forces medic, rushed to the aid of the wounded and was himself shot and instantly killed. The enemy disappeared without a trace. Medical evacuation for the wounded and dead was requested but was precluded by weather. As a result, one of the wounded Yards died. The rest of the unit, by now beginning to suffer from exhaustion and illness, as well as from very cold, wet weather and from being repeatedly set upon by an unseen enemy, did not want to bury their comrades along the trail, but they were persuaded to do so by a promise to return soon and recover their bodies.

The next day, the same platoon that had been hit on the 18th was at the rear of the column, where it once again received sniper fire. This time, about half the platoon, led by the platoon commander, took off running up the column, away from the VC. None of the Montagnards would follow one of the Special Forces NCOs who urged them to attack the ambushers.

During the last few days in the field, chaos ruled. The recon platoon had moved to a river valley (probably the Rao Nai) and the main body struggled to rejoin them. Straggling, fatigue, hunger, and vomiting imperiled the entire force. Captain Morris requested extraction on 20 November; this was denied. On the 21st, a resupply drop finally took place. On the ground, the guerrillas were beyond control, looting and hoarding food. Captain Morris again asked that the operation be terminated, as the physical condition of the unit was frightful. LTC Stein, for his part, "felt that CPT Morris was 'spooked' and had lost control." Stein sent two of his staff officers to fly over the patrol and talk to Morris. They did so and reported that the commander on the ground was not "spooked" and indeed had control difficulties, because of the failure of all of the short-range radios the unit carried for communication between platoons. Finally, the III MAF operations officer ordered, over pleas from LTC Stein for the mission to continue, that the Mobile Guerrilla Force be withdrawn. This was accomplished from 23–26 November.

The assigned objective for Operation BLACKJACK 11/LONGSTREET was the southern part of the A Shau/A Loui Valley, and that is certainly where the MGF was originally headed. It is not clear from the available after action report that A-100 actually entered the A Shau Valley. The only grid coordinate provided is where SGT Smith and the Montagnard KIAs were buried, YC593850. That location is on the east side of the Rao Nai Valley, which itself is about 6 km east of the A Shau. The recon platoon was 2–3 km from the main body, and as on 20 November a river crossing was made in order for the unit to be joined, it is likely that final extraction was completed near the Rao Nai. It is also surmised that if the unit had been in the A Shau Valley, much stronger VC contact would have been made. On the other hand, if the grid coordinate is incorrect, then LONGSTREET may have operated in the A Shau.

The Mobile Guerrilla Force had failed in its first full-force operation, in what might serve as a notable example of how "friction" can compromise a military plan. The weather may be blamed, but many shortcomings in concept, preparation, and implementation were also revealed.

Mobile Guerrilla Force Combined Reconnaissance Platoon

The diamond in the I Corps MGF coalbin was the combined reconnaissance platoon (CRP). Under the command of 1LT Raymond C. Morris and Platoon Sergeant Henry D. "Hank" Luthy (later Command Sergeant Major of the Special Operations Command), the platoon spent more than a month in the A Shau during the latter part of 1966. Among the enemy sightings reported was the presence of NVA tanks in the valley, an intelligence item "viewed with skepticism" by headquarters.[12]

Mobile Guerrilla Force – Last Operations and Disbanding

In January 1967, the I Corps Mobile Guerrilla Force was reorganized. Team A-100 at this point had only a single member who had served on OPN BLACKJACK 11, Australian Warrant Officer Barry Rust. According to Aaron Gritzmaker, an SF medic assigned to the reconstituted team:

> In late Jan 67 our recon platoon was sent back into A shau to develop targets for the rest of the MGF. They were trapped and surrounded within 2 to 3 hours. They were completely pinned down. . . . the Marines were able to land and get our guys out. . . . The CH-46 that picked our guys up was heavily damaged. The Marines pilots deserved medals for going into one of the hottest LZs I ever saw. We lost several of our Yards on this operation but no Americans.[13]

What happened to that MGF platoon in January 1967 would establish a pattern, repeated many, many times in the ensuing years, as small American reconnaisance units tried to pry open the enemy's secrets in the A Shau Valley.

The next foray by the 5th Special Forces Group into the A Shau Valley occurred in early February 1967. Operation 1-67 (AUSTIN) was able to remain in the northern part of the valley for only two days before being extracted while under heavy contact with a large enemy force. The quarterly *Operational Reports – Lessons Learned* for the 5th SF Group concluded that "the information [AUSTIN] collected added greatly to the overall intelligence picture in I CTZ."[14]* Most of the problems evident in BLACKJACK 11/LONGSTREET were addressed and overcome by the new MGF leadership. Training was much improved. From 31 March till 19 April 1967, the next MGF operation, BLACKJACK 12/OCONEE,

---

* I have been unable to identify the Special Forces unit assigned to Operation 1-67; it is possible that this refers to the MGF recon platoon action in January.

was directed to the vicinity of the former Special Forces camp at Nam Dong. This time, the reconnnaissance-in-force mission was carrried out and led to repeated and heavy contact with the enemy. An Australian WO, John Stone, was KIA and there were several wounded, but over 60 VC were killed.[15]

The third and final operation of the I Corps MGF, BLACKJACK 13, moved the GWOA back to the A Shau Valley. The mission was to be another reconnaissance-in-force, scheduled for 25 May–11 June 1967 (this period overlapped Operation PIROUS and must have been directed to an area of the A Shau Valley not assigned to Project DELTA's AO), and the troops involved were esentiallly the same as those who had participated in BLACKJACK 12. It was over in a matter of hours. Dropped off by truck some distance from the valley, the unit began moving toward the objective after dark had fallen. Around midnight, the point triggered a booby trap, wounding two strikers and Green Beret Jerry Knight. The unit formed a perimeter and waited till morning for medevac. One of the wounded CIDG soldiers died during the night. In the morning, the request for medical evacuation was denied, and the unit was ordered to bury the dead and continue the mission, carrying the other wounded striker, who was unable to walk. The CIDG troops balked, refusing to continue until their compatriot was evacuated. The plan changed. One platoon was to take the wounded and the body of the dead trooper back to the truck drop-off point, where the trucks would return to meet them.

While heading back down the trail they had traversed the previous day (never an action recommended for infantry in Vietnam), the platoon was ambushed. In the battle that followed, another Special Forces Soldier, Aaron Gritzmaker, was severly wounded, and several more Montagnards were also wounded, one of whom later died. An LZ was evenually secured and medical evacuation completed. The patrol commander decided to return to the drop-off point and recommence the operation via an alternate route into the A Shau. The Yards said no. Additional American officers arrived on the scene to try to persuade the Rhade troops to continue. They still refused, and the operation ended on the afternoon of 27 May. The CIDG units were soon disbanded.[16]

After one additional failed operation in the Nam Dong vicinity, Team A-100 would be folded back into A-113 and and the I Corps Mobile Guerrilla Force would cease to exist. The next American venture into the valley would be the largest and most successful ground reconnaissance operation launched into the A Shau Valley area during the Vietnam War.

5

BETTER KILLING THROUGH CHEMISTRY

Project POPEYE

On 18 March 1971, Jack Anderson's nationally syndicated column carried the subtitle "Rainmakers Proving Success Over Ho Chi Minh Trail."[1] Thus, four years to the day after the beginning of its operational phase, the "Top Secret Special Category" rainmaking program in Southeast Asia was revealed to the American public for the first time. Anderson gave the code name "Intermediary Compatriot" for the program, although that had long since been changed, after possible security leaks, to "POPEYE." The newspaper report was denied by the Pentagon at that time and would continue to be disclaimed for more than a year. Indeed, as late as April 1972, in an appearance before the Senate Foreign Relations Committee, then Secretary of Defense Melvin Laird stated: "We have never engaged in that type of activity over North Vietnam."[2] When he did finally 'fess up in January 1974, he claimed that the requisite information "was not available to [him]" at the time of his earlier testimony.[3]

On 3 July 1972, a story in the *New York Times* by investigative reporter Seymour Hersh provided new details on the use of cloud seeding to modify weather patterns during the Vietnam War. According to his source(s), the earliest experimental application of rainmaking in the war zone occurred in the early 1960s and was conducted by the CIA.

According to interviews, the Central Intelligence Agency initiated the use of cloud-seeding over Hue

. . . "when the Diem regime was having all that trouble with the Buddhists. They would just stand around during demonstrations when the police threw tear gas at them, but we noticed that when the rains came they wouldn't stay on," the former [CIA] agent said. "The agency got an Air America Beechcraft and had it rigged up with silver iodide," he said. "There was another demonstration and we seeded the area. It rained." A similar cloud-seeding was carried out by C.I.A. aircraft in Saigon at least once during the summer of 1964, the former agent said.[4]

Hersh noted that the CIA's 1963 experiment, which likely occurred during the period 13–16 August, at a time when the Diem regime was severely threatened by Buddhist demonstrations, including the public self-immolation of monks, was "the first confirmed use of meteorological warfare [by any nation]."[5] This was not, in reality, "warfare" use, but for the purpose of controlling political and religious opposition to the Catholic Diem. It was, however, an omen of future weather "militarization" in Laos, North Vietnam, and one small area of South Vietnam – the A Shau Valley.*

By the mid-1960s, the United States government had for some years been studying the use of cloud seeding to stimulate rainfall, with tests on the modification of hurricanes to perhaps reduce their impacts by altering the cloud structure near the eye of the storm (Project STORM-FURY). The experiments also included, by 1964, research into possible military uses of silver iodide smoke in rainmaking. These military trials were conducted at the secret Naval Ordnance Test Station at China Lake, California. The work at China Lake was then transferred for a trial under actual war zone conditions in Southeast Asia in 1966. Secretary of Defense Robert McNamara said later (18 March 1968) that the tests at China Lake had shown weather modification to be "a potentially powerful weapon against lines of communication in Southeast Asia." And so, on to Laos.[6]

The first full-scale experimental weather modification trial in the theater of war, under the codename Project POPEYE, and with supervision

---

* Project POPEYE was terminated permanently two days after the appearance of the Hersh article.

by personnel from the China Lake facility, was conducted in great secrecy in the Se Kong river valley in the eastern panhandle of Laos in October 1966 by the U.S. Marine all-weather fighter squadron VMFA-115, operating out of Da Nang air base. Using F-4B Phantom aircraft equipped with a modified photoflash ejector, three specially trained crews (pilot and radar intercept officer) dispensed silver iodide flares, called "Wimpy" in keeping with the POPEYE theme. Forty-eight of the 56 missions were deemed successful.† One of the flights seeded clouds that drifted over the border into South Vietnam while continuing to dump torrential showers, and a Special Forces A-Team camp (probably Kham Duc) reported nine inches of rain in four hours.[7] The trial was deemed a success, leading to subsequent implementation at the operational level under the modified codename Operation POPEYE.[8]

What did the U.S. hope to accomplish with POPEYE? On 20 March 1974, a full seven years after implementation and three years after it was first exposed by Jack Anderson, the United States Senate Subcommittee on Oceans and Environment of the Committee on Foreign Relations held a top-secret hearing on "Weather Modification." The Department of Defense presented a report titled *SEASIA RAINMAKING*. This report gave the objective of the operation as

increas[ing] rainfall sufficiently in carefully selected areas to deny the enemy the use of roads by:

(1) Softening road surfaces
(2) Causing landslides along roadways
(3) Washing out river crossings
(4) Maintaining saturated soil conditions beyond the normal time span[9]

The operation was approved on 25 February 1967 by the secretary of defense[10] and began on 20 March 1967. It was conducted annually during the southwest monsoon season, from March through November, with the

---

† The streams in the A Shau Valley are a part of the Se Kong watershed, and the test area was just over the mountains from South Vietnam.

goal of extending the worst effects of the rains on road traffic on the Ho Chi Minh Trail. Initially permitted only over the eastern side of the Laotian panhandle and a small area in North Vietnam, the target area over the A Shau Valley was added on 13 September 1967.[11] *The A Shau was the only part of South Vietnam included in the program until some additional areas along the South Vietnamese border with Cambodia and Laos were included in 1972*, during the final few months before POPEYE was terminated on 5 July 1972.

The DoD report to the Senate went on to list, by year, the sorties flown and units of rainmaking chemicals expended, with a total of 2.602 sorties and 47,408 units for the entire war. Selection of targets depended on their strategic role in NVA lines of communication and the likelihood that increased rainfall could reduce enemy road traffic. Target areas were designated drainage basins, and "it was usually possible to seed every suitable cloud within a drainage basin, but priority was given to seeding clouds directly over roads, intersections, and river crossings within each basin."[12] The Seventh Air Force at Tan Son Nhut AFB was the controlling headquarters.

Operation POPEYE used three WC-130 aircraft from the 54th Weather Reconnaissance Squadron and two RF-4C aircraft stationed at Udorn, Thailand, with crews rotating from duty on Guam. The 54th WRS used the code name MOTORPOOL for POPEYE operations and adopted the slogan "make mud, not war."[13] Apparently, there was an early success in the A Shau Valley, even before that area was added to the target list in September 1967. A memo to President Johnson from Walt Rostow on 10 July 1967 stated:

> There was probably substance behind your instinct about landslides.
> The incident, you will recall, concerned landslides in the A Shau valley trapping enemy forces which were then heavily bombed by B-52s.
> In the week of June 16–22, the area [Route 922 in Laos, just west of the A Shau] was seeded, which may well have contributed to the result.[14]

In reporting the results of the project, and with an ambiguity that all too often typified commanders' after-action evaluations of the American war effort, it was stated: "The results of the project cannot be precisely quantified." The best the Air Force could come up with was a Defense Intelligence Agency estimate "that rainfall was increased in limited areas up to 30%" and that "it is believed that this . . . did contribute to slowing the flow of supplies into South Vietnam along the Ho Chi Minh trail." Hardly a ringing endorsement. At the conclusion of his presentation to the Senate subcommittee, the DoD representative, LTC Ed Soyster from the Joint Chiefs of Staff, concluded: "While this program had an effect on the primitive road conditions in these areas the results were certainly limited and unverifiable. It was conducted because of its apparent contribution to the interdiction mission and the relatively low program costs." Additional remarks on the impact of the cloud seeding came from Dennis J. Doolin, Deputy Assistant Secretary for Defense (East Asia and Pacific Affairs). Mr. Doolin, having viewed the material provide by the JCS representative, opined, "It looks to me like when you are getting 21 inches in a given area, and we add 2 inches, if I was on the bottom, I do not think I would know the difference between 21 and 23."[15] The hearing transcript was made public on 19 May 1974.

Senator Claiborne Pell, chairman of the subcommittee that held the March 1974 hearing on "Weather Modification," was curious as to why POPEYE was so highly classified. "I know in my own experience here that this particular program was the only program about which the DOD did not feel able to respond to questions in either public or private session."[16] Dennis J. Doolin responded that the secrecy was likely due to the "perceived sensitivity of the operation" related to the possibility that a friendly country, Thailand, might have been deprived of water for agriculture, while denying that this was in fact the case.[17]

Project POPEYE was, truth be told, the first known operational use of weather modification for military purposes by any country. An outcome of the exposure of top-secret POPEYE was a prolonged national and international debate over the morality and "legality" of playing with Mother Nature. At the core of this debate is the inevitable fact that altering the weather in one location will certainly cause changes elsewhere. It is analogous to altering the flow of a river; what brings about fertile soil in one area may lead to persistent parched soil in another. By 1974, the United States Congress had banned the use of weather control for military purposes, and an international treaty, the Environmental Modification Convention, followed in 1977, although this document has yet to face a true "trial by fire."[18]

COMMANDO LAVA

In an effort to deter enemy movement on remote mountain roads, Dr. [William G.] McMillan [Westy's scientific adviser] and his scientists tried an ingenious gimmick. They found a chemical solvent that when mixed with water and soil would turn the soil to slush, which as long as it was wet would not stabilize. At the start of the rainy season, they dumped tons of the solvent from C-130s on a constricted road in the A Shau Valley, but no substantial evidence was ever found that it proved effective in deterring movement.[19]

Not long after the Project POPEYE trials in Laos commenced in March 1967, another experimental use of chemicals to inhibit vehicular travel on the Ho Chi Minh Trail was also being investigated in the southern part of that country. Researchers at Dow Chemical (the good folks who also made millions from their "Agent" products during the Vietnam War), had created a product called Calgon, a 50-50 mixture of trisodium nitrilo-triacetic acid and sodium tripolyphosphate that they believed could be applied directly to the soil, resulting, when the rain fell, in a gooey quagmire that would render vehicular travel impossible. It was estimated that about 20 tons of Calgon, at 1 pound per square yard, would be needed for the average choke point. "A C-130 aircraft began the experiments in Laos on 17 May, free dropping the chemicals (packed in palletized bags) on chokepoints along routes 92 and 110. Ten days later, both chokepoints appeared impassable to traffic. A sloping segment on route 110 oozed downhill like lava, inspiring the project's code name."[20]

The initial test results elated Ambassador Sullivan so much that on 27 May he had his "staff give Generals Westmoreland and Momyer a briefing on operation Commando Lava." In a follow-up telegram from his embassy to the Department of State, Sullivan's enthusiasm was unbounded. Excerpts:

2. . . . mud loses all consistency and becomes incapable of supporting vehicles or any other substantial weight. This not only vastly enhances nature's own mud making in quality, but also extends the effect of the rainy season because of the chemicals' persistency.

3. We believe this could prove a far more effective road interdiction device (at least in the rainy season) than iron bombs and infinitely less costly. . . .

7. I also feel that, if we could combine these techniques with techniques of Operation Popeye, perhaps within the concept of Practice Nine, we might be able to make enemy movement among the cordillera of the Annamite chain almost prohibitive. In short, chelation may prove better than escalation. Make mud, not war![21]

On 16 June, CG III MAF was studying what type of anti–infiltration methods could be used as part of the PRACTICE NINE system in the flood plain east of Gio Linh. "He . . . recommended installation of the SPOS [Strong Point Obstacle System] to each flank of the flood plain, interdicting forward of the trace with COMMANDO LAVA and observed artillery fire. . . ."[22]

With the endorsement of Sullivan, Westmoreland, Admiral Sharp (CINCPAC), the Joint Chiefs of Staff, and Secretary of Defense McNamara, Operation Plan 500-67, COMMANDO LAVA II, moved forward. In June, one choke point each in Laos and the A Shau Valley were targeted for the next round of tests. "Two Seventh Air Force C–130s flew their first chemical drops on July 20 and 21, 1967, along segments of Route 548 in the A Shau Valley. The airdrops succeeded even though small-arms fire struck both planes. The next day, three C–130s spread chemicals on a 200-yard segment of the same route." Several subsequent missions were flown in July and August, but "aerial observation showed that the chemical compounds did little to slow or halt communist traffic. The DRV still used the roads and made no unusual efforts to bypass seeded sectors. Deep rutting on four road segments was seen, but was corrected by road-repair crews. They cover the ruts with logs and bamboo matting. . . ."

The tests were abandoned for good on 21 October 1967. The final report on COMMANDO LAVA confirmed chelation, "but there was scant evidence of mud or mudslides, and enemy traffic was not impeded. The report blamed the route's soil texture."[23]

COMMANDO LAVA was never implemented.[24‡]

Project POPEYE and Operation COMMANDO LAVA were unique attempts to use science to close down the Ho Chi Minh Trail. They were also similar in that the A Shau Valley was the only site in South Vietnam where use was attempted (except for the last couple of months of POPEYE in 1972). There are three reasons why these experiments were limited to the valley. First, the Route 548 corridor, with its connections to Base Areas 607 and 611 in Laos and Routes 547 and 614, leading to Hue and Da Nang, was the most important crossroads and cache area of the Ho Chi Minh Trail in South Vietnam. Second, the remoteness of the valley made ground interdiction problematic and propelled the search for alternatives. Third, in 1966 and 1967, COMUSMACV knew that allied ground troops could not and/or would not be deployed into the A Shau, and the hoped-for successes of the rain- and mud-making would impact only *enemy* movement.

RANCH HAND – 1961–1964

The history of herbicide use by the United States during the Vietnam War is still being lived. The story of the employment of Agent Orange and other chemical defoliants during the war by the United States Air Force is told in *Operation Ranch Hand: The Air Force and Herbicides in Southeast Asia, 1961–1971.* In this chapter, the use, effectiveness, and impact of defoliation in the A Shau Valley during the war years will be discussed. The long-term effects of Agent Orange et al. are beyond the scope of study.

---

‡   LTG Cushman, former CG III MAF, had a somewhat different recollection of the effectiveness of COMMANDO LAVA in a 1982 oral history interview: "Just at that time [1967], a chemical was developed in conjunction with rain that was supposed to turn dirt into impassable mud. It actually worked to some extent and we really plastered the A Shau Valley with that. We had pictures of water stretching all the way across that valley. So, it slowed down the enemy and they had a difficult time with it."

The first application of defoliants in Vietnam was delivered by a Vietnamese Air Force C-47 on 29 December 1961.[25] The initial American missions began on 10–16 January 1962 near Vung Tau, east of Saigon. The program, code named RANCH HAND, expanded slowly on a test basis in the early months of 1962. On 18 July, General Harkins requested, among other new targets, that "a planned road construction project between the two outposts of A Shau and A Luoi in the later famous A Shau Valley" be approved. This request rattled around between Admiral Felt (CINCPAC) and the Joint Chiefs for several months, after initially being turned down by the admiral due to proximity to the Laotian border (fear that the chemicals might drift across the border and damage crops). On 3 October, a more emphatic request and justification was submitted, noting the importance of the A Shau road for patrols along an infiltration route from North Vietnam. This time around, the Joint Chiefs endorsed the request and forwarded it to Secretary of Defense McNamara. After further bureaucratic and political maneuvers, on 30 November 1962, President Kennedy accepted the proposal that A Shau and four other target areas be approved for treatment with herbicides.

In the proposal for expansion of the target list, "General Harkins had rated [defoliation] as 90% to 95% effective against mangrove forests, and 60% effective against tropical scrub. Then, he told the President that U.S. advisors located in the vicinity of spray operations had reported no reaction from the local population." President Kennedy delegated the authority to approve future herbicide targets to Ambassador Nolting and General Harkins, but that authority specifically excluded crop destruction operations. A few of the newly approved target areas, not including the A Shau Valley, were sprayed in December before the fixed-wing operation entered a pause until June of the following year.[26] The first herbicide spraying in the vicinity of A Shau was on 16 February 1963, when a helicopter dispensed 62 gallons of Agent Purple over Be Luong in the nearby Rao Nai Valley. Five additional helicopter Agent Purple missions in the Rao Nai followed, from 18–24 February 1963.[27]§ For unknown reasons, but likely having to do with a perceived drop in the priority of defoliation in the valley

---

§ Interestingly, this is the same time period when CPT Powell was stationed in A Shau.

compared to other targets, no action was taken to spray defoliants in the A Shau until over a year later. We pick up the story in 1964.

The Agent Orange Data Warehouse contains a wealth of information on Operation RANCH HAND, including flight dates and tracks, type and gallons of herbicides, and, in many cases, documentation of the mission history and after action reports on defoliation programs. The following section relies on that source. The first authorization of defoliation missions in the A Shau Valley begins with the *Psywar and Civil Affairs Annex to Plan 2T–20T* of the 11th Division Tactical Zone (Nam Hoa District, covering the entire western portion of Thua Thien Province) in January 1964. In response to increased VC activity, the Psywar and Civil Affair mission included "provid[ing] mutual support for the defoliation and destruction of enemy crops to cut off enemy supply sources among the people and destroy enemy installations." The Psywar concept of execution was to "use propaganda to explain to the people the usefulness of the defoliation and settlement of Highlanders [Montagnards]."(!?) The "propaganda topics" that were to be used specifically are as follows:

With Highlanders:

1) **If you live with the Communists, you will suffer misery and diseases. If you return to the Government, you will be well off.**

2) **The government and Armed Forces always help you to live in peace.**

3) **Enumerate VC crimes – forcing Highlanders to clear abandoned land, to pay taxes to them, and to enlist in their army.**

4) **Cooperate with the armed forces to destroy VC.**

5) **Tell them how happy those Highlanders are who have resettled in the resettlement areas of ASHAU and TABAT.**[28]

You can't make this stuff up.

Towards the goal of resettlement, a number of "resettlement areas" were to be established, including three in the valley: A Luoi (75 hectares, capacity 350), Ta Bat (35 hectares, capacity 200), and A Shau (50 hectares, capacity 250). There is no evidence that any consideration was given to the proximity of the "resettlement areas," where farming would be necessary, to the defoliation flight paths. Yet the center of the A Shau site (YC525815) would be just 300 meters from the termination point for spraying (YC522815), and the Ta Bat site (YC418942) was almost directly under the spray zone.

On 7 April, a meeting was held to discuss proposed defoliation targets in I Corps. The I Corps Senior U.S. Advisor and the ARVN I Corps Commander's representative agreed that the A Shau was "priority 1." The following day, a reconnaissance flight in a C-123 took place along the proposed target corridor. Participating were LTC Peter Olenchuk, MACV J3 staff, CPT Hagerty and crew, RANCH HAND Commander, and two I Corps chemical advisors. As a result, LTC Olenchuk, Chief of the MACV Chemical Operations Section, recommended that the "defoliation of the A Luoi-A Shau Valley by C-123 spray aircraft be approved for execution upon receipt of the JGS [Joint General Staff] request." The recon flight observed: "The habitation is essentially confined to the three outposts; although judging from the trails transients cross the area from the nearby hillsides where slash/burn areas would indicate some habitation." There is no indication of whether or not the aerial observers were aware of the planned resettlement areas.[29] The province chief endorsed the request on 11 April 1964, including a required "Province pledge to pay for any damages caused to civilian crops and property by the spraying of defoliants on the objective areas."[30]

It is important to understand the 1964 RANCH HAND missions in the A Shau Valley in the context of crop destruction and resettlement of villagers and not solely as an attempt to defoliate areas where the enemy might hide camps or from where they might ambush allied troops. There also appears to have been the assumption that the Pacoh were an incredibly gullible tribe that would be easily persuaded to concentrate in slums adjacent to allied installations. As discussed elsewhere, the concept of Special Forces border surveillance camps embraced the need for a local populace to provide both manpower and intelligence. However, by mid-1964 it was evident that the few remaining Montagnards in the vicinity of A Shau would never ally themselves with the government. Thus, the Agent Purple spraying merely helped to finalize the depopulation of the valley that had commenced much earlier with search and destroy missions such as that of CPT Powell and his advisees, and so Special Forces support for defoliation is not remarkable. I find it interesting, nonetheless, that there is no evidence of a single representative of Special Forces command participating in the decision-making process regarding the A Shau.

On 30 April, COL Leroy Collins, senior advisor to the ARVN 1st Infantry Division, added his endorsement to the bureaucratic back story of defoliation in the A Shau, apparently dismissing out-of-hand concerns about the local population:

2. Due to increased VC activities against aircraft utilizing the airstrips in the valley, I feel rapid execution of the defoliation plan is of the highest priority. . . .

5 . . . In the past the Psy/War and Civic Action backup plans have hindered the execution of defoliant operations. These plans will not affect the valley operations *because the area is uninhabited* [emphasis added].

6. Request that portions of the 20P&T Plan pertaining to the A Shau A Luoi valley be executed as soon as possible.[31]

COL Collins's mention of "increased VC activities against aircraft" is undoubtedly a reference to the incident at A Shau involving Deputy MACV Commander Westmoreland's CV-2 Caribou on 24 April (see chapter 1). Indeed, that incident was directly addressed at a meeting of the MACV 203 Committee on Defoliation Target 20-24 (the I Corps target list) on 4 May. The MACV 203 Committee, chaired by LTC Olenchuk, also recommended approval of the defoliation. It noted:

Habitation primarily at the 3 outposts. No crops except at outposts (approximately 5–10 hectares) which belong to the indigenous personnel settled around the outposts. These hamlet personnel are currently under 60% subsistence and should damage to their crops occur (a likely possibility in view of RVNAF and U.S. advisors' strong desires to defoliate the nearby shrubs), complete restitution and continued substance at 100% level will and can be provided by the U.S. Special Forces and Provincial authorities.[32]

The next day (5 May), MACJ315 Summary of Target Request 20–24 included the following comments:

The area is a valley that is essentially uninhabited except for the RVNAF outposts at A Luoi, Ta Bat, and A Shau. . . . [T]he need for additional chemicals should be determined approximately 2–3 weeks after the 3 gallon per acre spraying has been completed.

. . . Leaflets will be distributed to the people in the area to explain the harmlessness of the defoliants to the people and animals.

According to the 1965 Summary of Target Request 2–24, however, the Pacoh "do not have a written language," so leaflets could only be distributed in Vietnamese, which the Montagnards could not read! The chemical deemed "harmless to people and animals" was Agent Purple, which would be used in the 1964 spraying and "contained 16 times the mean dioxin content of formulations [e.g. Agent Orange] used during 1965–1970."[33]

On 16 May, citing a letter from BG Nguyen Van Thieu, ARVN chief of staff and future president, the ARVN added its endorsement, also noting the Westmoreland aircraft attack:[34]

It was at A Shau on April 24 that the plane carrying General Westmoreland . . . was hit. The purpose of the spraying will be to increase visibility along the road and at the airfield in order to increase the security of operating in and controlling the area. There are also plans to resettle Montagnards along the road at some later date when security has been improved. If any Montagnards should come over to the government side at the time of the spraying, there are adequate facilities to resettle and provide for them.

Leaflets will be distributed explaining the purpose of the spraying and its harmlessness to humans and animals [sic]. In order to obtain maximum effectiveness, it has been requested that the posts and airfields themselves be sprayed. This will entail the deliberate spraying of about 10 hectares of crops at A Luoi. Provision has been made, however, for complete reimbursement of this minor acreage and continuous subsistence for the Montagnards affected if necessary. U.S. Special Forces officers have confirmed that Special Forces funds and resources will be used. These Montagnards will be informed in advance of the reason for the spraying and that they will be provided with sufficient food to make up for their loss.

The Vietnamese endorsement reveals that the financial burden of supporting the locals had been moved from the province chief to Special Forces (all American funds, in any case), and the Montagnards would now be entirely dependent for their subsistence on the local Special Forces team and its tenuous air supply system.

Finally, on 24 May 1964, BG William Depuy, MACV assistant chief of staff, J-3, issued a message "to execute the defoliation of ASHAU – ALUOI road axis . . . from 27 May to 1 July 1964. . . ." The target area (Target 20-24) was to include 200 meters on each side of the road, as well as around the airfields and camps.

The first defoliation mission in the A Shau Valley was flown on 30 May 1964 by three C-123 aircraft. Two more missions were flown on the same path on 31 May and 1 June. The flight on 31 May reported receiving .30-cal fire and 2 hits. A total of 10 aircraft dispensed 9,375 gallons of Agent Purple during these missions, which flew a path from the A Luoi vicinity to A Shau.

The results of the 1964 spraying operation are documented in the Report on Herbicide Operations issued by the I Corps Army Advisory Group on 22 July 1964:

2. Aerial observation indicates that the spraying was very effective in killing the vegetation along the road. Ground observation, however, has shown that a second canopy exists in the target area. This canopy was

protected by the top canopy and will require [a] second spraying to kill all the vegetation. Visibility along [the] road has not been improved. This is due to the fact that although the top canopy has been killed, the leaves have not yet dropped off the trees.

3. Camp A Shau and its airstrip are considered to have been completed. Although there remain some small areas to be sprayed, these can be accomplished by hand spray. While the number of incidents has remained unchanged, the camp has improved its security. By killing vegetation around the camp, the field of observation and fields of fire have been greatly improved.

4. Camps Ta Bat and A Luoi did not meet with the same success as Camp A Shau. These two areas were sprayed again on 30 June and a report will be forwarded at a later date.

5. The only crops in the area belonged to a small settlement at Camp A Shau. Before the spray mission, this settlement received 50% of its provisions from Camp A Shau. Since the spray, Camp A Shau has provided the settlement all its provisions.[35]

The three sorties on 30 June marked the completion of the 1964 herbicide program in the valley. The contradictory language so often seen in Vietnam-era military reporting is evident here. The spraying was deemed "effective," although visibility along the road was not improved, and while fields of observation and fire around the A Shau camp improved security, "the number of incidents [of enemy fire] remained unchanged."

RANCH HAND – 1965–1969

The next appearance of the C-123s of Operation RANCH HAND in the valley took place from 7–12 September 1965, under the code name TRAIL DUST IV.** This group of herbicide flights was in response to the intelligence in August 1965 that two VC battalions were training for

---

** The TRAIL DUST code name would continue to be used until RANCH HAND operations ceased in 1971.

attacks on allied bases in the valley. "The requested defoliation includes an area with a radius of 1000 meters around each camp and an area extending 500 meters on each side of the road connecting the two camps [A Luoi and A Shau]. These requests are in effect a re-initiation of requests which were approved in May 1964 as Target 20-24. . . ."[36] Five crop destruction missions using a total of 14 sorties were flown, including three missions in the northern part of the valley from A Luoi to east of Tiger Mountain. Agent Purple had been replaced at the start of 1965 by Agent Orange, which was less volatile and also less toxic.

A typical spray application was made by three C-123s flying at an altitude of 150 feet and at a speed of 130–150 knots, covering an area 14 kilometers long and 240 meters wide.[37] Beginning in 1965, there were three primary herbicides in use. Agent Orange, the most prevalent, included the chemical compound 2,4,5-T, of which 2,3,7,8-tetrachlorodibenzo-p-dioxin is a contaminant. "Orange was the general-purpose herbicide for defoliation and crop destruction, with leaf fall in three to six weeks and control persisting for seven to twelve months." According to a 1967 Project CHECO report, "The chemical did not poison the soil, and as a result regrowth could occur in the form of new grass and scrub after a period of four to six months."[38]

Agent Blue, which did not contain 2,4,5-T "was used for crop destruction missions; it was the agent of choice for destruction of rice crops." It produced rapid defoliation, "causing browning or discoloration in one day, with maximum desiccation and leaf fall occurring within two to four weeks."

Agent White also did not contain 2,4,5-T. It was used for defoliation and "was most often used in areas where longer persistence rather than immediate defoliation was desired. . . . [it was] effective principally on broadleaf herbaceous and woody plants . . ., and full defoliation did not normally occur for several months."[39]

In February 1974, the National Academy of Sciences–National Research Council issued a report using aerial photographs and other sources titled "The Effects of Herbicides in South Vietnam." "Part B, Working Papers: Economic Stress and Settlement Changes" looked at 25 areas within the

Republic of Vietnam, including Study Area 16, the A Shau Valley. The following excerpt of that study (pages 49–50) summarizes the impact of herbicides through early 1970 (the last fixed-wing defoliation flight occurred on 12 February 1971; helicopter operations ceased on 31 October 1971):

In 1958, the territory from the A-Shau valley to the Laos border was almost completely speckled with clearings and the scars of formal clearings that denote swiddens. These were the basis for subsistence cultivation practiced by the Pacoh, a group of non-Vietnamese Highlanders. Clearing of vegetation throughout the valley and its margin has been going on for a long time to result in such extensive removal of the forest cover. Dr. Gerald Hickey visited the valley in 1964 and reported most of the villages gone; Pacoh refugees subsequently were reported to have moved to the vicinity of Hue, close to the tomb of Emperor Tu-Duc. In 1966 five hamlets were located at distant intervals in the central A-Shau Valley, with a total population of not more than 210. Between September 1965 and May 1966 nine crop destruction missions and three defoliation missions, all 12 with Agent Orange, are reported. A defoliation line is evident on the eastern margin of the valley in the January 1966 photos. Between August 1966 and April 1970, an intensive defoliation campaign [was] conducted (all but 21 of the 110 spray runs used Agent Orange), with over half of the spraying being done in 1969. Prior to June 1969, but continuing into 1970, very intensive carpet bombing of the area also occurred, destroying both airstrips that had been built in the valley. By June 1968, the area of new swiddens in the A Shau valley was nil and the area between the valley and the Laos border only one-tenth that of 1958. By 1969, all cultivation essentially had ceased. The combined effects of massive bombing and repeated spraying, together with ground action, had rendered the area untenable for its indigenous inhabitants. All of the hamlets occupied in 1966 were in the path of the spraying runs and by 1969 no longer existed. To the west and south, outside of the zone of greatest destruction owing to warfare, there appears to be an increase in the number of new swiddens; apparently those Highlanders who survived have shifted not only their cultivated fields but their homes as well.[40]

The report (Table III) listed a total of 121 spray sorties, with 100 Agent Orange, 10 Agent White, and 11 Agent Blue. Of 53 original settlements in the valley (Table IV), only 2 were never sprayed and 31 were sprayed four or more times. The report concluded (page 59): "Non-Vietnamese mountain people were the groups most seriously affected by the spraying. Since their swidden economy, being dependent on the forest as a renewable reservoir of plant nutrients, is the most fragile, they suffered the most by crop destruction and forest defoliation. In addition, once displaced from their established territories, these people have no viable alternative skills."

While the impact on the indigenous population is clear (it was driven from the valley, as intended), the effects of defoliation and crop destruction damages on the enemy were mixed. Below is a sampling of after action reports and other postdefoliation observations:

1) **Operation TRAIL DUST IV (sprayings in September 1965; AAR dated 17 November 65): While "coverage of cultivated areas [in use by enemy] . . . was good, . . . all areas are either replanted or being prepared for new crops."**

2) **For Project 1/20/3/66, the 1966 defoliation plan, it was noted: "Use of agent ORANGE during February 1966 in the immediate area of A Shau Special Forces camp resulted in the defoliation of trees and brush but produced no effect on the elephant grass. The predominant elephant grass areas are in the vicinity of the three old Special Forces camps of A Shau, A Luoi and Ta Bat."†† (Agent Blue would be more effective against elephant grass.)**

3) **During Operation DELAWARE (April–May 1968), the attempt by 5/7 Cavalry to flank the NVA line of resistance west of Tiger Mountain was stopped because the area**

†† After the fall of Camp A Shau in March 1966, the entire valley was considered to be a "VC cultivation area," similar in intent to a "free fire zone" for herbicide spraying.

between the American position and the enemy on higher ground had been defoliated, giving the NVA clear fields of fire.

4) On the other side of Tiger Mountain, the NVA bypass route was later partially exposed by defoliation.

5) On 26 August 1968, a "Herbicide Operation Evaluation" referring to Operation SOMERSET PLAIN noted, "Defoliation proved extremely effective in permitting increased surveillance of enemy infiltration routes and LOC's such as Route 547 out of the A Shau Valley."[41]

As with air power, the use of herbicides in the A Shau Valley does not readily lend itself to either a "pass/fail" grade or rating on any single scale. The early herbicide use did abet the exodus of the last Pacoh tribesmen from the valley (a dubious goal), but while VC/NVA crops were eliminated, the enemy never ceased to replant his own crop areas that were hit. Defoliation was useful in some areas and at some times, but I know of no cache or major base area that was exposed first by herbicides; they had to be found by ground or aerial reconnaissance. Allied forces continued to be ambushed by enemy units hidden in the jungle, and the enemy continued to build roads concealed by foliage for the duration of the war.

6

OPERATION PIROUS

Strategic Differences – COMUSMACV and III MAF, 1966–1967

The Project DELTA reconnaissance immediately after the fall of the A Shau SF Camp in March 1966 and the abortive Mobile Guerrilla Force operations in 1966–67 took place in an atmosphere of continuing tension between the overall war strategy of General Westmoreland and the priorities in the I Corps area as implemented by the Third Marine Amphibious Force (III MAF) commander, LTG Lewis Walt (through 31 May 1967) and his successor, LTG Cushman. GEN Westmoreland was on the strategic defensive but believed that the key to eventual assumption of an offensive strategy was to employ his U.S. infantry battalions to engage the enemy main forces vigorously and wear them down to a point where they could no longer replace their losses. This "strategy," really a tactic known as "search and destroy," meant using the American units to "locate, engage, and destroy the North Vietnamese and Viet Cong main-force units."[1] In practice, this usually meant dangling his infantry as bait to "locate" the enemy by being ambushed or attacked in fixed positions (firebases or night defensive positions [NDPs]), then applying air power, helicopter gunships, and artillery to assure a high "kill ratio."

Harry Summers, in his classic analysis of the U.S. in Vietnam, *On Strategy*, wrote:

> We thought we were pursuing a new strategy called counterinsurgency, but actually we were pursuing a defensive strategy in pursuit of a negative aim—a strategy familiar to Clausewitz in the early nineteenth century. In his chapter on purpose and means in war, Clausewitz discusses various

methods of obtaining the object of the war. One way is what Clausewitz calls "the negative aim." It is, he said, "the natural formula for outlasting the enemy, for wearing him down." In a later chapter, Clausewitz discusses the relationship between the negative aim and the strategic defensive. "The aim of the defense must embody the idea of waiting," he said. "The idea implies . . . that the situation may improve. . . . Gaining time is the only way [the defender] can achieve his aim." Basic to the success of a strategic defensive in pursuit of the negative aim, therefore, is the assumption that time is on your side. But the longer the war progressed the more obvious it became that time was *not* on our side [emphasis in original]. It was American rather than North Vietnamese will that was being eroded."[2]

Westmoreland's "strategy" was fatally flawed, but did the Marine approach offer a better outcome?

The Marine leadership favored the strategic precepts of LTG Victor H. Krulak, the commander of Fleet Marine Force, Pacific (FMFPAC). Krulak had extensive experience in counterguerrilla operations and had served as a special assistant for counterinsurgency to the Joint Chiefs of Staff when Kennedy was president. Although in the administrative (not operational) chain of command for III MAF, Krulak had the ear of ADM Ulysses S. Grant Sharp, the Commander in Chief, Pacific (COMCIN-CPAC), Westy's boss, and was co-located with him in Oahu. Krulak theorized

that there was no virtue at all in seeking out the NVA in the mountains and jungle; that so long as they stayed there they were a threat to nobody, that our efforts should be addressed to the rich, populous lowlands. . . . It is our conviction that if we can destroy the guerrilla fabric among the people, we will automatically deny the larger units the food and the intelligence and the taxes, and the other support they need. At the same time, if the big units want to sortie out of the mountains and come down where they can be cut up by our supporting arms, the Marines are glad to take them on, but the real war is among the people and not among these mountains.[3]

In retrospect, there were several serious drawbacks to the Marine approach. First, as the Battle for Hue so clearly revealed, enemy units in the mountains of I Corps *did* constitute a threat. Second, the idea that they could be easily destroyed by "supporting arms" once they came into the populous areas was false. Not only in Hue, but also in the Tet contests in the cities and district and provincial capitals, as well as in countless battles for well-fortified villages over the course of the war, the use of supporting arms did not ensure success and resulted in extensive destruction of towns and villages, creating hundreds of thousands of refugees and alienating the population. Third, the use of American forces in pacification operations was shown to be inferior for a number of reasons (e.g., culture, language, fostering unsustainable reliance on U.S. supporting arms) to the assignment of South Vietnamese forces to this task.

In February 1966, while President Lyndon Johnson met with Premier Nguyen Cao Ky in Honolulu, GEN Westmoreland met with Admiral Sharp, Chairman of the JCS General Earl Wheeler, Secretary of Defense Robert McNamara, Secretary of State Dean Rusk, and Maxwell Taylor. This meeting established the strategic guidance for allied operations for 1966. The dual goals of pacification and destroying "V.C./PAVN forces . . . at a rate as high as their capability to put men into the field [i.e., the elusive 'crossover point']" were reemphasized. Several specific tasks in support of the military goal were identified, among them "defend military bases, political population, and food-producing centers" and "intensify operations against PLAF/PAVN lines of communications by 'almost doubling the number of battalion-months of offensive operations from 40 to 75 a month.'"

In I Corps, the pacification and population defense goals would dominate for the next two years. During late 1966, under pressure from COMUSMACV, the Marines would launch an increasing number of battalion or larger operations, but virtually all of those took place within 25 miles of the coast or along the DMZ.[4] During 1966 and 1967, there would not be one single "battalion-month" of operations in the A Shau Valley.

GEN Westmoreland was familiar with the USMC counterinsurgency doctrine, but he perceived it to be the wrong approach.

I believed the marines should have been trying to find the enemy's main forces and bring them to battle. . . . As a first step, I wanted General Walt to create a two- or three-battalion force capable of moving quickly by helicopter into enemy-dominated regions and more and more to join with the ARVN in operations large enough in numbers to punish the enemy's big units and disrupt their bases and supply caches.[5]

Yet COMUSMACV was, according to his apologia, more concerned with not upsetting Army–Marine relations than with the implementation of his strategy, so that he chose to let the III MAF follow its own mind.

Although there was no question that since I was commander of all American military forces in Vietnam, the marines were under my over-all command, I had no wish to deal so abruptly with General Walt that I might precipitate an interservice imbroglio. . . . General Walt had a mission-type order which by custom afforded him considerable leeway in execution. Rather than start a controversy, I chose to issue orders for specific projects that as time passed would gradually get the marines out of their beachheads.[6]

None of those "projects," until the planning for Operation YORK in late 1967, would take allied troops into the valley.

In his autobiography, Westmoreland is transparently defensive in his remarks about his relationship with the Marine command. He says:

Their self-confidence also apparently made them reluctant to ask for help so that often I had almost to force them to accept U.S. Army support. Self-confidence and independence, however admirable, constitute a two-way street in command relationships. Whereas it is incumbent on the over-all commander to assure that his subordinates have what he discerns they need, it is also incumbent on the subordinate to ask for what he needs.[7]

The last remark is ironic, given that until the self-destruction of the VC during Tet 1968, COMUSMACV repeatedly asserted the adequacy of *his* forces, yet never could find the troops to clean out or occupy the A Shau.

It is also symptomatic of Westmoreland's failure to fully grasp the operational implications of doctrinal differences between the Marines and Army with respect to air mobility. Westmoreland acknowledges that "the marines in comparison to U.S. Army units had few helicopters. That restricted mobility."[8] Yet in the same commentary, the former COMUSMACV expressed his desire for a "two-or three-battalion" USMC airmobile force. At the peak of their deployment in Vietnam (1966–1968), the Marines had 15 medium transport helicopter squadrons with less than 400 aircraft including, until 1968, a majority of older UH-34s (the Army 1st Cavalry Division alone had over 400 helicopters, as would the 101st Airborne Division after its conversion to Airmobile status in 1969).[9] The newer twin-rotor CH-46 was not nearly as well suited as the Army UH-1 Huey for use into "hot" landing zones. As the planning for Operation CUMBERLAND/CLOUD in 1967 would reveal (see chapter 7), the Marine Corps never could mass the helicopter assets to conduct a major operation in the A Shau Valley, while maintaining the necessary support of the III MAF battalions engaged in northern I Corps. Thus, even had General Westmoreland "ordered" the Marines to go into the A Shau, that could not have happened unless additional helicopter battalions could have been provided by the U.S. Army. Spare or reserve helicopter units, however, did not exist, and as time would tell, it was only when U.S. Army divisions with their own *organic* helicopters, the 1st Cavalry and 101st Airborne, were sent north that incursions into the A Shau became feasible.

Underlying the never-ending debate between the "pacification" proponents and those who believe "the big unit war" was the path to victory is the acknowledgment that the United States and its allies never had sufficient forces to *fully* engage in both efforts simultaneously. And therein, perhaps, is a key to understanding why the war was lost. The NVA/VC always pursued both and could and did shift their strategic emphasis, in accordance with the theory of protracted war, from one to the other, and from military operations to political negotiation, as the conditions changed. The allies, with their preset strategic plans for each year, thereby ceded the initiative to the foe.

By the end of 1966, the Marines in I Corps were by all accounts over-stretched. While III MAF had doubled in size during the year, if there had been a time when it might have realigned its regiments and poked a nose into the valley, that time had passed. There was severe political turmoil in the spring, the NVA launched an offensive in Quang Tri province in the summer, and both the NVA and the VC were on the march in increasing numbers in southern I Corps. "III MAF's high hopes for pacifying and unifying its three enclaves during 1966 had been dashed" and pacification was at "a complete standstill."[10] It would be the second quarter of 1967 before army reinforcements started shifting north, and 1967 would be almost over when GEN Westmoreland began to think about using more than air power in the A Shau. In the meantime, he would try to keep track of what Victor Charlie and his uncle Ho were up to.

"Search and Destroy" vs. Pacification: Was There Another Alternative?

Had Westmoreland imposed his "strategy" earlier in I Corps and mandated offensive operations into the A Shau Valley, would it have made a difference in the long run? My hunch is no. The reason is that COMUSMACV never emphasized or prioritized destroying *and denying* enemy "base areas" vs. attacking the enemy main force units ("bringing them to battle"). Quite the contrary. When U.S. commanders executed any mission, they knew that body count and kill ratio were the key statistics. Capturing enemy war materials generated good press, but if you wanted to advance your career, dead gooks were the ticket. As any student of management knows, the way rewards are structured has everything to do with the performance that follows. Give a bonus for attendance and employees will show up for work. Period. Recognize high-volume sales, and you will likely be rewarded, but probably at the sacrifice of high margins. And so on. In Westy's War, attacking base areas meant attacking NVA or VC Main Force *units.* This is reflected in the fact that even in the many operations that targeted base areas (e.g., CEDAR FALLS and JUNCTION CITY), occupation was always very short-term, as it would be in the only A Shau operation under his command, DELAWARE. General Westmoreland's desire to "disrupt . . . bases and supply caches" could only have been effective if the bases were occupied (controlled) for extended periods of time and also if the enemy lines of communications could have been severed or at least substantially impeded, but by the mid-1960s the allies lacked the maneuver battalions to pursue this strategy.

Harry Summers, agreeing with a 1965 paper titled "The Strategy of Isolation" by General Cao Van Vien, RVN Defense Minister, reckoned that extension of the DMZ to the Thai border would have been necessary to achieve this goal.[11] GEN Vien estimated that eight divisions would be required to implement this plan. General Abrams, with his later desire to attack the enemy logistical "nose," hinted at the necessity of isolating the battlefield, but by then his rapidly diminishing forces lacked the capability to accomplish this.

An interesting question is whether, with a limited number of motorable crossing points from Laos into South Vietnam, it is possible that the battlefield could have been substantially isolated from inside RVN during the early years of the American War (1965–1966), starting with Khe Sanh and the A Shau Valley.* This brings us full circle to the Border Surveillance Camps. The North Vietnamese recognized the threat that they posed to their future logistical and infiltration system and made each and every one of them a key target. When A Shau fell (to an entire NVA division), they saw that the U.S. would not or could not commit the force necessary to defend or reestablish isolated border outposts. Thus, the long-term fate of all of the other camps was sealed and western South Vietnam ceded to the Communists.

Operation PIROUS in the A Shau Valley

The largest recon operation of the war to focus on the A Shau Valley and adjoining areas was launched by Project DELTA on 18 April 1967 under the code name PIROUS (Operation 5-67). Over the course of almost two

---

* Such a strategy would have necessitated the capability and willingness to carry out limited cross-border spoiling attacks.

months, recon (2 US SF and 4 ARVN LLDB) and roadrunner (6 LLDB) teams would spread out over the A Shau and Rao Nai valleys and adjoining mountains to "conduct reconnaissance, surveillance, and exploitation operations within the assigned area of operations (AO), and employ TAC AIR strikes on enemy installations as appropriate."[12] A total of over 40 team insertions would be made, although only 10 of those teams (5 recon and 5 roadrunner) actually entered the A Shau or its immediately surrounding highlands. Here is the story of those teams.

Recon Team 10 became the first team to insert near the valley floor, landing about 1.5 km southeast of the abandoned Camp A Shau at last light on 22 April. As they jumped from their helicopters into an LZ that had once been a corn field, SSG Herb Siugzda, on his first mission with DELTA, dropped directly onto a three-foot tall sharpened cornstalk, an improvised punji stake of immense proportion, that entered the left side of his groin and exited on the right, just below his abdomen. Siugzda was pinned by the stalk, unable to move and in severe pain.[13] The other American on the team, medic SFC Walter "Doc" Simpson, reported to the DELTA commander, Major Allen: "We have to medevac Herb immediately. He took a punji stake through his groin." Allen responded that the mission should be continued, but Simpson insisted: "You don't understand. This isn't one of those little sticks—this thing is the size of a freaking baseball bat!" According to R.C. Morris, the team spent a quiet night and the team was evacuated safely at first light.

It is interesting, but not unusual as memories become less clear with the passage of time, that R.C. Morris's account of this action makes no mention of enemy activity and indicates that extraction took place on 23 April. The official after action report for this same team states that several enemy rifle shots were heard upon landing, and that the team moved about 1 km to the southeast, where it reported receiving small arms fire from all sides and estimated it faced a force of 60–70 enemy troops, with additional enemies seen farther to the north and south. "Due to heavy ground fire during medevac, entire team extracted."[14] That history is supported by the story as related by Tom Yarborough, which relies also on the text of Siugzda's Army Commendation Medal for Heroism.[15] Yarborough's narrative indicates a much more perilous escape: "As a hail of bullets ripped through the leaves and branches around him, SSgt Siugzda, immobilized from his injury and in great pain, nevertheless directed the perimeter defense and then orchestrated the successful helicopter recovery of this team through a wall of hostile fire." SSG Siugzda survived the surgical removal of the corn stalk, returning to duty with Team B-52 in June. The final intelligence report on PIROUS cites a number of "concentrations of enemy activity," one of which, at YC5480 at the southern end of the A Shau Valley, lies directly at the location where Recon Team 10's action occurred.

Things got off to a slightly better start farther up the valley, where Recon Team 9 led by SFC "Robbie" Robinette inserted safely at last light on 23 April. They, too, heard enemy signal shots shortly after disembarking. Landing near Route 547, approximately 1.5 km from where it intersects Route 548 in the central valley, the team had no enemy contact or sightings for several days. Then, on the night of 26 April, about 1.5 km west of their LZ and the same distance from Route 548, they reported hearing "domestic animals, women, children, and chopping and digging sounds all night."[16] Early the following morning Team 9 was being pursued by an estimated enemy platoon and fired at an enemy trail watcher while crossing a well-used trail. The team requested extraction from an LZ on Hill 890.

At 0850 hours, when the first extraction helicopter was lifting off, by hoist, two of the ARVN team members, heavy enemy anti-aircraft fire was received from a string of emplacements in an arc along the west side of the east wall of the valley from northwest to south of the pickup zone. The second helicopter to approach was shot down, landing near the recon team, and the crew was able to link up without casualties. At this time, the 5th Company of the ARVN 91st Airborne Ranger Battalion (the reaction force for Operation PIROUS) was placed aboard four USMC CH-46 helicopters and sent in to reinforce the recon and helicopter troops at Hill 890.[17] Three of the CH-46s made successful landings in the draw east of Hill 890, and the fourth crash-landed several hundred meters to the west, with only a minor injury to one of the Rangers. All of the units now on the ground were able to link up and were successfully extracted, under fire, by a flight of five more CH–46s. Before extraction, the recon team ambushed an unsuspecting squad of NVA armed with AK–47s, and all 12 of the enemy were killed in a superb action by SFC Robinette, who received the Silver Star for his action on this day.[18]

Roadrunner teams, consisting solely of indigenous troops, were also part of Project DELTA. Team 105 was the first into action in the A Shau Valley during Operation PIROUS, inserting on 30 April about 2 km from the A Shau stream, near a trail connecting the A Shau and Rao Nai valleys. The team continued to explore westward, finding more evidence of a well-used trail connecting the adjacent watersheds, but spotted no enemy and was withdrawn as scheduled on 4 May.

Operating generally several kilometers east of Roadrunner Team 105, Recon Team 12 was in the field from 1–6 May. This team patrolled along several tributaries of the Rao Nai, up to the range of mountains separating the Rao Nai from the A Shau. They found a few huts and trails and were alerted on several occasions by enemy rifle shots but did not see any enemy troops or engage in any firefights.

To the north of Team 105, Roadrunner 101 was inserted at last light on 9 May. Operating in the area of the Khe Ta Li stream, a tributary of the Rao Nai, this team found a 3 m wide section of Route 547 at YC478952, some 5 km east of the location shown for this track on allied maps. They also found the tracks of a company-size enemy unit that had moved through the area perhaps a day earlier. After engaging in a brief firefight with four VC on 13 May, the team was extracted on schedule on 14 May.

Approximately 6 km north of the A Shau airstrip, Roadrunner Team 106 was inserted on the east side of the mountains bordering the A Shau Valley on 13 May. Soon after arriving at their LZ and discovering a freshly used trail, this unit was fired on by a small group of VC. There were no casualties, but the team was extracted as their mission had been compromised.

In mid-May, several teams were sent to the west of the A Shau Valley, into the rugged hills near the Laotian border. On 15 May, Roadrunner Team 108 was dropped off at an LZ just 4 km from the border, near a trail connecting Laos with the southern A Shau Valley. The LZ was hot, with heavy fire directed at the helicopters from both east and west, and one team member wounded. An enemy company was detected approaching from the northwest. The team was quickly extracted, and tac air and helicopter gunship fire was directed against the enemy positions with unknown results.

The northernmost patrol of Operation PIROUS was that of Recon Team 4, inserted near the northern end of the A Shau Valley (the south end of Quang Tri Province) on 10 May. The U.S. SF team members were LT Coulter and SFC Smyth. For six days, this unit covered an area of about 2 x 3 km, finding and observing small trails and elephant tracks, but no enemy. On the night of 13–14 May, an air strike was called in on several bright lights that were observed in the vicinity of Route 548 in the Da Krong Valley. The team was withdrawn on the morning of 17 May, having completed its mission. The AAR contains conflicting information whether the team's extraction was conducted under fire[19] or "without incident."[20]

Next, Roadrunner 107 was inserted into an area a few kilometers west of the A Shau airstrip where several tributary streams of the A Shau flow down from the mountains on the Laotian border into the valley to the northeast. The LZ was cold on 17 May, and the team's searches found nothing but an old encampment. On 21 May, gunshots were heard at 1000 and 1300, but no enemy was sighted. One of the team members became ill and medical evacuation was requested.

When the first extraction helicopter from the 281st Assault Helicopter Company (AHC) attempted to land, it was hit by automatic weapons fire, wounding the aircraft commander, WO Jerry Montoya, and the SF recovery NCO. The Huey was able to return to Phu Bai. As the second attempt at recovery was in progress, a UH-1C gunship from the 281st AHC was shot down, crashing into a hillside to the north of the LZ. The Huey caught fire, rolled into a ravine, and exploded. It was thought doubtful that there were any survivors. Fortunately, a C&C (command and control) helicopter spotted a signal panel, indicating that crew was alive and needing rescue. A USAF Search and Rescue (SAR) HH-3 Jolly Green Giant helicopter arrived but was unable to land near the signal panel and was used instead to successfully evacuate Recon Team 107. The next rescue attempt, by a USMC CH-46, successfully extracted SP5 Craig Szweg, the crew chief of the downed gunship. A new signal panel was sighted, and two Americans were seen on the ground, so yet another rescue try was undertaken. PFC Michael P. Gallagher of the 281st AHC, door gunner on the extraction ship, was hit by enemy fire and died before he could be returned to base. At this point, darkness fell and further flights were called off for the day. The three remaining crewmen of the downed UH-1, WO Donald "Corky" Corkran, WO Walter Wrobleski, and SP5 Gary Hall, were still missing, somewhere on the western side of the A Shau. Tactical air strikes were used throughout the afternoon to support the rescue operations.

At 0745 on 22 May, the SF reaction force, Company 1 of the 91st ARVN Ranger Battalion, was landed in the valley about 2.5 km west of Route 548 by USMC CH-46 helicopters. The LZ was hot and one of the Marine helicopters was shot down by .51-cal anti-aircraft fire, crash-landing in the LZ without casualties. At the same time, the C&C ship once again spotted two of the American crewmen on the ground. The Rangers moved to the downed helicopter while under continuous enemy contact. There, they found and rescued SP5 Hall. There were no human remains in the wreckage, and the search continued until nightfall for the two pilots.

Early on the morning of 23 May, the rangers finally found WO Corkran. The first attempt to evacuate him was defeated by heavy enemy ground fire. After heavy application of tactical air strikes, the company moved to a new LZ. En route, a firefight with an ambushing enemy unit cost the lives of three ARVN Rangers with three others wounded. The new LZ was secured and the reaction force and WO Corkran were evacuated. Although aircraft continued to search for him until Operation PIROUS ended in June, the remains of WO Walter F. Wrobleski have never been found.[21]

The final team slated to recon the area to the west of the valley was Recon Team 7. When they tried to land 5 km due west of A Luoi and about 2 km north of Dong Ap Bia (later better known as Hamburger Hill) on 17 May, a heavy stream of enemy fire greeted the helicopters. Seven enemy troops were seen on the LZ. The landing was aborted and the area was hit with radar–directed air strikes (Sky Spot).

Operation PIROUS – After Action Report and Intelligence Summary

Operation PIROUS represented the very best in interservice and interallied cooperation. Marine helicopters from HMM-265 carried Army and ARVN soldiers into and from the valley and rescued downed USAF pilots and Army air crew. The Army's 281st Assault Helicopter Company likewise carried teams of Project DELTA into action and supported them with gunships, and when a USMC F-8 Crusader crashed on 4 June, the 281st AHC was quickly on the scene to rescue the downed pilot. Air Force SAR helicopters helped to extract recon personnel, and to rescue downed pilots of all services. Air strikes were provided by fighters from the USAF, USMC, US Navy, and the Vietnamese Air Force (VNAF).

By "body count" standards, PIROUS wasn't much, but then it was a reconnaissance operation and not "search and destroy." Only one American was MIA and one American life was lost, both casualties coming during the extraction of Roadrunner Team 107. Three ARVN Rangers also gave their lives during that action. Given that most of the insertions and extractions were conducted under fire, these were incredibly low losses. Seventeen enemies were killed in ground actions, 12 of those by Robbie Robinette of RT 9. Another 29 enemy soldiers were estimated to have been killed by air strikes.

Of the 48 team and company insertions during the period 18 April to 13 June 1967, only 10 were in the A Shau watershed. The remaining patrols took place to the east and northeast, in and around the Rao Nai Valley or even farther to the north or east in the rugged hills between the A Shau Valley and Hue. The Rao Nai runs more or less due north from a couple of kilometers east of the south end of the A Shau Valley until it crosses Route 547 and then empties into the Song Bo. In addition to the intelligence gathered by troops of B-52, there were daily sightings of enemy troops, vehicles, anti-aircraft positions, and camps by helicopter crews and the forward air controllers who repeatedly risked their lives against NVA fire.[†22]

The final intelligence summary of Operation PIROUS summarizes the limits of allied knowledge about the enemy in the A Shau Valley between the fall of Camp A Shau and the 1968 Tet offensive:

(1) Route 922 entering RVN from Laos at the northern end of the ASHAU Valley and Route 548 running southeast through the ASHAU Valley are connected and have been improved to accommodate heavy truck traffic along their entire length to the southernmost portion of the ASHAU Valley. Evidence of truck traffic and road repair work has been reported daily. . . . Movement of material and supplies by truck is believed to be in a southeasterly direction throughout the ASHAU Valley, with empty trucks returned in a northwesterly direction along the same route towards Base Area 611 in LAOS for reloading. . . .

(2) . . . the following concentrations of enemy activity have been established: 1) The northern end of the ASHAU Valley, including surrounding high ground. 2) The southern end of the ASHAU Valley . . . mainly in vicinity of YC5480. 3) The rolling hills area west of A LUOI (YD3800). 4) Valley areas approximately 5–8 kilometers east of TA BAT (YC4295). 5) Along the RAO NAI River Valley between 80 and 90 east-west grid lines (YC). . . . 6) The

---

† Future Chairman of the Joint Chiefs of Staff, then CPT Hugh Shelton, in his biography, chronicles his participation as the leader of a reconnaissance team during Operation PIR-OUS. His account begins: "Our team of two Americans and four ARVN Special Forces deployed north to the old A-Team camp on the edge of the A Shau Valley. . . ." His story is repeated in Thomas Yarborough's book (pp. 75–77), "The team inserted at last light on the east wall of the A Shau opposite the abandoned Special Forces camp at A Luoi. . . ." The future four-star general's narrative is captivating and includes a graphic description of a typical recon patrol in western Thua Thien Province, as well as a not-so-typical encounter with a Bengal tiger! As with many "A Shau stories," however, this one did not take place in the valley. The official Debriefing Report for Recon Team 5 (USSF Advisors CPT Shelton and SFC Robinette) lists the patrol grid coordinates, which place the mission in the vicinity of tributary streams of the Rao Nho, approximately 10 km north and east of the A Shau Valley. The fact that the patrol did not enter the valley does not, of course, in any way diminish the risks or contributions of that mission.

mountain and valley area immediately southwest of the confluence of the RAO NAI and RAO NHO Rivers at YD5203. Areas 1 and 2 appear to contain the main facilities for movement of supplies by truck . . . to include ground and air defense systems and security and labor forces. Areas 3, 4, 5, and 6 are believed to be mainly food producing areas. . . . A possible continuation of the infiltration route . . . seems to originate in base area 607 leading east across the LAOS-SVN border and then continuing . . . deeper into South Vietnam. . . .

(6) No evidence of infiltration routes exiting the ASHAU Valley to the northeast into Base Area 114 or east towards HUE/PHU BAI was found. Route 547 connecting the ASHAU Valley with HUE does not show signs of being used as an infiltration route.[23]

While the magnitude of the enemy reliance on the A Shau Valley as a key element in its line of communication had begun to be exposed, nothing was revealed about the direction that future NVA operations based in the A Shau might take.

7

OPERATION CUMBERLAND/CLOUD

The Developing Threat in Early 1967

The MACV/RVNAF Combined Campaign Plan for 1967 differed from that of the preceding year primarily in an increased focus on Pacification, with the armed forces of the Republic of Vietnam to take on the primary responsibility. U.S. units were to go after the NVA/VC main forces. This, of course, was contrary to the Marine view of strategy. The application of the Plan was of limited relevance to the disposition of troops in I Corps, however, as the enemy was calling the tune in the northern provinces. In addition to the steadily increasing NVA presence in Quang Tri on the border with North Vietnam, which fully occupied the 3rd Marine Division, the enemy also was on the move elsewhere in I Corps.

In January 1967, a threat in Quang Ngai, the southernmost province in I Corps, suddenly materialized, with no less than three NVA regiments identified there, compounding III MAF and MACV concerns about the situation along the DMZ and around Khe Sanh. To make matters worse, the A Shau thorn in the side also became inflamed once again.

While this ominous activity [in Quang Ngai] continued, another, far more serious threat began to emerge farther north in the A Shau Valley.... Marine observation flights had revealed no significant evidence of enemy activity in the valley. Yet when Special Forces long-range reconnaissance patrols entered in February 1967 [OPN AUSTIN] they met heavy resistance and noted that the enemy was improving roads and trails, perhaps to accommodate rapid movement of troops and supplies for a major offensive. Possibly

more important, an intelligence report indicated that two North Vietnamese regiments were already gathering in Base Area 607, inside Laos, within
a few hours' march of the A Shau. MACV immediately sent B-52s to strike
the suspected locations of the two units, hoping that an air attack could
blunt this looming threat.

This increased enemy activity [in Quang Ngai] and in the A Shau, together
with growing trail activity during the 8-12 February Tet ceasefire, *came as
an unexpected and quite unpleasant surprise to Westmoreland and demanded
his immediate attention and response, if not a major change in his campaign
plan for 1967* [emphasis added]. The MACV commander remained convinced
that sooner or later Hanoi would launch a conventional offensive across
the Demilitarized Zone, supported by attacks from Laos [i.e., along Route 9
through Khe Sanh and Route 922/548 in the A Shau]."[*1]

So once again, as had happened a year earlier when the Special Forces
camp at A Shau was overrun, in early 1967 the American commander was
surprised and disconcerted by enemy moves in and around the A Shau
Valley. The Third Marine Amphibious Force Headquarters (III MAF)
was the senior American command in I Corps, and the official USMC
history for 1967 observes: "Infiltration from Laos and the influx of NVA
troops and supplies into the A Shau Valley and the mountains west of Hue
caused some very touchy and precise repositioning of units."[2] Westmoreland brought three U.S. Army brigades north (TF OREGON), primarily
stationed to protect major allied bases and the heavily populated coastal
lowlands, but could spare no Army maneuver battalions for an A Shau
operation.

The only tool immediately available to COMUSMACV to attack NVA
positions in the A Shau was to send in the heavy bombers,[3] although
such raids had no more than nuisance effect on the enemy given the lack
of timely intelligence on the location of his troops and supply dumps,

---

* Map 22 in the official history, which accompanies the narrative, incorrectly locates Base
Area 607 well to the north of A Shau, on the Quang Tri / Thua Thien border. Its actual
location was just to the south of the valley, on the Thua Thien / Quang Nam border. Base
Area 611 occupied the spot indicated for 607.

his practice of keeping units constantly shifting from one well-hidden camp to another, and the likelihood that ARC LIGHT targets were often known to the enemy in advance. From mid-April through mid-June, Project DELTA's reconnaissance teams were tasked with developing a much better picture of what the enemy was up to (Operation PIROUS – see preceding chapter). Once that intelligence was provided and indicated the magnitude of the NVA base development, it became imperative to think seriously about an attack into the valley itself. It was time to send in the Marines.

Operation CUMBERLAND/CLOUD

The III MAF plan for the incursion into the A Shau Valley in mid-1967 was created under the code name CUMBERLAND/CLOUD. The basic concept for the operation was to secure and improve over 20 kilometers of Route 547 between Hue and the A Shau, to emplace a heavy artillery unit that could fire in support of ground troops in the valley (CUMBER-LAND), and then to conduct a raid on enemy installations (CLOUD).

The forces initially assigned to the road clearing mission were three companies of the First Battalion, Fourth Marines, and some Marine engineers. The move down 547 began on 8 June, while the Project DELTA recon teams were still prowling the A Shau and Rao Nai Valleys. By 7 July, the forward construction point for the road had reached the area where Fire Support Base BASTOGNE (initially named CUMBER-LAND by the USMC) would later be located (YD620094), some 15 km from the start point and about 26 km from Hue by road. Contact with the enemy had been minimal, and although a number of Marines had been wounded, none had lost their lives.

On 7 July 1967, a major USMC/NVA battle, Operation BUFFALO, was in progress near the DMZ, and all of the line companies of 1/4 Marines except for A Company were ordered north as reinforcements. A Company remained in the Route 547 area to provide security for the engineers, who were building a base for the artillery and continuing to upgrade the road. Ten days later, when the rest of the battalion returned from Operation BUFFALO, Phase II of Operation CUMBERLAND resumed.

In early planning with the ARVN for the CLOUD phase of the operation, two options for the size of force to send into the A Shau were presented.

> CG 1st ARVN DIV [GEN Truong] stated at planning conference [on July 29] his concept that ARVN forces should participate with one of two alternatives. Alternative one: that JGS [ARVN joint general staff] provide complete ARVN ABN DIV {7 BNS} for operations, lasting at least 1 month, including three ABN BN currently OPCON to CG I corps. Alternative two: provide one company (Black Panther) for OPN (approx strength 270). These alternatives were also expressed by GEN Lam at conference 29 July.[4]

General Truong was probably the most highly esteemed division commander in the ARVN, and General Lam was his chief, the I Corps commander. LTG Cushman (III MAF) continued this memorandum to his boss, GEN Westmoreland, with the comment that the first option was "obviously impractical at this time" for several reasons:

1) The weather: there were only about four hours of marginal flying weather each day;

2) III MAF did not have enough helicopters to support the larger operation.

The III MAF commander went on to note that in his conception, the USMC battalion would establish a base and the ARVN Black Panther Company "would provide an excellent rcn in depth capability." Further remarks revealed that intelligence reports, as well as recent enemy action along the DMZ, would require limiting the USMC contribution to a single battalion, and both the weather and lack of any readily identifiable reserves made the whole plan "risky." LTG Cushman nevertheless concluded on an optimistic note: "I believe that one USMC BN reinf plus one ARVN co will actually be more mobile, more easily inserted and extracted and more easily supported than original concept. Therefore in accordance with your verbal orders I intend to execute A Shau raid with one USMC BN (reinf) in conjunction with one ARVN Black Panther Company."

While the 1st Battalion, 4th Marines was winding its way down Route 547, the plans for the second part of the operation, the incursion into the A Shau Valley (Operation CLOUD), were still evolving. This raid into the heart of the A Shau Valley would be the first allied venture there in force since Camp A Shau's demise. LTC James "Wes" Hammond, Jr. had just arrived in the country to take over the helm of 2/4 Marines in mid–July. LTC Hammond was quickly drawn into a round of briefings and planning sessions for Operation CLOUD. In a letter to his son, Jimmie, in February 1999, he recalled:

> Initially, it was to be a combined operation with the ARVN and there were several conferences with the Army advisers to the 1st Arvn Div. The Army colonel was named Kelly, I think. One of the earlier schemes of maneuver was for ARVN Rangers to parachute into the northern part of the objective area near the old Special Forces fort which had been the scene of a big fiasco in 1965 or 66. I balked at not having command of all the forces on the ground. Subsequently, the ARVN Rangers were withdrawn from the operation and the Huk Bao [sic] or Black Panther Co. substituted. They had a New Zealand CWO as adviser. The Rangers were to have an air-dropped 105mm btry which I thought to be tactically unsound inasmuch as both the paratroops and the battery would impinge on the helo assets for extraction. One of the "rumors" fed back through Col. Kelly was that the paratroops felt that there was a heavy concentration of AA MGs around the drop zone. That raised my suspicions that the ARVN involvement had already gotten back to Charlie [the enemy].[5]

LTC Hammond had other reservations. The Public Affairs Office wanted to have the press included in the first wave into the valley, in order to be sure that the operation would be correctly reported as a "raid" and not a defeat, when the troops were pulled out after just two days. But press seats on helicopters meant fewer slots for Marines, and the battalion commander knew he would need every man possible. Then there was talk about including a battery of 106mm recoilless rifles, but that could only be done if all of the few available Marine heavy helicopters (CH–53s) were used, and the Marine commanding general in the Pacific, "Brute" Krulak, had "said he'd kill anyone" who lost one of those aircraft. Because of the need for a large commitment of Marine helicopters and fixed-wing air, the force planned for CLOUD evolved into an air-ground task force ("KILO"), with the regimental commander of the 4th Marines, Bill Dick, as overall ground commander. All in all, Operation CLOUD seemed to be shaping up to be another A Shau Camp or Mobile Guerrilla Force fiasco, with troops ordered to proceed on an ill-conceived mission that the operational-level leaders knew was far more likely to end in disaster than success, but with senior command adamant that the plan be carried out.

The mission statement for Operation CLOUD reads: "2d Bn, 4th Marines (Rein) will conduct a combined heliborne assault with the Black Panther Co. (HB) into the Southern portion of A SHAU VALLEY to conduct a raid to destroy all enemy installations, lines of communications, base camps, and to kill or capture all enemy within assigned AO."[6] The Concept of Operation ordered that after securing the LZ (SPARROW), "moves one company in Northwesterly direction to conduct screening mission perpendicular to Valley while three companies attack in Southeasterly direction conducting search and destroy mission in Southern portion of A SHAU VALLEY." Company F, 2/3 Marines was attached and would secure the LZ while acting as reserve.

The last few hours before commitment of the Marine battalion were counting down. As the attack force assembled at the Phu Bai Marine base and bivouacked for the night of 3–4 August, prepared to fly into the valley the following morning, there was another component to the plan that was unknown to the CO of the 2nd Battalion, 4th Marines: "What further bothers me is that there was a Recon patrol that went in there before us and I was never told about it as the CO of the operation. I certainly would have wanted to know they were going and . . . to debrief [them] after."[7] Yet in this peculiar case, what LTC Hammond didn't know may have saved his command. There were not one but two Marine patrols scouting the valley.

Teams Party Line One and Mono Type II

On 1 August at 1130 hours, Team Party Line One from A Company, 3rd Reconnaissance Battalion, inserted into an LZ 5 km due east of the abandoned Special Forces camp at A Shau. They would be patrolling directly between two major enemy activity concentrations noted by the Project Delta teams just weeks earlier: the Rao Nai Valley and the southern end of the A Shau Valley (YC5480).[8] The patrol was led by 1LT Bill McBride and included six enlisted Marines and a Navy medical corpsman. The members of the patrol had little time to prepare; they were a last-minute replacement for the team originally assigned. Their mission was to investigate the ridgeline between the A Shau and Rao Nai valleys, as this had been an area occupied by NVA antiaircraft batteries during the Camp A Shau battle, and to look for other signs and camps of the enemy, particularly along trails connecting the A Shau Valley with the village of Be Luong in the Rao Nai Valley, just to the east.

The insertion was cold (no enemy fire) but difficult nonetheless, as the elephant grass on the landing zone was 10–12 feet tall and had to be battered down by the helicopter's ramp before the men could disembark. Shortly after landing, two shots were heard from nearby. This was the usual means by which enemy trail watchers and trackers signaled that a recon team was spotted on the ground. The team moved to the southwest, discovering a well-used trail about one meter wide showing no evidence of recent use. Walking an enemy trail was not a favored course of action when deep in "Indian country," but in this case it seemed to be a better option than trying to blaze a trail through the high, sharp elephant grass that was all around. The patrol moved, with extreme caution, along the trail toward Hill 745. At about 1800 hours, they reached the hill, where they found an abandoned Montagnard camp and set up a harbor site for the night. A half hour later, five or six additional gunshots were heard from the valley to their west, indicating that the enemy was nearby and seeking to converge on and engage the Marines. The patrol settled in for the night, calling in spotting rounds to register on-call defensive fire from the Army 175mm guns located at the firebase set up on Route 547 during Operation CUMBERLAND.

A little before 0500 hours, the men were awakened (two of the eight were on watch at all times through the night) by a B-52 ARC LIGHT strike, about 5 km away in the valley. "The whole hillside shook like an earthquake."[9] This was part of the preparation for the impending air assault by the 2/4 Marines. This day (2 August), the team had intermittent problems maintaining communication with a C-130 aircraft overhead, their only link to outside the valley. They began moving downhill and had only gone about 200 meters when wood chopping was heard a short way ahead. A pause – they stopped, waited, and observed, but saw nothing, and the chopping ceased. They moved ahead cautiously and soon heard more chopping and voices. The point team was sent ahead to scout and soon reported that they had encountered Montagnards – two women and two boys with axes, who had run away from them down the trail.

A quandary. It was well known that all of the Montagnards in this area were pro-VC, and so it was expected that the presence of the Americans would quickly be transmitted to the enemy. LT McBride took the team in a circle of about 300 meters diameter, returning to a site about 40 meters from their original position. All was quiet; it was now about 1400 hours. LT McBride takes up the story:

> I took three people, without our packs, and moved back to the trail and cautiously retraced our steps downhill to try to find out what was happening. Then my point spotted two men in uniform. They wore light-colored uniforms, bush hats, and had bandoliers of ammo across their chests, but there were no weapons visible. The point carefully returned to the rest of us, but then we saw the two run down the trail. I decided to get the rest of the patrol and move uphill. By the time we got back, and the point man was down the trail a short distance, watching from there, we started taking fire.[10]

The point man had been shot in the hand, and the team fought a brief firefight with an estimated three enemies with AK-47s. The firing stopped and LT McBride received word via the C-130 radio link that he was to find an LZ so that the team, which was "compromised," could be extracted. He started the team moving to the north but had gone only a short distance when it began to receive fire from the front (uphill), and two more Marines received minor wounds. The Marines returned fire and threw grenades until at about 1630 hours, the enemy withdrew. While fending off the enemy, the patrol had attempted to clear an LZ at their location, as the enemy was to the north, the trail (east or west) was certainly ambushed, and the slope downhill to the valley floor offered little chance to evade the enemy or find a flat, cleared area.

At around 1700 hours, the Marine helicopters and fighter-bombers arrived and immediately encountered enemy automatic weapons fire from east, south, and especially from the valley floor west of the patrol. In spite of the team's effort to provide a cleared LZ, the extraction chopper pilot reported that he could not get into the LZ they had worked to prepare. The pilot identified an alternative clearing where he thought he could get on the ground and directed LT McBride to take his team to that location.

Moving through extremely difficult terrain at a pace of only about 100 meters per hour, Team Party Line One arrived at the new pickup point at 1910 hours. While the team had made their move without enemy interference, the helicopters and fighters continued to be fired upon. Although the Marines on the ground could not see them, the airmen reported lights all around the patrol, and as darkness fell the men were told that their rescue would have to wait until morning.

While Party Line One sought to evade its pursuers and find an LZ for extraction, the other 3rd Recon team, under 2LT Al Weh, was so far maneuvering without enemy contact. Team Mono Type II was inserted on the afternoon of 2 August about 2000 meters to the east of LT McBride. The two units were expected to accomplish their tasks independently and not to work together. The mission of Mono Type II was to look for suspected NVA units and base camps and also to keep a careful record of weather conditions.

LT Weh and his team (also six enlisted Marines and one Navy medic), who had more time to prepare for the mission than Party Line One, had rappelled into the LZ, in a plan devised to minimize the chance of their being spotted. Intelligence indicated that Montagnards would be keeping an eye on all clearings large enough for a helicopter to land. Weh said: "My objective was to find a location in the jungle some distance from my objective and rappel into it. But what I really wanted to do was to find a 'depression' in the jungle canopy that was lesser in height than the surrounding jungle. If I could find a place like that, then the jungle would help mask the chopper as it hovered within that bowl and let us drop down through the trees."[11] It worked. The insertion took longer, but the team was on the ground unsighted by the enemy. While Mono Type II was moving toward their objective, they heard in the distance the firefight between Party Line One and the enemy. They harbored about 600 meters from their designated observation point (OP) and, except for an unidentified animal that landed squarely on LT Weh's face in the darkness, passed a quiet night.

On the morning of 3 August, LT Weh and his men moved into their OP, still without enemy contact. The jungle around that location was so heavy that he did not expect to be able to see any enemy activity and could only carry out the "weather reporting" part of his mission.

LT McBride and his team also passed the night safely. During the night, sounds of the enemy moving were heard all around, and 29 rounds were fired in support by the 175mm guns, with unknown results.

With morning came the return of the Marine fighter bombers and helicopter gunships, strafing the area around Party Line One. The men on the ground tried for a second time to clear an LZ at the spot the pilot had directed them to, but once again the site was unacceptable due to remaining obstructions. The team was told to move about 200 meters further down the slope to a third pickup point. There they found a large clearing of about 100 x 150 meters that had been cleared by the Montagnards. Wasting no time, a Marine CH-46 piloted by LT J.L. Connolly came quickly down into the field, while the recon team moved from concealment to board. LT McBride describes what happened next:

I was the last to jump in, but as I was going to the chopper, it took a hit and a two-inch hole blasted open above the left wheel. Then I was being pulled aboard as the chopper was trying to take off. It got about ten to fifteen feet in the air and started dancing around side to side, and then plunged back to the earth. It hit on its right side and rolled over one and a half times. It had taken other hits. . . . We had seriously wounded [passengers and crew members], and people were trying to help each other out. The crew chief was killed. I had to cut the corpsman's bag off him to get him out of the tangled mess. I was concerned about fire and explosion.[12]

CPL Thomas A. Gopp of HMM-164 was the crew chief killed in this action.

In the clearing, chaos briefly reigned. There were now 11 men alive on the ground, including the 3 surviving crewmen from the helicopter, and several of them were severely injured. LT McBride had to set up a perimeter, ensure that the wounded were cared for, and establish contact with the second CH-46 that was overhead as backup. Soon that helicopter, too, was settling into the LZ under heavy fire. The team radio operator was badly wounded, but all the men got aboard and the CH-46, piloted by Captain Al Kirk, lifted off. At about 20 feet from the ground, this aircraft was struck again by enemy fire. It shuddered, then gained momentum and continued to climb, making the flight out of the valley and back to the Marine base. Three of the members of Team Party Line One died of their wounds: HM3 James "Doc" McGrath, USN, and LCPLs John B. Nahan III and Jack Wolpe. All of the other team members were wounded: Bill McBride, James McCarty, Cletus Kitchens, Bobby Latham, and Dan Heckathorne. LT McBride received a Bronze Star for his actions as leader of Party Line One and CPT Kirk was awarded the Distinguished Flying Cross.

As Team Party Line One and the helicopter crews of HMM-164 were fighting to complete the extraction, LT Weh and his group passed an uneventful day, hunkering down in trenches in an old Special Forces outpost. That night (3–4 August), the patrol was notified that it would be extracted at first light the next morning. The night passed peacefully until between 0330 and 0530 in the morning, when sounds of enemy movement were heard around the position.

At 0730 on 4 August, the CH-46 was inbound. Al Weh instructed the pilot on where and how to land so that the team could move quickly to embark. As the chopper came in, "the jungle erupted with small arms fire," including rocket propelled grenades. Once the helicopter was on the ground, six members of the team soon moved up the ramp. LT Weh and his radioman, PFC Dodd, brought up the rear. Just as they got on board, both were seriously wounded. Two other team members received minor wounds, but all would survive. LT Allen Weh, who would receive the Silver Star for his courage and leadership of the patrol, "later said about his men, 'I have a lot of respect for those Marines, Sergeant Goodwin, who was my APL, Lance Corporal Gerdan, my rear point; Pfc. Kenck, my radio operator; Pfc. Gauthier; Pfc. Carter; Doc Earl; and Pfc. Dodd.'"[13]

The Operation that Never Was – Operation CLOUD Is Cancelled

Operation CLOUD was cancelled during the late evening hours of 3 August 1967. Most of those who were scheduled to take part did not find out until early the following morning.

In a top-secret memo to LTG Krulak dated 1242Z(ulu) (1942 hours local time) on the evening of 3 August 1967, LTG Cushman passed on the message he had just sent to General Westmoreland, "reluctantly" cancelling Operation CLOUD. He cited three reasons:

1) **A forecast for deteriorating weather conditions in the A Shau Valley on 6 August**

2) **An NVA buildup in the DMZ / Con Thien area, which might require helicopters or infantry tied up in the A Shau**

3) **Fear that a military "reverse" might negatively impact the impending SVN elections.**

All three reasons mirror discussion points in LTG Cushman's memo to Westmoreland of just three days earlier, at which time he had still supported the plan. So what changed? First, it would seem that the weather forecast had become even more unfavorable, which may be attributed in part to the reports submitted by LT Weh's team then operating in the valley.

Second, his concern about the DMZ area now emphasized the possible need for helicopters even more than infantry reserves: "Ability to move troops to and within the DMZ-Rte 9 area may become a critical factor." This concern was driven by both weather issues and possible helicopter losses during the A Shau operation, a concern elevated by the opposition encountered during the extractions of Party Line I and Monotype II. Third, the election worries – which were primarily a concern of the ARVN commanders – related directly to the perceived riskiness of OPN CLOUD, and that, too, seems to have increased. It is therefore important to recognize that the intelligence about the level of enemy activity, especially the intensity of the anti-aircraft fire, and the observations about changing weather patterns that came from the 3rd Reconnaissance Battalion patrols in the A Shau during the critical three days between his 31 July endorsement and his 3 August cancellation of Operation CLOUD probably played a pivotal role in convincing the III MAF commander not to launch the air assault into the A Shau.

Many years later, on learning about the fate of the recon teams, Wes Hammond opined: "Had we gone, I think that we would have taken our casualties during insertion and extraction. Once on the ground, I think we would have been all right." Others who were primed to participate were less confident. Robert Bliss USMC, RET, was with G/2/4 Marines at the Phu Bai staging area. In an e-mail to Bill McBride in 1998, he remembered "being pretty damned scared about my prospects. . . . We were all aware that this could be our last night together for many of us. However, the next morning we were told the operation had been canceled, much to our collective relief."[14] LT (later COL) Michael Gavlick was a young platoon commander in F/2/4:

At first light, the Bn (Rein) was to helo in, with our company as one of the leads. . . . The enemy situation was briefed as very grim, including 50-cal machineguns on the sides of the valley, which would take a toll on us trying to land on the valley floor. We, at my level, believed our casualties in the initial waves could reach 50%. And, with numerous roundtrips per helo assumed, even getting everyone in seemed questionable. We were to walk out! Everyone was quite subdued throughout the night, but as Marines, we were all pretty much resigned to our fate. I cannot adequately describe how relieved we were when the Op was called off in the pre-dawn hours.[†115]

The two patrols from the 3rd Reconnaissance Battalion triggered a response from the NVA that resulted in the tragic loss of three Marines and one Sailor on 3 August 1967. But the unheralded bravery of the men who served on Party Line One and Mono Type II and the helicopter crews who saved them, risking their lives in the Valley of Death to gain the knowledge that led to the cancellation of Operation CLOUD, likely prevented heavy Marine casualties had it proceeded.

Notwithstanding LTG Cushman's statement to GEN Westmoreland when he cancelled CLOUD that "the artillery and air interdiction campaigns are underway in the valley and will continue to be pushed hard,"[16] the big guns on Route 547 were displaced for movement back to II Corps just a few days later, on 9 August.[17] There was no one left to spot for them in the A Shau. Only bombs and chemicals would impede the NVA in the valley for the next eight months.

---

† The Operation Order did specify that extraction was to be by helicopter, not on foot!

8

ENEMY CONSTRUCTION AND MACV
COGITATION

Evolution of the Enemy Road Network Into and Through the A Shau Valley – 1966 and Early 1967

With the successful elimination of the last American/"Puppet" base in the A Shau Valley in March 1966 by the 325th NVA Division, the time had come to begin the development of the valley as the single most important enemy base in northern South Vietnam. For two full years, the development of that complex of caves, camps, and caches would be harassed by American air power but unopposed by ground troops.

The popular historical images of the Ho Chi Minh Trail are either that of a long column of NVA infantry hiking for months through the jungle, suffering malaria and other diseases and constantly on the lookout for American fighter-bombers, or porters pushing specially constructed bicycles carrying a couple of hundred pounds of rice. In reality, the continuing supply of those troops was as critical as the infiltration of the men themselves, and once the People's Army of Vietnam (PAVN) regulars in divisional strength were operating in South Vietnam, manpower alone could not suffice to provide for the logistical needs of regiments of infantry supported by many battalions of artillery, anti-aircraft weapons, engineers, and finally tanks. As early as 1964, the supply function was largely performed not by human carriers, but by truck. In mid-1966 the CIA estimated that North Vietnam had 11,000–12,000 trucks in inventory.[1] In considering the role of the Ho Chi Minh Trail as the main supply route for Communist troops in the northern half of the Republic of Vietnam, two points should be borne in mind:

1) **It is true that a majority of the total logistical tonnage needed to sustain the NVA and VC was generated in-country. That majority, however, was almost exclusively food. Rice and other rations were farmed by VC-controlled villages, by soldiers detailed from combat units, and was collected through taxation of contested and government-controlled hamlets. For all of the weapons and ammunition requirements, however, the enemy was almost entirely dependent on what came down the Trail. This was increasingly true with each passing year, as numbers of artillery, rocket, anti-aircraft, and even tank battalions were added to the order of battle. Also deliverable only from the north were all of the POL products (petroleum, oil, and lubricants) to keep the vehicular armada running – delivered first by truck, but later augmented by pipelines.**

2) **Although vast amounts of supplies were destroyed in transit by allied air power, in the long run, the enemy was able to keep hundreds of battalions in the field and also construct an extensive network of stockpiled caches inside the RVN and in its Cambodian and Laotian sanctuaries. Those caches near battle areas were essential when monsoon rains reduced the flow of supplies from the north to a trickle, and vital when offensives placed high demand on munitions expenditures.**

The increased reliance on trucks meant the need for all-weather roads with good drainage, concealed river crossings, vehicle repair facilities, etc. Over time, the People's Army of Vietnam was able not only to maintain an ever-increasing level of supply for its forces in the South, but also to expand the road network mile by mile until it spread throughout southern

Laos into Cambodia and deep into South Vietnam. Tracing the history of the expansion of that "limited all-weather" road network is analogous to following the enemy's slow but certain progress toward victory in the "protracted war" to reunite Vietnam. The A Shau Valley occupied a key position in that network. It served as a supply hub and staging area for all of the ARVN I Corps provinces and was positioned to support threats to the second and fourth largest cities in the Republic of Vietnam, Da Nang and Hue.

By 1 June 1966 the Communists' limited all-weather road system within Laos extended south to the Cambodian border and included several miles off the main trunk line down Rte 922 toward the A Shau Valley.[2] Unimproved Route 922 continued eastward, intersecting with Route 548 at the northern end of the valley. For the next six months or so, allied intelligence did not detect any road work in or near the valley. Then, in late 1966, a top-secret CIA Memorandum reported on "intensified Communist activity in A Shau Valley area":

12. According to U.S. aerial observers, repair and construction work extending Route 922 . . . northeast to a point approximately 10 kilometers across the Laos/Thua Thien Province border into the A Shau Valley has been in progress since at least mid-December 1966. There is evidence of considerable recent supply movements along this route, which is reportedly also being used for the infiltration of North Vietnamese Army (NVA) troops.[3]

The report went on to mention "a US/CIDG reconnaissance patrol" (likely OPN AUSTIN) that on 23 February 1967 had discovered that roads were being improved and were being guarded by many two- or three-man defensive positions that were elaborately camouflaged. The patrol was pulled out after engaging in a battle with an estimated two NVA companies.

Additional aerial photography was conducted and described by the National Photographic Interpretation Center (NPIC) in a report on 24 March 1967: "This is the first time a road, capable of being used by the

Communist forces, has been constructed connecting the road network of Laos to existing roads in South Vietnam."[4] *It would be difficult to overstate the importance of this development, given all that was to follow.* The NPIC memo detailed the construction work, which connected Route 922 all the way to Route 548 at the northern end of the A Shau Valley, and then as far south as the A Luoi airfield area. A follow-up memo from the NPIC on April 6 disclosed new vehicle tracks "following the alignment of Route 548 . . . passing near Ta Bat Airfield (abandoned) . . . and terminating at the SE end of the abandoned A Shau runway. Road clearing is observed SSE from the terminus of vehicle tracks to [several km south of the A Shau airfield]."[5] Just one week later, another NPIC message advanced the farthest point of enemy road construction by an additional 1.2 nautical miles.[6]

One year after taking Camp A Shau, the enemy had a road connecting the south end of the valley with North Vietnam. A CIA Memorandum on Communist road construction later summarized, by season, the construction of roads in Laos and adjacent areas. For the 1966–67 dry season (November 1966–March 1967), only 50 total miles were added to the network, while for the following six months, overall construction was negligible. Comparing those figures with just what was being done in the valley, it would appear that the enemy engineers' primary objective for an entire year was the work on Route 548.[7]

Central Intelligence Agency reporting in April 1967 disclosed growing concern with the A Shau. A cable on April 17 noted: "The new VC road into the A Shau Valley is reported to be replete with fortifications and anti-aircraft gun emplacements."[8] By April 24, when the first reports from Project DELTA's OPN PIROUS began to come in, "Trucks, trailers, and quantities of oil and/or gasoline drums have been sighted at both A Luoi and A Shau airfields" (see Map 3).[9]

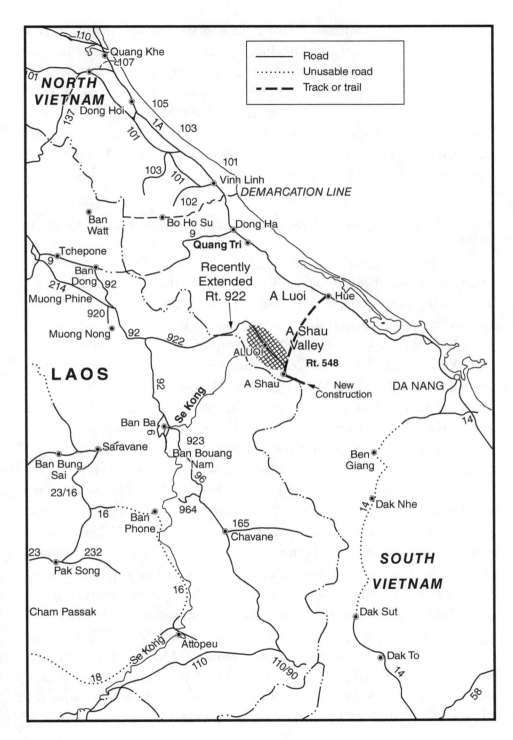

MAP 3

Source: CIA MEMORANDUM: *The Situation in Vietnam, 24 April 1967.*

The movement of enemy combat units into and through the valley also attracted the attention of allied intelligence. In June 1967, "at least one regimental-size unit has been reported in the area."[10] By November, "elements trained by the 304th Division have recently moved through the A Shau Valley area . . . where they became part of the 6th Regiment." The 31st Regiment (one battalion of which would take part in the attack on Hue) had been located by signals intercepts (SIGINT) just north of Route 922 in Laos and was believed to be heading for either A Shau or Khe Sanh.[11]

The Evolution of General Westmoreland's Plan for Operations in the Valley

For months, intelligence reports had shown the slow but undeniable construction of the enemy road network into and through the valley and the growing power of the NVA in the A Shau, and "Westy" had repeatedly expressed his concern. At a debriefing at HQ, USMC, on 1 August 1967, BG John R. Chaisson, USMC, Director, Joint Operations Center, MACV, remarked: "The two outstanding tactical issues that are involved between MACV and III MAF right now, I'd say are the A Shau Valley [and Khe Sanh]. General Westmoreland is still worried about the A Shau Valley. We've been plastering, as you all know, with B–52 strikes and tactical air, but he has not got out of his craw the fact that they are operating in the A Shau Valley and they are using it as a line of communication."[12] Five months later, speaking to a group at HQ, USMC, Chaisson reiterated COMUSMACV's concern: "Now, coincident with this threat in the north [re: Khe Sanh] is the age-old problem of the A Shau Valley. I guess, again, for those who've been out there, you know you can ring a bell and General Westmoreland will come out of the corner like a . . . like a pug. And two or three of the bells you can ring that get this reaction are A Shau Valley, the Do Xa [an enemy stronghold in southern I Corps], and, as I say, the Highlands. I know he hasn't quit on A Shau Valley."[13] Yet the only raid into the valley, by a single reinforced Marine battalion, had been cancelled in early August (OPN CLOUD), in large part as either the helicopters or the troops would likely be needed elsewhere. How did COMUSMACV plan to address his A Shau worries?

An important conference took place on 21 November, when President Lyndon Johnson met with his Vietnam advisors. The CIA prepared a top-secret summary of that meeting. Several items on the agenda pertained, directly or indirectly, to the A Shau. General Westmoreland hoped to be able to attack NVA Base Area 607 with "two Vietnamese battalions and a raid of three to four days." Base Area 607 is located in an area of Laos that protrudes into South Vietnam between the southern end of A Shau Valley in Thua Thien Province and Quang Nam Province to the south. The proximate reason for the desired raid was that a recent heavy rocket attack on Da Nang had been launched from BA607. General Wheeler (Chairman of the Joint Chiefs of Staff) remarked on the recent building of a "truck road through this area," referring to Route 548, with the implication that the rockets had come through the valley. The ARVN raid would never happen.

Near the end of the meeting, the President "said his main concern was that General Westmoreland get what he wants as soon as possible. General Westmoreland said from a practical standpoint *he had all he needed at this stage*" (emphasis added).[14] Westmoreland did not have "all he needed" in I Corps in November, but with the planned reinforcement by two additional army divisions (1st Cavalry and 101st Airborne), he expected to within the next several months.

Operation YORK[15]

In mid-August 1967 (i.e., almost immediately after the cancellation of Operation CLOUD), COMUSMACV informed his corps-level commanders of his intention to employ the 1st Cavalry Division "as a theater exploitation force," once enough of the battalions authorized in the latest increase in his troop "ceiling" were at hand. That increment would include, in I Corps, seven battalions of the 101st Airborne Division and the 11th Light Infantry Brigade. The plans for the First Cav were developed under the code name YORK. In the first version of YORK, units other than the Cav would carry out preliminary missions in II and III Corps. These were completed with little opposition by the end of November 1967.

In early December, the YORK plan was revised. In the modified plan, Phase II was to take the 1st Cavalry Division into the A Shau Valley sometime in April.[16] III MAF was instructed to prepare a detailed plan for this operation; that was submitted on 6 February, a week after the enemy Tet Offensive began. In the meantime, the siege of Khe Sanh began in January and the 1st Cavalry was sent north to support the Marines in Quang Tri Province. MACV formed a new headquarters, MACV (FWD), to plan for future operations in I Corps, and on 24 February transferred the planning for YORK II to that HQ.

The plans for the relief of Khe Sanh now took precedence over the earlier YORK concept of operations, and on 23 March YORK was cancelled permanently. It would be replaced by Operations PEGASUS (relief of Khe Sanh) and DELAWARE (A Shau reconnaissance in force). The logistical plans developed for YORK and helicopter and port facilities being prepared for the 1st Cavalry would be useful in the new planning cycle.[17]

A key takeaway about the YORK plans is that as army units were fed into I Corps, they were rapidly tasked with meeting imminent enemy threats and therefore were unavailable for offensive initiatives. Some 13 battalions (11 LIB, 198 LIB, & 2 brigades of the 101st Abn Div) did not arrive in-country until December 1967 or later, and by the time their base camps and logistical structures were in place, the Tet Offensive had already occurred. This is why Westmoreland's offensive in the A Shau was so late in coming. When the time for the reconnaissance in force into that "so long . . . unchallenged enemy domain" finally arrived, in April 1968, it was because "sufficient troops were at last available to deny the enemy undisputed use of the valley as a route to Hue."[18] Did the fierce fighting at Hue in February 1968 delay the A Shau operation or impel it? Not so much either. The April 1968 time frame carried over from YORK planning, and even after Hue, Khe Sanh was the commanding general's top priority for the 1st Cavalry Division.

One final quotation from Westmoreland's autobiography: "Having lost the lone CIDG Special Forces camp in the A Shau Valley in 1966 and *having lacked the forces to move back in* [emphasis added], we possessed no block such as Khe Sanh against enemy infiltration through the A Shau

Valley toward Hue. . . . at least eight VC and North Vietnamese battalions, equivalent to a division, infiltrated. . . . into the old imperial city."*[19] Yet while Khe Sanh, the importance of which to the enemy's line of communication paled in comparison to that of A Shau, was defended at all costs, A Shau had not even a solitary allied outpost.

The Combined Campaign Plan for 1968 in I Corps[20]

The planning for YORK went hand-in-hand with the issuance of the Combined Campaign Plan for 1968. That document, distributed on 11 November 1967, described the I Corps Scheme of Maneuver: "Military forces will be primarily committed in the anti-main force war to contain the enemy along the DMZ and the Laotian border and to destroy the 2nd NVA Div and other enemy forces and bases in I CTZ." Curiously, Inclosure I of that plan, which mapped "Areas for Priority of Military Offensive Operations, "excluded the A Shau Valley and much of the mountainous western part of I Corps.

Annex K of the Plan covered "Employment of Civilian Irregular Defense Group." Appendix 2 to that Annex provided for the (re)establishment of three (!) CIDG camps to interdict infiltration and supply routes through the A Shau Valley:

1. I CORPS TACTIAL ZONE

(a) Establish Nam Dong Camp (YC8783) during 1st Quarter CY68, to interdict the infiltration routes from southern A Shau Valley north to Hue and east to Danang

(b) Establish Ta Bat Camp (YC4295) during 2nd Quarter, CY68, to interdict the infiltration routes from Laos, the northwest-southwest route in A Shau Valley and the old road east to Hue [Route 547]

(c) Establish Thon Luu Camp (YD3828) during 3rd Quarter, CY68, to interdict routes out of the northern A Shau Valley to Hue and Quang Tri

---

* The final estimate is that 16–18 enemy battalions were involved in the battle for Hue.

How three CIDG camps some 25 miles apart were to materially impede the enemy main force units and how the camp at Ta Bat was to be protected from the NVA units just across the border, including artillery, are open questions. There were only two other CIDG camps in the three northernmost provinces of I Corps, and one of those, Lang Vei, would soon disappear under the treads of NVA tanks.

Annex L, "Neutralization of VC/NVA Base Areas," was equally audacious: "Base area goals for CY68 are 80% of the identified base areas in RVN *neutralized on a continuing basis* by the end of the year" (emphasis added). There were 46 base areas, so about 37 were to be permanently put out of action. This annex contains several paragraphs that display astonishing hubris, if not complete ignorance of the preceding two years of the American war experience.

Para 3.(3) "Maximum use will be made of B-52 bomber strikes and other Air Force and Naval firepower, with rapid follow-up by ground forces to complete the neutralization." Bombers had never and would never achieve anything close to "neutralization," and the notion that a bit of "rapid follow-up" could complete, let alone sustain, neutralization is absurd.

Para 3.a.(4) "Base areas determined to be habitual safe havens for the enemy will be selected for total destruction when warranted. Jungle growth which provides the cover and concealment necessary for these areas to be of value to the enemy will be destroyed when feasible by cutting, clearing, or other appropriate measures." Having seen these tactics repeatedly fail to be sustainable in a host of enemy bases in III Corps (War Zones C & D, Iron Triangle, Ho Bo Woods, etc.), it is incredible that they are proposed here as a panacea.

And again in Para 3.(3), "For those bases where complete neutralization and permanent denial to the enemy is not possible, repeated air strikes and random pattern ground operations will be conducted. . . ." Since all but nine base areas were to be leveled (80%), we may presume that in no more than that number would it be "not possible" to achieve "permanent denial."

What were they thinking?! How many "battalion-months" would be needed to "permanently" close 37 base areas and randomly interdict 9 more? Did this plan assume that the enemy would meekly crawl across the border into Cambodia or Laos when in-country bases were attacked and remain there when the allied troops moved on?

As the years 1968–1969 would show, only a continuing presence of allied ground troops could keep the enemy from rapidly rebuilding installations and reinstituting supply lines through the A Shau Valley. Air power could and did contribute much to the allied war effort, especially when used in remote areas with identified targets (including road "choke points") or in close support of troops in contact, but it was not effective as an area denial weapon in unobserved, heavily jungled terrain.

The Communist Line of Communication Developments – November 1967 through February 1968

By early 1968, evidence was observed of a development in the A Shau that was unique at this stage of the war in South Vietnam: an apparent effort to prepare the abandoned A Luoi and A Shau airstrips for use as drop zones, helicopter pads, or even runways for North Vietnamese fixed–wing aircraft.[21] Sound like fiction? It was not. On 6 February 1968 the CIA intercepted an enemy radio message from Laos stating that "our propeller aircraft would be landing in the A Shau Valley on 6 February."[22] A partially redacted follow–up report noted,

These planes probably succeeded in parachuting some supplies into the A Shau Valley area. On the 7 February flight the aircraft took off from Gia Lam [in NVN], passed over Bai Thuong and crossed the Demilitarized Zone in the vicinity of the Laos border. As the IL-14s flew south they were sighted near Khe Sanh by USAF fighter pilots who described them as 'unmarked, silver, twin-engine, C-47 type transports.' The precise location of the IL-14s' objective is not known, but it may have been the probable position of the North Vietnamese Army rear services headquarters in the A Shau Valley. [i.e., Binh Tram 7 in the Punchbowl] . . . at least two planes could not find the probable drop zone because of bad weather. The cargo delivered to the A Shau Valley area by the IL-14s is not known. . . . It is possible that the IL-14s were involved in replenishing essential items such as communication gear or specially trained individuals.[23]

The CIA report does not explain why the USAF failed to engage the enemy transports, but the U.S. pilots may have believed they were observing Air America or CIA "black ops" aircraft.[†]

In reading the Central Intelligence Agency reports from November 1967 through March 1968 on the enemy progress in developing all-weather lines of communication into and through the A Shau Valley, one cannot fail to perceive growing alarm at the rapid pace of the NVA engineering effort (see Map 4).[24] There were two primary focal points for new road construction. In the central valley, a new route from the Ta Bat vicinity to connect with Route 547 at Ta Luong, 22 miles southwest of Hue, was the major supply line for the enemy troops in Hue during the Tet Offensive. While construction continued on the peripheral arteries, Route 548 was open for traffic, and traffic was good. "Aerial photography of A SHAU Valley indicates the valley road to be passable and continuously used by 1½ and 2 ton trucks."[25] Ominously, by early March, "the North Vietnamese have introduced tanks in the A Shau Valley and on the newly completed road between A Shau and Hue. At least two tanks were detected on this route at a point about ten air miles southwest of Hue."[26]

At the southern end of the valley, improvements in the road network connecting Route 548 with Laos Route 923 and RVN Route 14, leading to Da Nang, would be a major expansion of the Communist logistical network threatening the second largest city in South Vietnam and the second largest allied base complex (next to Saigon / Long Binh / Bien Hoa) in the country. This road would become known to the allies as the "Yellow Brick Road." It traversed Base Area 607 in Laos, joining Route 548 in Thua Thien Province to Route 14 in Quang Nam Province. After

---

† An incident in November 1967 involving another type of North Vietnamese aircraft operating over the A Shau is reported on the internet but not confirmed by any available official record. U.S. Army CPT Ken Lee of the 131st Aviation Company (Surveillance Aircraft), piloting an OV-1 Mohawk en route to Laos, was attacked by a MIG-17 fighter. The Mohawk was peppered with 23mm cannon fire, but in return was able to hit the MIG multiple times with its .50 cal and 2.75" rockets. When last seen, the enemy fighter was "well ablaze" and headed into a cloud bank over the A Shau; the Mohawk was able to return to Phu Bai. This is claimed as the only Army air-to-air victory after WWII (when the Army Air Corps was the predecessor to the USAF). See https://www.avgeekery.com/avgeekery-exclusive-mig-killing-army-ov-1-mohawk-pilot-ken-lee/.

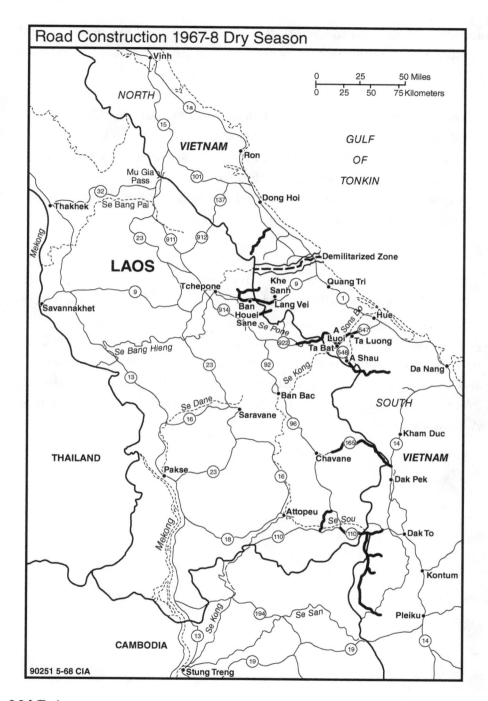

MAP 4

Source: *Intelligence Memorandum: Road Construction in the Laotian Panhandle and Adjacent Areas of South Vietnam 1967-1968. May 1968.*

abandoning work on this artery at the end of the 1967 dry season (April), construction was resumed, reaching a feverish pace just prior to Operation DELAWARE:

> The Communists have continued Route 548 through the valley into a salient of Laos southeast of A Shau [Base Area 607]. Between 28 March and 9 April the enemy built 18 miles [!] of road through dense jungle and diffi- cult mountainous terrain east-southeast in the general direction of South Vietnamese Route 14. The last reported terminus of the road was in South Vietnam, some 10 miles northwest of an enemy-controlled town on Route 14 west of Da Nang."[27]

The North Vietnamese were also in the process of establishing an exten- sive landline network that would improve the reliability of their com- munications and protect them from allied signals intelligence (SIGINT) penetration. By March 1968, "Branch 2A extends east from Line 2 . . . near the junction of Routes 92 and 922, generally parallels Route 922 . . . and crosses the Laos/South Vietnam border. . . . From this point, Branch 2A extends southeast approximately four nm toward Ta Bat Airfield in the A Shau Valley."[28]

9

OPERATION DELAWARE – PRELIMINARIES

The First Cavalry Receives Marching Orders

Planning by COMUSMACV (GEN Westmoreland) for an incursion into the A Shau Valley was initiated in August 1967, with the YORK-series planning documents. Once the prerequisite deployment of three divisions of United States Army troops into the coastal areas of I Corps was completed (1st Cavalry Division, 101st Airborne Division, and the American Division [formerly Task Force OREGON]), the allied commander finally expected to have enough troops at hand to consider an offensive move in western Thua Thien Province. When the communist Tet Offensive struck at the end of January 1968, those plans had been shelved but by no means forgotten. Rather, the allies recognized that the enemy's tenacious fight for Hue was supplied and reinforced via the A Shau, making it a more important objective than ever.[*1] Once the Tet battles in the northern part of South Vietnam, including the prolonged battle for Hue, had subsided, the allied command turned immediately to the relief of the besieged Khe Sanh Combat Base, carried out by the First Cavalry Division in Operation PEGASUS in coordination with the Marines in SCOTLAND II.

---

* See cable from Rostow to LBJ 22 Feb 68 reporting on GEN Johnson's conversation with GEN Westmoreland that morning: "General Westmoreland expressed the view that Hue may have been the enemy's primary objective from the outset of the Tet Offensive. Reports indicate an enemy effort to reinforce Hue through the A Shau Valley and he is watching the A Shau Valley very carefully."

In mid–March, GEN Westmoreland prepared a top-secret eyes-only memo for LBJ on the subject of "I Corps Operations":

1) **I have reviewed Gen. Cushman's [CG III MAF] plans for the next six–eight weeks in Northern I Corps. Priority will be given to destroying enemy forces in coastal areas of Thua Thien and Quang Tri; blocking enemy use of Route 547 and interdicting his line of communication in the A Shau valley; and mounting offensive operations against enemy forces in Khe Sanh areas; to include opening of Route 9. The latter is planned to commence about 1 April [which became the start date for PEGASUS].**

2) **As a result of [the] above priorities, with extreme demands on logistic and helicopter support, Gen. Cushman does not plan to execute the assault and occupation of A Shau valley and raids into the associated base areas 607 and 611, as had been planned and approved for approximately 1 April. In lieu of assault operations into the A Shau valley, extensive reconnaissance operations by U.S. and Vietnamese special forces and organic assets will continue along Route 547 to the valley. The reconnaissance will be backed up by at least one brigade of the 101st Airborne, operating out of Cumberland [Bastogne] fire base. Targets in the valley will be developed for air and artillery attack. Raids of short duration by air mobile forces may be conducted to harass and destroy enemy logistical installations.[2]**

Several key points about COMUSMACV's strategic thinking in relation to the A Shau Valley as of 15 March 1968 are revealed. First, although he continued to indicate satisfaction with the overall level of forces available to him, a paucity of "logistic and helicopter support" in I Corps was a clear constraint on operations, even with the shift of the three army divisions. That shift had caused the overall logistical structure in the northern part of the country to be temporarily stretched beyond capacity.† Second, despite Westy's professed keen concern about the enemy's logistical buildup in the A Shau, any major ground operation there would have to come after (1) enemy forces in the coastal areas were destroyed, and (2) Khe Sanh was relieved and Route 9 opened. Third, nothing more than "raids of short duration by airmobile forces" could be expected for the A Shau earlier than late April to mid-May. Since we know that the mid-April to mid-May period in early 1968 was assumed to be the best, if not the only, time weather would permit a major operation into the A Shau (an assumption that would be invalidated in April, when it was learned that low clouds and fog were a greater impediment to air operations than the monsoon rains), it would therefore appear that "the assault and occupation of A Shau valley and raids into the associated base areas 607 and 611" that Gen Cushman "does not plan to execute" would be postponed indefinitely. From an NVA perspective, we may conclude that the Tet Offensive had successfully, if inadvertently, acted as a spoiling attack to delay or prevent the American move into the A Shau, originally scheduled for "about 1 April," but not by much.

Once PEGASUS began, the expectation of significant NVA defense during PEGASUS was quickly dispelled, as the enemy chose once again to fade away in the face of a major commitment of force (see Mao Tse Tung: "enemy advances, we retreat"). The NVA's rapid withdrawal from the Khe Sanh plateau opened the door, in turn, for the First Team to leave the "mopping up" around Khe Sanh to just two of its battalions (the 2nd Brigade, with the 1/5 and 2/5 Cavalry), while the rest of the division moved south to Thua Thien Province.

As early as the fall of 1967, 1st Cavalry Division planners had considered "possible operations [YORK]" into the A Shau Valley. But it came as a considerable surprise to Division Commander Major General John J. Tolson, III, when, at noon on 10 April at LZ STUD, he was told by

---

† There is a bit of irony here: allied operations in I Corps were *constrained* by a limited logistical base; the enemy offensive at Hue was *empowered* by the logistical base in the A Shau Valley.

his commanding officer, LTG William B. Rosson of Provisional Corps Vietnam (PCV), that he was to commence a division-sized "reconnaissance in force" into that valley less than one week hence, with preliminary recon to start in less than 72 hours.[3] This was particularly unsettling news as the First Cav was at that time still heavily committed to Operation PEGASUS. Key to this sudden major operational shift in the northernmost part of South Vietnam was the weather in the A Shau Valley. It was expected that April and May would offer the best (a very relative term given the A Shau climate) opportunity all year to go into the valley with a large air assault, air support, and aerial resupply.

The following day, MG Tolson began the withdrawal of his battalions from the Khe Sanh plateau. For the attack into the A Shau, he would have his 1st and 3rd Brigades, with the 2/12 Cavalry in division reserve and the 3rd Regiment of the 1st ARVN Infantry Division attached. Simultaneous with the First Team's "cavalry raid" into the valley, the 101st Airborne Division would move down Route 547 to eliminate the enemy units and bases along that corridor and establish long-range artillery outposts to support the cavalrymen. Artillery support was a critical concern, as this would be the first time the cavalry division had conducted a major air assault beyond the range of its own 105mm and 155mm howitzers.[4] Early delivery of the big guns to a group of mutually supporting firebases was important; the delay in establishing one of those positions (VICKI/GOODMAN) would create fire support problems for another (LZ TIGER).

On 14 April, with aerial recon in the valley already underway, MG Tolson presented his concept for the operation to LTG Rosson, LTG Cushman (the USMC III MAF commander), and GEN Westmoreland. The plan was for "the 1st Brigade leading the assault into the center of the valley around A Luoi Airstrip. This brigade was to be followed subsequently by the 3rd Brigade to the north and the regiment from the First ARVN

Division to the south."‡5 The senior commanders concurred completely with Tolson's plan; the actual start date (D–Day) would depend on how soon the preliminary scouting and air attacks could be completed. The aerial reconnaissance was of critical importance, as so little was known about the extent of the enemy buildup in the A Shau over the preceding two years.

Binh Tram Seven

American intelligence, relying almost exclusively on aerial reconnaissance, knew very little about the composition or missions of enemy forces in the A Shau Valley at the commencement of Operation DELAWARE in April 1968. It was estimated that there were about 3,000 NVA (seven battalions) in the A Shau, including the 280th Antiaircraft Artillery Regiment and part of the 3rd Battalion, 203rd Armored Regiment. There were no known line infantry units present; engineers and support troops comprised the rest of the expected enemy order of battle.[6] Subsequently, it would become known that the NVA command element for the area, Binh Tram 7, was located in the "Punch Bowl" area west of A Luoi. Binh Tram 7 was the largest subordinate of the 559th Transportation Group, the overall command for the Ho Chi Minh Trail, and while most of the binh trams (military stations) were considered by allied headquarters to be regimental-sized units, the A Shau Valley force was in excess of divisional strength, consisting primarily of engineer,

---

‡ MG Tolson had expected to also have an ARVN airborne task force and his own 2nd Brigade under command. GEN Westmoreland denied him the use of either. However, as the ARVN TF mission was in Base Area 114, some distance north of the valley, and the role of guarding the First Cavalry's base areas outside the valley was assumed by the 196th Light Infantry Brigade, these changes had no impact on the plan of operations in the A Shau Valley.

truck, and anti–aircraft battalions. In early 1968, Binh Tram 7 included the following:[7]

Anti-aircraft – 280th Regiment with two 37mm battalions (12th & 23rd), one 23mm battalion, & 2 companies (4th & 22nd)

Bulldozers – 1 battalion

Counter-reconnaissance – 1 battalion[8] (may have been the battalion from the 312th Infantry Division)

Infantry – 1 battalion (4th Bn/ 312th NVA division) & 1 security company (643rd)

Engineers – 4 battalions (K3, K4, K5 & 52nd) & 1 female platoon (B)

Medical – 1 company (2 dispensaries [hospitals], first-aid stations, & treatment teams)

Tanks – 1 battalion (18 x PT-76)

Trucks – 3–4 battalions (29th, 53rd, 55th [?] & 90th) & 1 company (71st) (mostly 2-ton GAZ-63)

The report "Organization and Activities of the 559th Transportation Group"[9] was issued on 13 June 1968 by BG Phillip Davidson, commander of the Intelligence Section (J2) of MACV. With the capture of a treasure trove of NVA documents during DELAWARE, the magnitude of the NVA effort in the A Shau was revealed (see Maps 5 and 6). The logistical base was "massive" and included "a road system 125 kilometers in length. During the [Tet] Offensive, truck convoys used this road system to supply the Thua Thien and Hue battle areas." Not only did Station 7 in the valley support the Hue area, it also included a road network to the southeast, connecting with Highway 14 in Quang Nam province and the Da Nang area. Farther to the south, a route from A Shau led to the B3 Front in the Kontum area of the Central Highlands.

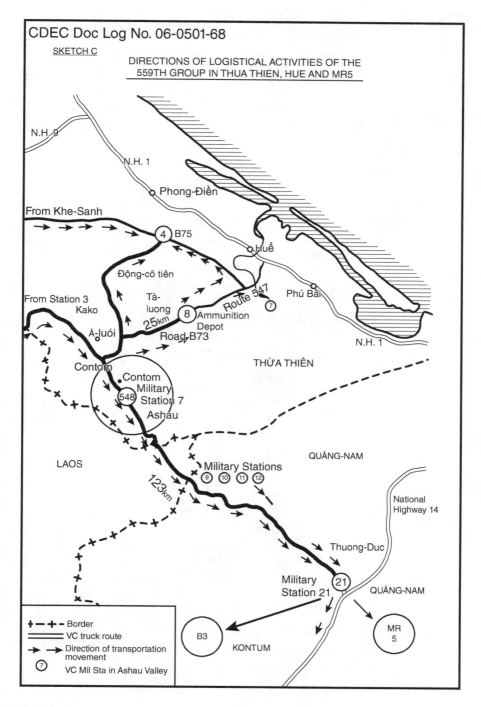

MAP 5

Source: Department of Defense Intelligence Information Report: *Organization and activities of the 559th Transportation Group, 13 June 1968.*

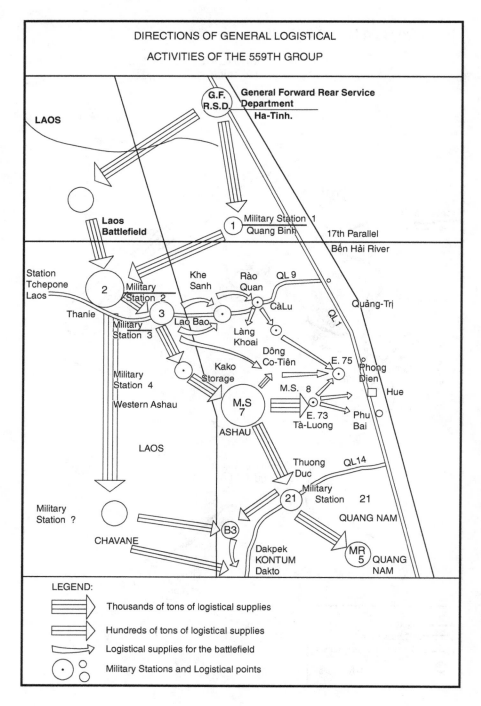

DIRECTIONS OF GENERAL LOGISTICAL

ACTIVITIES OF THE 559TH GROUP

MAP 6

Source: Department of Defense Intelligence Information Report: *Organization and activities of the 559th Transportation Group, 13 June 1968.*

The engineers assigned to Military Station 7, beginning in early December 1967 and finishing on 21 February 1968, had built a new road 25 km long connecting Route 548 with the Song Bo River Valley (the Song Bo flows northward and is 12–15 km west of Hue) and constructed an ammunition dump just 30 km from Hue. By this route, known to the NVA as B.73 and later to the allies as 547A, the enemy forces in the Hue and Phu Bai battlefields were supplied during Tet. LTG Cushman, CG III MAF, later facetiously remarked: "Lo and behold, they started building their share of the rural development here, and apparently, they're coming to meet the road we had built."[10] "During the last four months of 1967, and in early 1968, BT [Military Station] 7 depot at A Shau delivered thousands of tons of food and munitions supplies to the Hue Front, 1,200 tons to Military Region 5 [in the direction of Da Nang], using hundreds of vehicles, moving in convoys every second night, each convoy carrying hundreds of tons of war supplies." In spite of allied air attacks, Binh Tram 7 "met 162 percent of the planned requirements."

Female units were noted for their contributions in the A Shau area. The female engineer platoon "broke stones, paved roads, felled trees, etc." during 18-hour workdays. One of the two medical dispensaries was primarily staffed by women, and the BT 7 magazine reported that "all wounded and sick soldiers were pleased with this dispensary."

Special Forces Reconnaissance – Early 1968

As to date no original records pertaining to MACVSOG participation in Operation GRAND CANYON, which was fundamentally an air campaign, have been unearthed, it is unclear how early in 1968 MACVSOG Command and Control North (CCN), FOB-1 at Phu Bai was tasked with sending recon teams into the A Shau Valley. According to Robert Gillespie, the "teams . . . were promptly chewed up by PAVN," who employed counter-recon teams with devastating effect. A new FOB commander, LTC Robert Lopez, was assigned. "He was quickly irritated by the number of failed and aborted missions in the A Shau and was infuriated when team leaders refused operations in what had come to be called the 'Valley of Death.' In an effort to motivate his men, Lopez accompanied a team into the A Shau on 4 March."[11] On 6 March he was killed in action during a team insertion, when the Marine CH-46 helicopter from HMM-165, piloted by

MAJ William H. Seward, was shot down near Ta Bat. MAJ Seward was also killed, but the remaining crew members and passengers survived.

FOB-1 continued to attempt to place recon teams on the ground in the A Shau, weather permitting. On 20 March, one team was exfiltrated. An attempt to insert another team was aborted under enemy fire. A third team operating near A Luoi reported on suspected NVA caves and tunnels.[12] Two teams were dropped in the valley on 22 March.[13] On 25 March, a wiretap mission resulted in heavy enemy contact and was also aborted. Two teams were in the A Shau at that time; one called for an air strike, resulting in the destruction of a cache of about 150 artillery rounds.[14]

When the teams were extracted on 27 March, one was under fire.[15] That team of two Americans and four indigenous soldiers had completed their patrol on the western side of the valley, about 3 km south of Ta Bat and just 5 km southwest of where Major Lopez was shot down. While awaiting extraction, the team came under attack by a large enemy force. In order to permit the rest of his unit to commence escape and evasion (E&E) from the elephant-grass-covered pickup zone, the team leader, SSG Johnny C. Calhoun, provided covering fire until he was hit several times in the chest and stomach by enemy fire. He then pulled a pin from a grenade and clutched it to his body, taking many of the overrunning enemy with him into eternity. Johnny Calhoun was posthumously awarded the Distinguished Service Cross.[16] The remainder of the team was able to escape the enemy pursuers, engaging in several firefights along the way, and was rescued some 20 hours later.

On 28 March, yet another Studies and Observations Group (SOG) mission was inserted without attracting enemy attention.[17] The next mention of the SOG reconnaissance comes in Westmoreland's report to GEN Wheeler on 9 April: "Two ground reconnaissance teams inserted into A Shau during 7 to 9 April reported contact with enemy armed with automatic weapons in both missions. A third team remaining on road and river watch reports no contact."[18] There is no further mention of the SOG operation in Westmoreland's reporting, which ended on 12 April.

While the Green Berets of MACVSOG who sought to patrol in and around the A Shau were dodging (with limited success) NVA counter-reconnaissance elements, another Special Forces operation to the east was carried out, also under frequent enemy contact but with notable success, paving the way for the 101st Airborne Division to approach the A Shau down Route 547 as part of Operation DELAWARE. A full narrative of Project Delta's Operation SAMURAI IV, which took place entirely outside the A Shau, is beyond the scope of this book. However, as the intelligence gathered there illuminates how much the development of the road network into and through the A Shau Valley enabled heavy NVA use of both wheeled and tracked vehicles deep into South Vietnam, a brief consideration of SAMURAI IV is warranted.[19] Commencing on 4 March 1968 and ending on 4 April 1968, Detachment B-52 (Project DELTA) of the 5th Special Forces Group was assigned a mission "to conduct reconnaissance in force, surveillance and interdiction missions primarily along highway 547 in Western Thua Thien Province and to determine if alternate routes exist linking ASHAU Valley and HUE."[20] The patrols of SAMURAI IV covered new Route 547A (built by the enemy!) along the Rao Nho Valley from within a few miles of its junction with Route 548 in the A Shau Valley to its junction with Route 547 in the Rao Nai Valley. A large number of trucks were seen and/or heard and some were destroyed by air strikes, including at least four B-52 ARC LIGHT missions on 21–27 March. The trucks were observed along Route 547A in convoys as large as 14 vehicles. Several tanks, including at least one T-34 main battle tank, were also detected, and enemy anti-aircraft fire of 12.7mm and/or 14.5mm was regularly encountered, as were some heavier 37mm guns. In a total of 30 CCN insertions, including 10 of company size, there were 30 contacts with the enemy, who also fielded units as large as a reinforced company. In evaluating the condition of the roads, the after action report concluded: "547 and 547A are both improved to the point where they can be considered all weather secondary roads capable of handling traffic up to and including light to medium tanks." There was ample evidence that the enemy was constructing a "rear logistical and distribution area for future operations against the Hue/Phu Bai area" in the vicinity of the junction of the Rao Nho and Rao Nai Rivers, and of Routes 547 and 547A (see Map 7). This area would be known as Delta Junction (after Project DELTA) to the American troops soon to be entering in force. None of these developments, of course, would have been possible without the establishment in the first instance of a strong base in the A Shau Valley, the enemy's "back garden" gateway to northern I Corps.

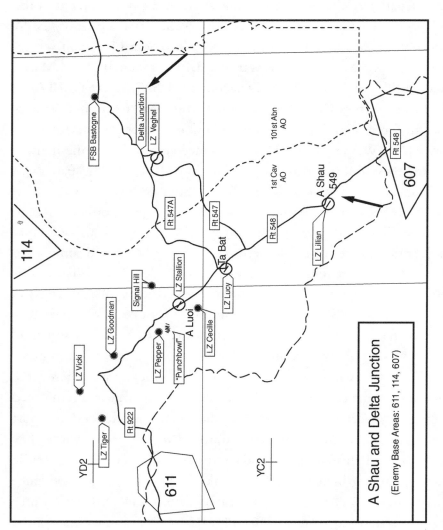

MAP 7

A Shau and Delta Junction
(Enemy Base Areas: 611, 114, 607)

Source: *Project CHECO Southeast Asia Report #91 - Special Report: Operation DELAWARE - 19 April to 17 May 1968, 02 September 1968.*

The First of the Ninth Cavalry Reconnoiters the A Shau (13–18 April)

The First Cavalry Division (Airmobile), aka "the Cav," the First Cav, or the First Team, was built around nine battalions of airmobile infantry. Those units are often mistaken for "air cavalry," when in fact they were light infantry that could operate either as "ground pounders" or, with the assistance of UH-1 Huey and CH-47 Chinook helicopters from the division's three helicopter battalions, could be transported by air, even into protected ("hot" or "red") landing zones if necessary. The UH-1 Huey transports were referred to as "slicks," while heavily armed Huey gunships were called "hogs" (later replaced by the AH-1G Huey Cobra).

The First Team's divisional reconnaissance squadron, the 1st Squadron, 9th Cavalry (Aerial Reconnaissance), was a true air cavalry unit with its own organic helicopter assets, including armed gunships and enough slicks to transport its Blue (rifle) platoons. The 1/9 Cavalry, under the command of LTC Richard W. Diller, was the ideal unit for recon in the A Shau Valley.

General John J. Tolson, the division commander, had scheduled three days of preliminary aerial reconnaissance and air strikes in the valley prior to the original D-Day of 17 April, with the primary objective of neutralizing the enemy anti-aircraft batteries that could wreak havoc on the helicopters that would deliver the airmobile troopers to the valley. The first day of the recon, 13 April, the weather disrupted the best-laid plans of the allies, and the 1/9 was unable to identify any enemy positions.[21] Bad weather continued for the next two days, with only four AAA sites identified, one of which was destroyed by an ARC LIGHT (B-52) strike.

General Tolson had made the entire operation conditional on his having three full days for the preliminary scouting and air strikes.[22] With the approval of his boss, Provisional Corps Vietnam commander General Rosson, D-Day was pushed back to 19 April, and the 1/9 Cav was then able to scour the valley during 16–18 April for enemy anti-aircraft guns, to direct air strikes against anti-aircraft artillery (AAA) and automatic weapons (AW) positions, and to identify air routes to the designated landing zones that would minimize exposure for the troop-carrying helicopters on D-Day.[23] On 15 April, the air cavalry helicopters conducted high-level surveillance, becoming familiar with the geography in and around the valley.

On the 16th, "the helicopters started getting down into the valley, utilizing reconnaissance by fire, better known as 'peeping and snooping.' They would go down to tree top level and locate a gun emplacement or other enemy position by an actual eyeball-to-eyeball sighting."[24] It was, in the modern vernacular, a "target-rich environment." "On D-3 [16 April] the anti-aircraft targets were not very hard to find. Everywhere the helicopters flew that day *it seemed to explode on us* [emphasis added]. They were mostly to the high ground on the northern side of the valley and also in the southwestern portion. . . ."[25] Tactical air strikes were scheduled every half hour, with additional fighter bombers available as needed. ARC LIGHT strikes were plotted for areas of especially high concentrations of enemy installations. During the three days of preparatory activity, the 1/9 Cavalry identified over 30 enemy anti-aircraft gun (up to 57mm) and .50 cal automatic weapons emplacements. Many of those were engaged and eliminated by tac air and the AH-1G Huey Cobra gunships of the air cavalry squadron.[26] Air strikes also resulted in a large number of secondary explosions, the largest result coming from a strike on an NVA command post and vehicle park on the evening of 16 April.

The NVA AAA fire was particularly strong in the vicinity of A Luoi in the central valley. The initial plan for DELAWARE was that the 1st Brigade, under COL John F. Stannard, would air assault onto the A Loui airstrip on D-Day to establish a base where aerial resupply could be maintained. When the anti-aircraft fire remained intense in this area on D minus 2 (17 April), MG Tolson decided to begin the operation in the northern part of the valley, with a 3rd Brigade (COL Hubert S. Campbell) air assault onto Tiger Mountain.[27] The general explained this decision as follows:

During the latter days of the reconnaissance by the 1/9, I could not get any assurance that an assault into the A Luoi area would not be very costly because the heaviest anti-aircraft encountered was in that area; and although a lot of it had been neutralized, there were still new positions that appeared each day. . . . [The alternative plan] I selected was going into the northern part of the area first because the anti-aircraft had not been as intense there. . . . By going up north first, I would immediately cut off the entrance into the valley of the new highway coming from Laos. . . .[28]

The 1/9 Cav "was tasked with choosing landing zones for the brigade and developing landing zone preparations. . . ."[29] The reconnaissance of the approaches to the northern valley determined that the safest route for the assault helicopters would be to approach from the northwest and depart to the southeast (more on this later).

The 1/9 Cavalry's opening role in Operation DELAWARE was costly. According to Shelby Stanton, 50 helicopters were hit by enemy fire, with 5 destroyed and 18 damaged too badly to be brought back into service.[30] Yet for all the damage to the machines, not a single life was lost by the cavalry squadron—truly a testament to the highly skilled army air crewmen of the First Cav! Likewise, neither the air force nor the marine airmen suffered any casualties during the run-up to D-Day, although 209 tactical airstrikes and 21 by B-52s were called in. One still unresolved question is whether the NVA employed radar-directed anti-aircraft fire in the valley in 1968. The Air Force concluded that "there was no evidence that either 23-mm or 37-mm weapons were radar-controlled or could accept radar data." There is disagreement about this from some of the helicopter crews, fighter pilots and forward air controllers (FAC), who encountered anti-aircraft explosions (flak) at up to 6,000 feet.[31]

In spite of the intense air recon and attacks of 16–18 April, most enemy AAA and AW guns remained undetected. The NVA used a variety of tactics to protect their anti-aircraft weapons until the time when they could be used to maximum effect:

1) **The NVA and VC were absolute masters of camouflage, and even the heavier guns were rarely observed. "A big problem we had with the 37mm was that our scout birds could not locate them." "Even the 23mm-gun that shot down several helicopters around LZ TIGER could not be located, although it was within a mile of the landing zone."[32]**

2) **The NVA built multiple positions for each weapon, frequently moving them around during the night. This greatly reduced the effectiveness of B-52 strikes, in particular, as by the time the planes arrived, the enemy was gone.**

3) **The enemy anti-aircraft gun crews were highly trained and seldom revealed their locations by firing until they had a sure and worthwhile target.**

The preliminaries were nearing an end. The morning of 19 April 1968 would mark the first American regular infantry troops ever entering the A Shau Valley.

Signal Hill – The First Outpost

One consistent problem for the allied recon teams that previously entered the A Shau was the difficulty of establishing communications with their bases outside the valley, located far away over many miles of deeply jungled hills and valleys. The patrols relied on helicopters or planes overhead to relay their messages, but with the much larger number of units to be operating in the valley during Operation DELAWARE, a fixed relay site would be needed to ensure a reliable radio net. The location selected, which would be known as Signal Hill, was atop the almost 5,000-foot-high Dong Re Lao Mountain, about 4 km northeast of the A Luoi airstrip.

The unit given the mission of occupying and protecting the signal facility was the First Cav's Ranger company, Company E, 52nd Infantry (Long-Range Patrol), with two engineer squads and some signalmen attached. A platoon of the "lurps," as they were known, would rappel from helicopters onto the hilltop, blast a landing zone out with explosives, and guard the site until more troops, including artillery, could be brought in. The D-Day mission was assigned to the 2nd Platoon under 1LT Joe Dilger.[33]

Staging out of Camp Evans, the major First Cav base 30 km to the east, the first wave of troopers, just 20 men in 4 UH-1 Hueys from the 227th Assault Helicopter Battalion[34] arrived over the hill on the sunny morning of 19 April 1968 and began to descend by rope 100 feet to the peak. The thin air at the high altitude barely provided sufficient lift to keep the choppers flying, and one of the Hueys during the insertion, with two lurps dangling 50 feet in the air, lost power and crashed into the jungle below. One of the rangers, SGT Larry Curtis, was concussed and trapped under one of the helicopter's skids. Those still on the aircraft, including the crew, were all unconscious but alive after the crash. With the fast-thinking work of the men from the other helicopters, who descended safely onto the LZ, the injured men were all saved and the equipment and explosives removed from the destroyed chopper before a major disaster could occur. The soldiers quickly commenced work with Bangalore torpedoes (explosives), chain saws, and machetes to clear the landing zone for reinforcements. By the end of the day, 14 helicopter-loads of troops had arrived and the crew of the downed helicopter was evacuated, without encountering any enemy fire.[35] Maybe it was too easy.

On the morning of 20 April, the Signal Hill LZ was still not ready to accommodate a helicopter, so SGT Curtis was evacuated by McGuire Rig, as the troopers continued to toil without rest to clear the landing area. While the Americans worked, the enemy climbed. NVA soldiers, ascending rapidly from the valley far below and unheard over the din of the land clearing work underway, began to harass the toiling troopers from unseen positions with accurate sniper fire at about 1130 hours. The First Cavalrymen responded with grenades, unable to actually see any of their besiegers. Soon, CPL Dick Turbitt and PFC Bob Noto were dead and SGT William Lambert and engineer PFC James McManus mortally wounded. LT Dilger was severely wounded in the chest. His survival was doubtful (he made it), and six other GIs were also wounded.[36] Late in the day, the LZ was reinforced with a platoon from the ground troop of the 1/9 Cavalry (D/1/9).[37] The heavy demand for slicks throughout the valley would deny the wounded evacuation that day.

On the morning of 21 April (D+2), the wounded were finally evacuated, and the company commander of E/52 Infantry, CPT Michael Gooding, arrived with more rangers. CPT Gooding assigned SGT Doug Parkinson to take his team and clear the area around the LZ. SGT Parkinson and his men, slogging through mud and the dense, tangled, impenetrable undergrowth, made two patrols around the hilltop. On the first, the team's lead scout, a Montagnard tribesman named Dish, was approached by a lone NVA soldier who mistook him for a comrade. It was his last mistake. In the late afternoon, two 105mm howitzers from Battery A, 1/21 Artillery arrived with a load of ammo. They would soon be sighted in and would support B/2/7 Cav at LZ PEPPER that night.[38] After three grueling days, Signal Hill was fully operational. For the next four weeks, the garrison on Dong Re Lao would fulfill its artillery and signals missions, while occupying a ringside seat (although sometimes from above the cloud line) for Operation DELAWARE.

The Signal Hill outpost would function as a "silent partner" to the infantry searching and skirmishing in the A Shau. It would also need to be resupplied when the weather permitted. On 25 April, a UH-1C gunship escorting a supply run to Signal Hill had an engine failure at high altitude en route. The attempted autorotation failed and the helicopter crashed about 14 km north of A Luoi, killing WO1 James E. Hoyt, 1LT Kenneth G Spencer, SP4 Michael P. Makuck, and SP4 James Kendall from D/227th AHB.[39]

A CH-47 Chinook was soon sent in to retrieve the Huey by sling load. With the sling rigged, the Chinook was hit by .50 cal automatic weapons fire as it hovered over the damaged Huey and crashed. There were nine men aboard the CH-47 from the 228th ASHB on that mission. Four of them, all from C/228 ASHB, perished in the crash: flight engineer SP5 Bolen P. "Mac" McGee, Gunners PFC Kenneth H. Delp, PFC Linden "Dale" Eiler, Jr., and PFC Jerry D. McManus. The pilot, MAJ Tony Vickers, was the only one not hurt, and would remain at the crash site until almost dark, when MG Tolson himself picked him up on his way back from the A Shau in his CC (Command/Control) bird. The aircraft commander and commanding officer of C/228 ASHB, MAJ Jerry Mathews, and Crew Chief Dan Dazell were medevaced to the Hospital Ship Sanctuary. Two cargo specialists from the 561st Transportation Company may or may not have also been on board the Chinook on its final mission. SP4 Joseph Burkes and PFC Ronald J. Campbell may have been present to assist with rigging the sling and may have already been on the ground in that role when the CH-47 was hit. In any case, they were not killed in this incident. Later that same day, both men were killed when a USMC Huey (BuNo 154761) was shot down elsewhere in I Corps, apparently having hitched a ride to Quang Tri Airfield. It will likely never be known with certainty whether they had been evacuated from the Chinook crash site, only to lose their lives in a second incident later that day, or whether they were never part of the 1st Cav retrieval mission.[40]

A Shau Anxieties

As the hour when several thousand "grunts" and helicopter crewmen of the 1st Cavalry Division (Airmobile) faced what might be their greatest challenge neared, there was deep concern, driven by rumors (true or false?) of lost patrols, masses of anti-aircraft weapons waiting in ambush, swarms of NVA on the ground, and an overwhelming sense of uncertainty. CPT John W. Taylor, commander of A Company, 5/7 Cavalry, one of the first two companies to enter the valley, recalled:

> The feeling the majority of the men had upon first coming into the valley was a sort of fear; distinctly different from that felt during Hue or Khe Sanh. We had heard so many stories about A Shau . . . the possibilities of running into large concentrations. We had a fear of the unknown. We thought that just around any corner we would run into a battalion of North Vietnamese.[41]

This was the norm for the men in the six infantry and three helicopter battalions scheduled to take part. Their commander, MG John J. Tolson, was, on the other hand, "not overly concerned by the risks."[42]

In some sense, both views were correct. The helicopters participating in the D-Day air assault would incur the division's worst losses on a single day of the war on 19 April 1968. Taylor's battalion, 5/7 Cav, would engage in the most sustained ground combat of any during DELAWARE and would have 16 men KIA and MIA – more than any other in the valley. Yet overall, after the enemy withdrew most of its anti-aircraft units, engineers, and rear-area troops during the first few days of the operation, casualties were remarkably low. Had the NVA elected to stay and fight, and/or brought in first-line infantry units to oppose the allied raid, it would undoubtedly have been quite a different story.

10

OPERATION DELAWARE: 3RD BRIGADE, 1ST AIR CAVALRY DIVISION

I n describing the course of Operation DELAWARE, the next three chapters will digress from a straight chronological narrative in favor of treating separately with each infantry battalion and its assigned landing zone (see Map 8). The U.S. brigades and the ARVN regiment and their respective areas of operation will be taken up in the order in which they entered the valley: the 3rd Brigade of the 1st Cav, then the 1st Brigade of the Cav, and finally the 3rd Regiment of the 1st ARVN Division.

5/7 Cavalry at LZ TIGER

The 5th Battalion, 7th Cavalry was the last infantry battalion to join the division, arriving in Vietnam in August 1966, over nine months after its sister units, the 1/7 and 2/7 Cavalry, were bloodied in the infamous Ia Drang Valley. Because the 3rd Brigade had led the way in Operation PEGASUS at the beginning of April, MG Tolson had not planned for them to go first into the A Shau. But when the 1/9 Air Cavalry reconnaissance reported that the central valley was too dangerous for the initial wave of the attack and a decision was reached to launch the operation in the northern valley, it was easier to change the order of insertion than the assignment of objectives, and so the 3rd Brigade would go in on D-Day. The 5/7 drew the short straw, going in first and going to the objective deepest in "Indian Country."

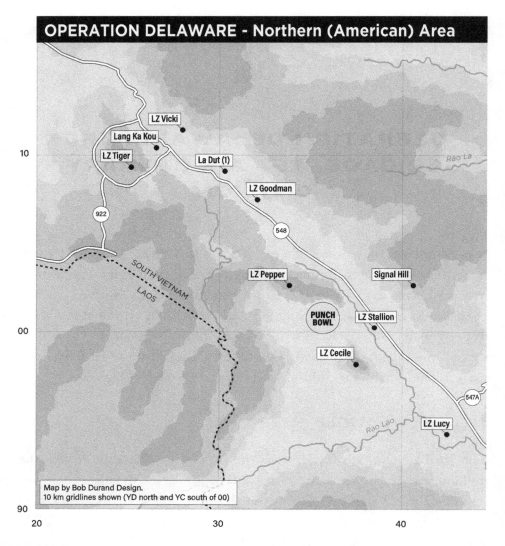

MAP 8

Source: Bob Durand Design

The first wave into LZ TIGER on the morning of 19 April 1968 departed the Camp Evans area at 0830, the same time as the LRP Platoon headed for Signal Hill, and comprised 8 helicopter gunships and 40 UH-1 slicks, each carrying 6 sky troopers.[1] The sunny conditions to the east did not prevail over the A Shau. Due to heavy clouds, the first lift came in over the valley at 6,000 feet and then had to descend one at a time through the clouds to the landing zones on Hill 1228, Co A Nong, known to the Americans as Tiger Mountain.* There were two designated landing areas. LZ TIGER (High) (YD253090) was atop the mountain and LZ TIGER (Low) (YD255088) sat about 300 meters down a barren ridge finger running toward the valley floor, which lay 2,000 meters farther to the east (see Map 9). The upper site

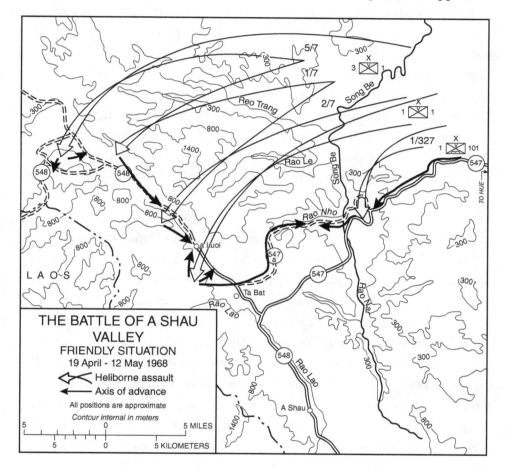

THE BATTLE OF A SHAU VALLEY
FRIENDLY SITUATION
19 April - 12 May 1968
Heliborne assault
Axis of advance
All positions are approximate
Contour internal in meters

MAP 9

Source: Pearson, Willard *The War in the Northern Provinces, 1966-1968* (Fort McNair, DC: Department of the Army, 1991), p. 90.

---

* The cloud cover cleared later in the morning and subsequent flights were able to execute a standard, straight-in approach. The name "Tiger Mountain" was a derivative of the local Pacoh Tribe's name for the hill. To the North Vietnamese, for unknown reasons, it was known as "McNamara Base." Phone conversation with Mike Sprayberry, 30 May 2020.

was nothing more than a one-ship LZ in a large bomb crater at cliff's edge (a B-52 ARC LIGHT strike had centered on the LZ early that morning).[2] The lower LZ had room for two helicopters to land simultaneously and had been selected because it afforded an unimpeded view of the valley floor, while sitting back from the military crest so as not to be exposed to fire from below[3]

Contrary to the recommendation of the 1/9 Air Cavalry scouts, the attack force approached Tiger Mountain from the east, flying across the valley, which soon came alive with enemy heavy machine-gun fire and exploding 37mm flak. The pilots took evasive action by constantly shifting altitude and, miraculously, only one of the helicopters in the first lift was shot down, although many were hit. Initially, the landings at TIGER High received the most intense fire, as the LZ there was more exposed to the heavy enemy AAA concentrated 2 km to the northeast in a jungled area on the valley floor. For both LZs, some of the heaviest anti-aircraft fire throughout the day came from a string of .51-cal heavy machine guns located along the ridge to the south and southwest of Tiger.

The lead slick from the 229th Assault Helicopter Battalion went down, hit by an RPG in the tail boom, which then separated as the pilot aborted the landing and attempted to gain altitude. Several passengers and the crew chief fell or jumped from the helicopter and survived, but 1LT David J. Nesset, WO Charles J. Harrington, and SGT Walter F. E. Matis, Jr., were killed in the ensuing crash.[4] With about 50% of Bravo Company's troops unloaded on the mountaintop, and company commander CPT Ralph Miles severely wounded in the abdomen by .51-cal fire (he would survive), it was decided to shift the remaining slicks to the lower LZ, where Alpha Company was debarking.[5] Once this was done, the rest of B Company landed safely, in spite of the enemy's redirection of the fire of all available weapons to the lower LZ.[†]

---

† A Fleet Marine Force message on 26 April summarizing Operation DELAWARE to date included the following text: "At 190905H, the 5th Battalion, 7th Cavalry, 1st Cavalry Division (Airmobile) commenced an unopposed helo assault into LZ TIGER (YD253090). One helicopter was shot down while departing the LZ." How helicopters were shot up and shot down during an "unopposed" assault is a mystery.

The helicopters trying to land at LZ TIGER faced an additional challenge not previously encountered by many of the pilots. The high-density altitude in the mountains around the A Shau caused reduced engine RPM, making it impossible to hover while dropping off troops and risking the rotor blades striking trees near the small landing areas. As a result, while the first lift carried six men per helicopter (the aircraft load, or ACL), the second wave ACL was reduced to five and later to only four for the trips from Camp Evans to the A Shau area. One Huey gunship from D/227 AHC was shot down to the south, but the crew was immediately rescued by another helicopter.

The cavalrymen quickly began consolidating their defenses on the landing zones. From TIGER Low, the approximate position of the heaviest concentration of anti-aircraft guns was identified as the vicinity of the abandoned village of Lang Ka Kou on Route 548.[6] This dangerous enemy AAA/AW battalion was almost midway between TIGER and the 1/7th Cavalry's LZ VICKI on the opposite side of the valley four kilometers to the northeast, and it seriously threatened helicopters approaching either location. The gunships of the 1/9 Cavalry were called in, but heavier ordnance was needed. Throughout the afternoon, repeated air strikes were made against this heavily camouflaged and defended area, to little avail.[7] The enemy seemed to be able to disappear like a turtle into its shell whenever bombs were dropped, only to reemerge soon after.

The second wave of the 5/7 Cavalry, Companies C and D, was directed to land at the lower LZ, but the NVA by this time had focused all of their aim on that target. Another gunship and one of the slicks were shot down, with PFCs Michael G. Lipsius and Curtis R. Riley of D Company killed by 12.7mm fire. The other crew members and passengers survived. This time around, the landings at the lower location became so hot that most of C Company was sent instead to the LZ at the top of Tiger Mountain.[8] By 1100, the entire battalion was ready to commence ground operations.

LTC James Vaught's company commanders began the missions they had been assigned. While Charlie Company consolidated TIGER High and Bravo Company did the same at TIGER Low, Alpha Company patrolled eastward toward the center of the valley. PFC Clifford Sell became the first ground fatality of Operation DELAWARE, when A Company was ambushed on the trail down into the A Shau.[9] Delta Company moved several hundred meters to the south, where a major branch of route 548 circled around the lower part of Tiger Mountain on its way to connect with Route 922 in Laos. Although familiar to the aero-scouts and forward air controllers (FACs), neither the original track of Route 548 across the south side of Co A Nong, nor the recently built bypass road circling around the north side was marked on the tactical maps used by the ground troops.[10] When D Company reached and tried to cross the road, it ran into the first enemy ambush. Thereafter, almost every bend of the road going west was defended by a well-concealed and entrenched enemy with automatic weapons.[11]

Meanwhile, although the heavy 37mm flak trap in the valley had not been neutralized, the CH-47 Chinook medium helicopters of the 228th Assault Support Helicopter Battalion (ASHB) had been ordered to proceed to LZ TIGER with artillery, ammunition, and other supplies. The greatest one-day loss of medium and heavy helicopters that was inflicted on the 1st Cavalry Division during the entire Vietnam War was about to occur. MG Tolson later conceded to the press, "By far it's the hottest place we've ever gone into, and the most [helicopter] losses we've taken in a single day."[12]

As the stream of the larger, less maneuverable and more vulnerable supply choppers approached LZ TIGER, the enemy below opened a ferocious barrage. A B/228 ASHB Chinook, with a 105mm howitzer and a sling load of ammunition dangling beneath, was hit in the aft pylon by 37mm and/or 12.7mm fire. After releasing its cargo, and with the rear of the aircraft engulfed in flames, the crew there, flight engineer SSG Anthony F. Housh and crew chief SGT Michael J. Wallace, attempted to jump to safety but apparently did not survive the fall; their bodies were never found. The rest of the crew survived the aircraft's crash in the valley and were rescued.[13] A CH-54 "flying crane" from the 478th Heavy Helicopter Company was carrying a bulldozer in a sling when it was struck by 37mm flak. The helicopter crashed into the side of Tiger Mountain not far from Lang Ka Kou. The crew of CPT Arthur J. Lord, CW3 Charles W. Millard, SSG Michael R. Werdehof, and SP4 Philip R. Shafer was missing in action.[14] In the afternoon, another Chinook from A/228 ASHB hovered over LZ TIGER Low as it dropped a sling load of mortar ammunition into a crater at the edge of the LZ. There was an explosion in the aft pylon and the CH-47 settled to the ground off the western side of the LZ on its nose, while a fire raged in the rear of the fuselage. The crew bailed out and about 20 minutes later the ammunition started what would be several hours of "cooking off."[15] In the confusion and smoke of the crew evacuation, badly burned tail-gunner Kenny Sager was left behind. 1LT James "Mike" Sprayberry, who earned the Medal of Honor a few days later, the D Company executive officer, medic "Doc" Norman McBride, and several other men went into the burning aircraft to recover what they thought was the body of the crewman. Only years later would they learn that Doc McBride's emergency treatment and the subsequent attentions of battalion surgeon Dr. Jeffrey Kahler were able to keep Ken Sager alive on the LZ and enable his evacuation and survival.

Late in the afternoon, a third CH-47, also from A/228 ASHB, went down while delivering ammunition to LZ VICKI. Both pilots survived the aircraft crash some distance from the LZ and were able to walk to safety the next day, but three crewmen, CPL Douglas R. Blodgett, SGT Jesus A. Gonzalez, and SP4 William R. Dennis were missing in action.[16]

According to Wikipedia, 132 CH-47 Chinooks were lost during the Vietnam War. Thirty-two of those were in the 228th ASHB of the 1st Air Cavalry Division, of which 18 were lost to accidents. Of the 14 CH-47s lost in combat by the First Cav, 5 went down during Operation DELAWARE, including 3 on 19 April 1968.[17] It was a heavy price to pay. The CH-54 Flying Crane shot down on 19 April was the only one of those giant helicopters to be lost in combat during the Vietnam War.

Five of the planned six 105mm howitzers and their crews from C Battery, 1/21 Artillery were successfully inserted at TIGER. But a large part of their first-day ammunition supply went up in fire and thunder when the supply dump near the landing zone was detonated in the sequel to the crash of the last Chinook nearby. This had a lasting impact when a very plum target could not be adequately attacked later that first night in the valley.[18]

While the late afternoon supply lift was underway at the LZ on the ridge above, D Company continued to use fire and maneuver to gradually overcome the enemy ambushes along the road around Tiger Mountain. LTC Vaught, who had traveled by "mule" (a small, light, flat-bed vehicle usually used to haul supplies around the LZ) to deliver much-needed water and ammunition to D Company, incurred a broken back in an accident on his way back to the landing zone.[19] By the time he was brought back to the LZ, it was pitch dark with broken clouds, and the medevac unit declined to fly to the A Shau on such a risky mission. MAJ Bob Frix and WO Al Eason courageously volunteered to fly the battalion C&C (command and control) helicopter on the mission. After a hair-raising flight, including some time searching for the LZ, Frix and Eason were able to pick up LTC Vaught and fly him to Phu Bai, where he was able to receive timely hospital care and survive his injuries.[20] The battalion was taken over temporarily by MAJ Joseph V. Arnold and then by LTC Thomas W. Stockton on 21 April.[21]

Not long after dark, C Company observed a convoy of at least 50 trucks on the valley floor in the vicinity of the NVA anti-aircraft bastion at Lang Ka Kou. Forward observers attempted to adjust fire from the 5/7's own 81mm mortars (at maximum range), the 105mm howitzers at TIGER (with little ammunition), and 175mm guns from LZ BASTOGNE (also operating at maximum range). Air Force Spooky gunship support was requested, but the convoy had gone dark and dispersed before the airborne artillery could arrive. It is highly probable that this convoy, which moved southeast down Route 548, was the departure of the NVA antiaircraft unit(s).[22]

Dawn on 20 April brought ground fog, which burned off only to be replaced by heavy cloud cover later in the morning that would persist for two days. Aerial resupply would be limited to 14 sling loads delivered by Huey on the afternoon of the 22nd.[23] On the slope below LZ TIGER, the enemy blocking force had disappeared when D Company moved out on its second day in the A Shau. Moving west along the road, at 1400 hours the point found an overturned truck containing about 200 "old and rusty" rifles, the first significant equipment capture of the operation.[24] B Company patrolled about 1 km to the south and C Company sent patrols to the northwest and southwest as occasional sniper fire struck the landing zones. During the night of 20–21 April, A Company heard vehicles again in the valley but was unable to identify their location.[25]

On 23 April at 0600 hours, the NVA launched an attack on B Company at LZ TIGER (Low) with mortar and small-arms fire. PFC Andrew L. McDaniel was KIA and 11 were wounded. PFC Charles S. Cox died later that day of his wounds. Counter-battery fire reportedly killed one enemy.[26] Soon thereafter, A Company, on the trail between TIGER High and Low, was mortared and then attacked by an NVA platoon. Two platoons of A company were already in movement toward the valley. They were able to hit the enemy assault force with a surprise flank attack, killing six NVA with no American losses.[27] Alpha Company then continued moving north on the road, skirmishing constantly with NVA snipers and losing several wounded before the day was out, while eliminating a number of the enemy. D Company, on the other side of Tiger Mountain, also advanced slowly along the road in the opposite direction throughout the day, with no casualties and six NVA KIA.[28] Enemy snipers also continued to be active around the perimeters at both TIGER LZs.

24 April brought with it a significant improvement in the weather, and D Company continued for the fifth straight day to work through scattered NVA contact in the direction of Laos. Ahead of them lay a saddle between Tiger Mountain and a north–south ridge along which ran the road. This area was considered to be a key enemy strongpoint, as 12.7mm anti–aircraft fire originated there. By noon, the cavalrymen were within 500 meters of the saddle, but enemy resistance stiffened as the afternoon wore on. The cavalrymen were advancing along the narrow road cut into the steep side of Tiger Mountain, with a front of scarcely 10 meters of level ground. The battle continued until dark, with a number of D Company wounded in exposed positions. Under heavy fire, SGT Henry Thomas used his helmet to scrape out trenches just deep enough to protect two wounded men.[29] In 2004, he finally received the Silver Star for his bravery.[30]

On 25 April, with no advance preparation by air or gunship strikes, D/5/7 Cavalry was ordered to attack the enemy position in the saddle head-on. The first attempt was stopped cold by heavy automatic rifle and machine-gun fire. The new battalion commander, LTC Stockton, instructed CPT Frank Lambert, the D company commander, to take one platoon to the north to try to flank the enemy bunkers. It took several hours to climb the cliff on the right (Tiger Mountain side) of the road and reach a position about 500 meters north of the rest of the company. At about 1330, the 1st Platoon commenced moving southwest toward its objective, through broken and heavily jungled terrain. Finally reaching the vicinity of the saddle at around 1800 hours and with daylight about to wane, the platoon launched an immediate assault and was immediately ambushed and surrounded as the sun went down. The platoon commander, 2LT David Barber, was one of the first to fall, mortally wounded. CPT Lambert was also down, severely wounded. PFC Hubia J. Guillory, SP4 Daniel M. Kelley, and SP4 David L. Scott had been killed in the initial burst of enemy fire, and there were eight badly wounded men needing medevac. But the platoon was caught in the kill zone and could not escape.[31] The artillery FO was unable to bring fire to bear directly on the enemy location, as Tiger Mountain masked the saddle area from the howitzers on the reverse slope at TIGER Low. The platoon sergeant, SSG Billy Bayne, and RTO George Mitchell protected the wounded, organized a defensive position, and drew into a tight perimeter. It would be a long, pitch-black night. Back on the road, the rest of D Company tried one more time to advance down the narrow shelf along which ran the road, only to be stymied yet again.[32]

While the maneuvering of D Company was in progress, the NVA hit LZ TIGER again, this time with mortars and rockets, at 1700 on 25 April. One GI was killed and 16 wounded.[33]

The executive officer of D Company, in command after the wounding of CPT Lambert, was 1LT James "Mike" Sprayberry. Back on the road, he was undaunted by the NVA entrenched to his front and recruited a dozen volunteers to join him in a nighttime attack down the road to link up with the surrounded platoon. 1LT James M. Sprayberry would be awarded the Medal of Honor for action on the night of 25–26 April 1968. His citation, for an action that lasted over seven hours, follows:

Capt. [then 1LT] Sprayberry, Armor, U.S. Army, distinguished himself by exceptional bravery while serving as executive officer of Company D. His company commander and a great number of the men were wounded and separated from the main body of the company. A daylight attempt to rescue them was driven back by the well-entrenched enemy's heavy fire. Capt. Sprayberry then organized and led a volunteer night patrol to eliminate the intervening enemy bunkers and to relieve the surrounded element. The patrol soon began receiving enemy machinegun fire. Capt. Sprayberry quickly moved the men to protective cover, and without regard for his own safety, crawled within close range of the bunker from which the fire was coming. He silenced the machinegun with a hand grenade. Identifying several 1-man enemy positions nearby, Capt. Sprayberry immediately attacked them with the rest of his grenades. He crawled back for more grenades and when 2 grenades were thrown at his men from a position to the front, Capt. Sprayberry, without hesitation, again exposed himself and charged the enemy-held bunker, killing its occupants with a grenade. Placing 2 men to cover his advance, he crawled forward and neutralized 3 more bunkers with grenades. Immediately thereafter, Capt. Sprayberry was surprised by an enemy soldier who charged from a concealed position. He killed the soldier with his pistol, and with continuing disregard for the danger, neutralized another enemy emplacement. Capt. Sprayberry then established radio contact with the isolated men, directing them toward his position. When the 2 elements made contact, he organized his men into litter parties to evacuate the wounded. As the evacuation was nearing completion, he observed an enemy machinegun position which he silenced with a grenade. Capt. Sprayberry returned to the rescue party, established security, and moved to friendly lines with the wounded. This rescue operation, which lasted approximately 7½ hours, saved the lives of many of his fellow soldiers. Capt. Sprayberry personally killed 12 enemy soldiers, eliminated 2 machineguns, and destroyed numerous enemy bunkers.

At about 0400, the attack force and the survivors of the surrounded platoon returned to the company lines. The night attack almost certainly saved the lives of many from the 1st Platoon who would likely have perished under attack from the NVA come morning. In addition to 1LT Sprayberry's medal, four men received Silver Stars for their heroism in this action: SSG Bayne of the ambushed 1st Platoon, Dave Bielski, 1LT Sprayberry's RTO (radioman), SP4 Barry Tranchetti, and SGT Delbert Mack (who killed one enemy soldier with his bayonet in hand-to-hand combat[34]) of the rescue force. LT Barber's body was retrieved, but the bodies of the others killed in the ambush could not be retrieved due to heavy enemy fire and remain missing to this day.[35]

26 April was a quiet day for both the Americans and the NVA around LZ TIGER, with no contact and no casualties. On 27 April, A Company resumed its patrol toward the floor of the valley, B Company moved to the south toward another east-west road 2 km south of Tiger Mountain, and D Company looked for alternative ways to flank the enemy bastion in the saddle.[36]

Fire and maneuver, air strikes, and artillery marked the renewal of the D Company attack on 28 April. A Company, nearing Route 548 in the valley, lost two men KIA, PFC Billy K. Ford and PFC Henry Wunderlich, but overran an enemy base camp and 82mm mortar position.[37]

The 29th was another day of progress for the 5/7 Cavalry. D Company obtained a position from which they could better observe air strikes, and a 2,000 lb. bomb seemed to obliterate one of the knolls at the heart of the NVA defenses. B Company reached the area of the abandoned village of Hu on the road to the south, uncovering a cache of 12.7mm anti-aircraft ammunition and other supplies. A Company was almost at its objective. Early on the morning of 30 April, B Company relieved the exhausted D Company on the road to the west, as D Company returned to guard TIGER Low.[38]

On 1 May, two platoons of B Company would make a wide circle to the south and then west, coming up to the road along the eastern side of the ridge about 1 km south of the saddle area. Engineers had already blown a large hole in the road around the south side of the mountain to interdict NVA vehicular movement. In a search for the missing soldiers from Delta Company, a scout helicopter and gunship (a "Pink Team") from the 1/9 Air Cavalry supported Bravo Company. As the OH-6A with a three-man crew approached the road along the western ridge about midway between the NVA on the road and the B Company flankers to the north, it was blown out of the sky. Missing in action were WO Warren

T. Whitmire, Jr., SP4 Richard D. Martin and SGT Donald P. Gervais.‡ After the "Loach" was hit, the company commander ordered his three platoons to move directly toward the crash in the hope of rescuing survivors. Around 1345, the platoon from the north (road) was able to observe the downed helicopter with binoculars from about 300 meters away. No survivors were visible, but NVA soldiers were at the crash site. The southern maneuvering force was then ordered to resume its assault toward the ridgeline to the west. As they cleared a tree line and headed up the slope, they came under intense fire, including the ubiquitous and deadly 12.7mm heavy machine guns. One infantryman was KIA, another seriously wounded by a 12.7mm round. The ridge in front of B Company was steep and barren, offering no opportunity for an assault. Reluctantly, the attack was called off, the wounded were medevaced at around 1700, and the entire company returned to TIGER Low.[39] The bodies of the scout helicopter crew remain missing.

The attack to the west, commenced when D Company walked off the LZ on 19 April, was over. The "stalemate in the saddle" between the 5/7th Cavalry on Tiger Mountain and the NVA occupying the road along the ridge to the west, marked the end of a battle that was costly for both sides. The Americans had succeeded in blocking the connecting road between Laos and the A Shau Valley; the NVA in guarding a base that would not be penetrated until almost a year later.

Down in the A Shau Valley on 1 May, Alpha Company at long last reached the area where the anti-aircraft fire had been centered on D-Day. The company found a truck, 200 rounds of 37mm AA ammunition, and a cave that had been used as an aid station.§

---

‡ The VHPA website [66-07810] states that this occurred around 1800 hours, but the narrative by Charles Baker, former operations officer of the 5/7 Cav, makes clear it happened fairly early in the morning.

§ The cave was U-shaped and large enough to drive a truck through, for drop-off ambulance service! A year later, when the 9th Marines swept into the area during DEWEY CANYON, it had been converted to a vehicle maintenance facility. Mike Sprayberry phone conversation 16 June 2020.

At the beginning of May, the 5/7 Cavalry did some reshuffling of company assignments. D company would remain as LZ security, less one platoon that together with a platoon from C Company would man a roadblock and observation point on the road south of TIGER High to prevent any enemy attack from the saddle area and the direction of Laos. The rest of C Company would patrol to the south, while A and B Companies would operate on the valley floor. TIGER Low was abandoned due to the threat of a tank attack from Laos. The road ran right through the edge of the lower LZ, and signals intelligence (SIGINT) indicated the presence of as many as three "groups" of NVA tanks in the area, while the U.S. infantry had few light anti-tank weapons in the event of an armor-supported enemy push.[40] The search for caches was on, as the end of DELAWARE was in sight.

On 2 May, six rounds of NVA artillery (!) fire were accurately targeted at LZ TIGER High, killing artillerymen CPL Steve Butorovic and PFC Edward L. Munson from C/1/21 Artillery, whose guns had been moved there from the Low LZ several days previously. Artillery, assisted by fire from at least one NVA tank on the ridgeline to the west (where the assault had ended on 1 May) also struck the roadblock, with two WIA.[41]

In the valley, both companies repeatedly found NVA equipment, including the remains of a 57mm AA gun, the first of that caliber detected in the A Shau. Just west of Lang Ka Kou, B Company discovered three 37mm AA guns on 3 May.[42] As dusk neared, the company moved to a knoll that had been selected for their NDP (night defensive position). The enemy was waiting in ambush. In the ensuing firefight, one Bravo infantryman, PFC Jesse Carmona, Jr., and medic SP4 Billy W. Bridgeman were KIA, and two NVA bodies and a machine gun were left when the enemy retreated. They didn't go far. A squad-sized probe was launched against Bravo's NDP at 2000. Artillery was called in, followed by the moans of many NVA wounded, and enemy movement was heard around the position throughout the night. On the morning of 4 May, a large number of blood trails and drag marks were found, indicating that the NVA had received a number of casualties. A "body count" of 7 KIA and 27 WIA was estimated. The enemy was gone.[43]

On 4 May, LZ TIGER High received two rounds of 122mm artillery fire, wounding two D Company troopers.[44] These were spotting rounds for the barrage two days later.

The next (and penultimate) contact for the 5/7 Cav came on 6 May, when heavy enemy fire hit 5/7 Cavalry positions far and wide. At TIGER High, fire was received on and off for two hours from 130mm artillery and 82mm mortars, resulting in six artillerymen WIA. At TIGER Low, the NVA used direct fire from B-40 rocket-propelled grenades, recoilless rifles, and a 37mm AA gun. C Company had one KIA, SP4 Edward L. Brock, and four wounded at that location. At the roadblock, tank fire from the ridge to the west was incoming. Overall, 250–300 rounds of rocket, mortar, and artillery fire were received.[45] Artillery from other LZs around the valley as well as aerial rocket artillery from the Cav's 1/20 ARA was called in on the enemy with unknown results. The expected ground attacks fortunately never came, but it was becoming clear that the enemy was beginning to return to the A Shau, and not just with engineers and anti-aircraft units.

In the following days, the search of the valley floor and adjacent areas continued to turn up enemy equipment and supplies. Enemy sightings, including several of reported NVA tanks, were numerous, and enemy indirect fire was occasionally received. The NVA buildup on the road running along the west wall of the valley just west of Hill 1228 also continued to be observed, until an ARC LIGHT strike was called on that area shortly before LZ TIGER was shut down.[46]

On 10 May, commencing at 0930 and ending at 1220, the Fifth Battalion, Seventh Cavalry and its attached units were withdrawn from the A Shau Valley. There was no reported enemy fire.[47] The experience of the battalion, the first American regular infantry to enter the A Shau Valley, was marked by three characteristics that would be repeated many times during forays into the valley over the next 18 months: (1) the operation came as no surprise to the People's Army of North Vietnam, (2) the landing zone was "simply too small to insert a battalion of troops, especially for a hot LZ,"[48] and (3) there were no ground reserves available in the event that a major enemy strongpoint was encountered.[**]

[**] These observations of Mike Sprayberry re: the 5/7th Cavalry are common to all U.S. operations in the valley, as subsequent chapters will show.

1/7 Cavalry at LZ VICKI and LZ GOODMAN

The second battalion to "Charlie Alpha" (combat assault) into the valley was the 1/7 Cavalry under LTC Joseph E. Wasiak. The helicopters departed Camp Evans on 19 April at 1330 hours and were soon descending onto the LZ, located on a ridge finger running to the southeast from Hill 1021, less than 1.5 km from Route 548 on the north side of the A Shau River. The LZ was small, with only enough room for two ships to unload at the same time.[49] It had been blanketed with a B–52 strike on 16 April. Numerous tac air strikes were targeted at likely enemy positions before the airmobile infantry arrived.[††]

In spite of the preparatory fires, however, just as at LZ TIGER, the AAA/AW fire was heavy. Between the two landing zones, some 23 helicopters in total were shot out of the sky on 19 April, 10 of which were completely destroyed (1 CH–54, 3 CH–47, and 6 UH–1). Eight aircrewmen were killed and nine missing on the downed aircraft from the Cav's 11th Aviation Brigade. At VICKI (YD282114), the designated approach route was from the west, across the valley floor, but NVA fire on that path was still intense. Sp4 John E. Wilburn of A/1/7 Cav was a passenger on one of the UH–1D slicks from B/229 Assault Helicopter Battalion. On approach to the LZ, the helicopter was hit in the tail section by enemy fire and suddenly began to spin. SP4 Wilburn was thrown out of the helicopter; his remains were never located. The pilot, WO Larry J. Branaugh, was able to crash-land the damaged aircraft, but was himself mortally wounded by enemy .50 cal machine-gun fire after he had landed. He died in the hospital the following day. The other crew members and passengers were picked up and survived.[50]

After the chopper from the 229th was downed, "a relatively secure route farther north was found with the aircraft skirting the open valley and hugging the eastern wall. The final approach was made from a dog leg to the southwest."[51]

———————

†† The short-lived LZ is variously cited as "VICKI," "VICKY," and "VICKEY" in the operational records. For consistency, I have elected to use "Vicki," per the 1st Cavalry Division AAR.

A howitzer battery was due to arrive later in the day so that the three bases of the 3rd Brigade could provide mutually supporting artillery coverage. But by the time the infantry lifts to VICKI were completed at 1515 hours, the weather and air losses combined to cause the temporary postponement of the big guns' delivery.[52]

On the morning of 20 April, the weather remained bleak. "Unbroken cloud layers blanketed the entire length of the valley, with morning fog and afternoon torrential rain limiting visibility to practically zero."[53] Determining that LZ VICKI was unsuitable as an artillery location[54] and would not be able to be resupplied by air (nor could its troops be evacuated that way), LTC Wasiak abandoned that base and marched his unit 6 km south to Hill 875, to be called LZ GOODMAN (YD324078). C and D Companies would complete that tortuous march on 22 April at 1330,[55] through drenching rains, dense jungle, and razor-sharp elephant grass. Shelby Stanton observed: "The drenched marchers became chilled and sick. They subsisted on ration tins and went without sufficient sleep."[56] MG Tolson's description of this trek in his *Vietnam Studies* monograph as "an overland attack" is a bit of a stretch, given that the terrain and weather, rather than NVA soldiers, were the primary "hostile forces."[57]

En route, on 21 April, the cavalrymen engaged in a battle with a small enemy force, killing three NVA, and then found two damaged Russian-made bulldozers and two trucks near the abandoned village of La Dut (1). Those were destroyed in place.‡‡[58] On 23 April, the 105mm howitzer battery was flown into GOODMAN, and the LZs were at last able to support each other with indirect fire.[59] A heavy battery of 155mm howitzers would also be based at GOODMAN.[60]

After its inauspicious start, the 1/7 Cavalry would lead the pack in the capture of NVA heavy equipment during the A Shau Valley campaign.

---

‡‡ The *Air Cavalry Division Magazine* reported in its September 1968 issue that the 1/7 "took the bulldozers along," but the AAR and the FMFPAC CC Msg both state that they were destroyed in place. "During the period, the 1st Battalion, 7th Cavalry found two the [sic] remains of enemy bulldozers and trucks, two kilometers southeast of LZ Vicky (YD557032). All of the enemy equipment found, had previously been destroyed by air strikes." It seems most likely that the dozers were blown up, given the terrain and that other reports indicate prior damage.

On 25 April, operating 1 km west of LZ GOODMAN, A Company captured three flat-bed trucks mounted with 37mm anti-aircraft guns just about 100 meters from Route 548 on the valley floor. C Company lost one man to enemy fire when SP4 Jack Biting was KIA on the 25th.[61] Continuing to search to the north and west, on 30 April four more 37mm guns in good condition were captured by C Company, 500 m south of Route 548 and 4 km west of GOODMAN.[62]

While patrols to the west were unveiling NVA equipment in the valley with alacrity, other patrols of the 1/7 Cav began to come up against some of the strongest resistance that would be met during DELAWARE as they moved just a short distance to the southeast. On 29 April, while working in the area of a small box canyon about 1 km off Route 548 and less than 2 km from the LZ, D Company received heavy enemy fire, killing PFC James E. Walker, Jr.[63]

On 1 May, moving deeper into the box canyon, Alpha Company got into a firefight with an enemy unit that hit from all directions with automatic weapons and a command-detonated mine, but the GIs got the better of the enemy in this battle, killing three NVA with no US losses. Another 1/7 trooper fell to enemy fire in the same vicinity on 1 May, when PFC Antonio Garcia of B Company was killed by small arms fire.[64] Contact with the enemy continued over the next two days, with 15–20 NVA in bunkers holding up the advance of Bravo Company. Two B Company troopers, PFCs Dervin J. Keisling and John E. Manson, lost their lives in the firefight on 3 May. On the 2nd, another cache was discovered, this one containing two trucks and 500 rounds of 37mm AAA ammunition.[65]

On 4 May, B Company found enemy communications wire leading to an NVA camp along the same trail where contact was made on the preceding days and launched an attack against the entrenched enemy. As the attack unfolded, the 2nd Platoon was suddenly exposed to impending disaster. Once again, an infantry lieutenant would risk his life, and in this case sacrifice it, in a Medal of Honor action in the A Shau Valley:

> While advancing uphill against fortified enemy positions in the A Shau Valley, the platoon encountered intense sniper fire, making movement very difficult. The right flank man suddenly discovered an enemy claymore

mine covering the route of advance and shouted a warning to his comrades. Realizing that the enemy would also be alerted, 1st Lt. [Douglas B.] Fournet ordered his men to take cover and ran uphill toward the mine, drawing a sheath knife as he approached it. With complete disregard for his safety, and realizing the imminent danger to members of his command, he used his body as a shield in front of the mine as he attempted to slash the control wires leading from the enemy. As he reached for the wire, the mine was detonated, killing him instantly. 5 men nearest the mine were slightly wounded, but 1st Lt. Fournet's heroic and unselfish act spared his men of serious injury or death.[66]

This would, however, mark the farthest that the First of the Seventh would be able to advance into this area. "A study . . . of documents captured in the A Shau Valley [later] revealed the existence of a large storage area/supply point [in this vicinity]. The importance of this cache was attested to by the stiff resistance deployed along the trails, preventing elements of the 1 ACD from discovering the cache."[67] Thus, the NVA once again demonstrated that it would stand up and fight when the prize was valuable and the conditions favorable.

Contact with NVA in the LZ GOODMAN area continued. On 6 May, SGT James L. Clark and SP4 Brian J. Cannada of H&H Company were KIA by small arms.[68] At about 2000 hours on the evening of 7 May, B Company received small arms and mortar fire from an unknown–size enemy unit, with six WIA needing evacuation.[69] This was part of a general increase in enemy indirect fire attacks against First Team LZs throughout the valley commencing on 5 May.[70]

On 10 May, after less than three weeks in operation, LZ Goodman was closed and the 1/7 Cavalry was extracted.[71]

2/7 Cavalry at LZ PEPPER

The third battalion of the 3rd Brigade, the 2/7 Cavalry, was scheduled to be the final unit into the valley on D-Day, 19 April. By 1630 hours on that day, the lift of 1/7 Cav to LZ VICKI had been completed, but the weather was closing in, visibility was dropping, and heavy losses had cut the availability of slicks by 50%. The insertion of 2/7 was postponed until the next day.[72]

The first UH-1 troop-carrying helicopter from C/227th AHB crashed while offloading at the LZ on 20 April. The crew and passengers had only minor injuries, but there could be no further insertions until troops from the 1/9 Air Cavalry rappelled in and used chain saws to enlarge the landing area.[73] When the LZ reopened, once again the first helicopter to land crashed. One company of 2/7 was finally on the ground on the morning of 20 April, establishing LZ PEPPER (Dong So Ridge) (YD339026), but thunderstorms then closed down the LZ for that day and the next, with the remainder of the 2/7 Cavalry and the forward 3rd Brigade command post of COL Campbell reaching the AO on the 22nd.[74] On 21 April, already patrolling well beyond the landing zone, B/2/7 got into a battle with a platoon of NVA about 6 km east of the LZ on the eastern wall of the valley.[75] SSG Cecil J. Moser and PFC Richard C. Klinkenberg were the first men from 2/7 Cavalry to die in the A Shau Valley. Eleven more would follow before the operation ended on 17 May. At around 1800 hours on 23 April, a CH-47 Chinook was hit by fire on approach, overshot LZ PEPPER, crashed, and burned. No lives were lost, but a number of crewmen and passengers were injured and had to be evacuated.[76] On the ground on the 23rd, PFC Lour LeDesma of C Company was shot and killed.[77]

The patrols of the 2/7 Cavalry continued to search the AO around LZ PEPPER. On 26 April, two men from C Company, SP4 Donald K. Dudley and PFC Earl P. Summersill, were killed by enemy small arms fire.[78]

28 April saw C Company engaged in a sharp firefight 1 km east of the LZ, on the top of the ridgeline running for some 6 km from northwest to southeast along the northern rim of what would soon become known as the "Punchbowl." Before the combat ended, two Americans were KIA, PFC Coyte D. Campbell and SP4 David T. Barnes, recipient of the Silver Star and Bronze Star Medals for Valor. Thirteen NVA were KIA.[79] The 2/7 contributed to the haul of enemy anti-aircraft guns on 28 April, with the capture of two 37mm weapons with 500 rounds of ammunition just to the south of Route 548 on the valley floor.[80] Anthony L. Moore of A/2/7 Cav was KIA on 29 April, as the sniping and skirmishing went on.[81]

On 7 May, D/2/7 Cav received 15 rounds of light (60mm) mortar fire, wounding eight.[82] The same day, A/2/7 was ambushed in a rugged draw on the western side of the western wall of the A Shau. Hit with heavy enemy automatic weapons fire from a concealed enemy platoon, the official recap states, "organic weapons fire was returned, resulting in three enemy KIA. According to the AAR, "eight USA were wounded and evacuated, and one USA was missing."[83] Unfortunately, that report must have been premature, for in fact four Air Cavalrymen were killed in that action and a fifth died of his wounds the following day, the largest loss in a single-day battle for the 3rd Brigade ground troops during Operation DELAWARE. The casualties were SGT Rick H. Fosburg, SP4 Berlin Shumate, PFC (medic) Isidro B. Jimenez, SP4 Roland C. Crosby (DOW), and PFC John R. Thornhill, III.[84] Contact with a platoon-size NVA element continued until 9 May, with one additional enemy KIA but none in A Company.

The infantry and artillery at LZ PEPPER were the last elements of the 3rd Brigade to be extracted, returning to Camp Evans on 11 May.[85]

OPERATION DELAWARE: 1ST BRIGADE, 1ST AIR CAVALRY DIVISION

2/8 Cavalry at LZ CECILE

On D+5, with the 3rd Brigade, including its supporting artillery, in position to the north, the 1st Brigade (COL John F. Stannard) commenced its role in Operation DELAWARE. First up would be the 2nd Battalion, 8th Cavalry, under LTC John V. Gibney. B Company arrived at LZ CECILE (Hill 674) (YC380981) between noon and 1400 on 24 April.[1] While no aircraft were lost in getting 2/8 Cavalry on the ground, weather and enemy fire continued to challenge the Air Cav pilots. En route to the valley, clouds forced the helicopters to fly as high as 11,000 feet:

> Once over the valley and through the holes in the overcast, the aircraft daisy-chained from LZ TIGER to the new LZ, covered en route by the gun [ship]s of both the 229th AHB and 227th AHB. CECILE was a two ship LZ at the southern end of a ridge about 2200 feet high. Although all aircraft were exposed to sniping fire on the approach, the main threat came from an automatic weapons position about 500 meters down the ridge to the southwest. Since the approach was made to the south, the enemy gunners got a crack at each aircraft as it departed no matter which way it broke. Very few hits were sustained, however, and no aircraft were lost. . . .[2]

B Company arrived at LZ CECILE between noon and 1400 on 24 April.[3] While no casualties—friendly or enemy—were reported on the LZ during the assault, clearly there was some ammunition expended, as an urgent resupply was requested in the late afternoon. Six Hueys from the 227th Assault Helicopter Battalion once again had to descend through a hole in the clouds near LZ TIGER, traverse the valley south to CECILE, and depart the same way. The trip in was not a problem, "but the return up the valley low level was another story. The flight received heavy automatic weapons fire during the entire traversing of the valley floor."[4] Fortunately, only one helicopter was hit, and it was able to complete the mission. Later in the day, a CH-47 attempting to drop a sling load of 105mm howitzer ammo was hit by .51-cal AAA, killing the left waist gunner, PFC David J. Caballero.[5] Before the day was out, the entire battalion was on hand, along with three 105mm howitzers.[6] That night, B Company marched about 2 km north to the A Luoi airstrip ("we had to touch the guy in front to make sure we would not get lost"), to occupy what would become LZ STALLION (YD385002) the following day.[7]

While B Company was headed north, D Company went south on 25 April. About 600 meters to the southeast of CECILE, they found a large cache of NVA electronics equipment.[8] The GIs would wonder at the variety and ingenuity of the always well-hidden enemy installations.

Once on the ground, the discovery of what was determined to be an NVA R&R area of sorts. Wet NVA laundry still hanging on clotheslines, caches of rifles and munitions, a cave with all kinds of electronic gear, radios, amplifiers, microphones, huge piles of rice sacks and supplies. How about the elaborate water filtration system that the NVA had built out of the side of a bomb crater? The green slimy water seen at the top of the crater, after having flowed down and through a 50 ft long trough of reeds, rocks, and sand, now ready as potable drinking water.[9]

For the next couple of days (4/26 and 4/27), things were quiet around CECILE, at least according to the official reports. It is likely, however, that one action not reflected in any of the higher headquarters' narratives occurred during this period. According to "one of [the] troopers" in B/2/8 Cavalry: "The night they fired the tanks up we were all standing around waiting to be interviewed by some famous newsman. . . . Those [NVA] tanks cranked up and when I turned back around the famous news guy was hauling a— for the chopper. . . . I believe there were five of them. The tanks stopped when our artillery illuminated the area and cranked up again when dark. We could hear their voices as they shouted orders and directions. The following day, Blackfoot (B Company) was sent down to follow the tank tracks that led towards Laos."[10] This account would appear to be confirmed by what followed.

On 28 April, a 2/8 Cavalry patrol found an abandoned PT-76 tank belonging to the 3rd Company, 3rd Battalion, 203rd NVA Armored Regiment. As an example of the difficulty that sometimes arises in reconciling contemporaneous reports, the 1st Cavalry Division AAR (11 July 1968), page 6, reports the location of the tank at YC392969, Tab C of the same AAR reports the event as occurring on 27 April at YC389971, and the FMFPAC SITREP #1,128 records the location as YD394970! In any event, it would appear that the PT-76, which had been destroyed by an earlier air strike, was about 1.5 km southeast of LZ CECILE, near a road diverging from Route 548 and going south to the Laotian border.

To the northwest of CECILE on the 28th, D/1/8 Cav operating from LZ STALLION made contact with an NVA company near the base of the ridge pointing down from Hill 975 (the eastern peak of Dong So Ridge) toward the airstrip (YD365013).* Moving on down a well-hidden, mile-long corduroyed road, D Company ran into an NVA company, supported by a dug-in PT-76 amphibious tank and using riot gas. The tank had the Americans stymied until SGT Hillery Craig was able to gain a firing position and knock it out with two M-72 antitank rockets. One Cav trooper, Fredrick E. Bauerle, III was KIA in this battle.[11] The North Vietnamese force was defending a large cache, including 122mm artillery ammunition

---

* Interestingly, it is a testament to the roughness of the terrain and the enemy's superb camouflage skills that one of the *last* enemy installations to be found during Operation DELAWARE, the dispensary discovered on 12 May by A/1/8 Cavalry, was located just 100 meters away at YD364013.

and AK-47 assault rifles.[†12] Thus began five days of continuous enemy contact for D/1/8.

C/2/8 Cavalry, moving to the support of D/1/8, became engaged in one of the toughest battles of the campaign, losing 5 men on 28 April while accounting for 18 NVA KIA.[13] The KIA were 1LT Larry C. Bolyard (HHC), PFC Jackie L. Elston (HHC medic and Bronze Star for Valor recipient), SP6 Edgar D. Page (C Co), SP4 Michael R. Lucas (C Co), and PFC John Duffy (C Co).[14] D/1/8 was at this time placed OPCON to (under operational control of) 2/8 Cavalry, to assist in exploitation of the "Punchbowl," as the rest of the 1/8 Cavalry was tasked with base defense around LZ STALLION.[15] The skirmishes in the Punchbowl would continue until 3 May, as the enemy sought to deny access to the large aggregation of installations and caches in the area of the Binh Tram headquarters.

A large storage area and truck park (vic YD 362012) were seized by 1st Brigade elements on 29 April as enemy forces continued to avoid decisive contact. The discovery of this large logistical complex confirmed intelligence reports processed and received prior to the operation. The continuing discoveries of hospitals, administrative, and logistical sites in the general vicinity of YD3601 lent further support to the theory that this was a major support complex for the A Shau Valley command and control element.[16]

Air strikes and helicopter gunship attacks were frequently called against targets identified by the Eighth Cavalry in the Punchbowl area. For example, in connection with the contact at YD362012 on 29 April, support from a gunship from the 1/9 Air Cavalry was requested. When it fired on the designated target, large secondary explosions resulted, reaching and downing the chopper. The crew was rescued and six destroyed trucks were later found at the site.[17] The helicopter was a loss.

----

† The *Air Cavalry Division Magazine* for September 1968 (page 7) reported that this cache included "five cases of 122mm rockets . . . and 187 cases of Russian assault rifles." This is doubtful, as the final First Cavalry Division AAR on *Operation Delaware* reports 33 semi-automatic rifles, 17 automatic rifles, and no 122mm rockets captured.

An enemy squad was in contact with D/1/8 on 1 May in the vicinity of the clash two days earlier. In a battle that took most of the day and in pouring rain,[18] small arms and artillery killed four of the enemy and caused 11 secondary explosions.[19] On the western fringe of the Punchbowl on the same day, B Company engaged an enemy unit of unknown size. Neither side took any casualties. On 2 May, D/1/8 again was in contact, this time with 15–20 NVA; SGT Evans B. Crocker, Jr., was KIA and several troopers were wounded; enemy losses were unknown.[20] Finally, on 3 May, D Company entered the complex the enemy had been so stoutly defending. They captured 400–500 bolt action rifles and a large munitions supply. Two men from D/1/8 had been killed and 24 wounded in the battles to reach the cache.[21]

On 4 May, Bravo Company departed CECILE and headed back to the western side of the Punchbowl, where an NVA supply route ran via a PVC-planked road west into Laos. The company discovered a trail running parallel to and near the road and advanced on a wide front, bracketing both routes. The NVA were waiting in at least six bunkers in a U-shaped ambush, with machine guns on both flanks. At around 1630, the dispersed American front caused the ambush to be triggered prematurely, and the NVA soon retreated into the bush before their lines could be flanked. They were not more than 5 km from sanctuary in Laos.[22] B Company lost two men, including medic Dempsey W. Parrott, a recipient of the Silver Star, and PFC David C. Schultz.[23] Ten to twelve NVA KIA were reported.[24]

5 May brought the discovery of a large weapons cache on the southwest side of the Punchbowl. Included were 27 submachine guns, six 75mm recoilless rifles, two .50 cal machine guns, and thirteen 122mm rockets.[25] A second cache located on 6 or 7 May contained over 800 rifles. Then on 8 May, the largest truck park was found by A/2/8 Cavalry to the northwest. There were "50 damaged enemy two and one-half ton and three ton trucks."[26]

In spite of the continuing American occupation of the central A Shau Valley, the enemy had by no means abandoned the area. Twice on 6 May, C/1/9 Air Cav observed and engaged enemy troops, with 6 NVA KIA in the first contact and 9 more later in the day.[27] On 10 May, a C/1/9 Cavalry gunship spotted an enemy unit 5 km west of LZ CECILE, killing 5. That same day, another 1/9 gunship caught the enemy in the open about 7 km southwest of A Luoi, claiming 11 KIA.[28] At about 1700 on the evening of 10 May, it was a 229th AHB armed helicopter that found the enemy on the move late in the day, 4–5 km southwest of LZ Cecile. Eleven NVA were KIA.[29] The number of these contacts to the southwest of the 1st Brigade LZ's indicates the probability that the NVA were finally beginning to mass for attacks on the American bases.

D/1/8 Cavalry, the troops who started the Punchbowl action on 28 April, had the last word on 13 May, when they engaged a reinforced enemy platoon to the northwest.[30] After falling back and hitting the enemy positions with heavy air and artillery shelling, the company overran the position in the afternoon.[31] The 2nd Battalion, 8th Cavalry was the last unit to be extracted from the valley, departing with its artillery on 16 May. It had been a very, very active operation, and 2/8 Cavalry was fortunate to have suffered only 10 KIA.[32]

1/8 and 1/12 Cavalry at LZ STALLION

Landing Zone STALLION, near the north end of the A Luoi airstrip, was slated to be the focal point for 1st Cavalry Division operations in the A Shau Valley. Occupied by B/2/8 Cavalry on the night of 25–26 April after a night march from LZ CECILE, two infantry battalions plus a host of engineer, artillery, and other troops would occupy and quickly construct the firebase, while preparing the long-abandoned airfield to receive transport aircraft.

At 0945 on 25 April, the 1/12 Cavalry completed insertion into the valley. By 1030, the forward headquarters of the 1st Brigade was in operation. Two companies of the 1/8 Cavalry arrived in the morning, and the remainder of the unit was on hand before the day was done.[33] By 1300, the 1/12 and 1/8 had linked up.[34]

Almost immediately, patrols from LZ STALLION began to find abandoned enemy equipment and supply caches. Before noon, A/1/8 Cav found 275 37mm AA rounds in cases. Late on the afternoon of the 25th,

D/1/8 Cavalry took possession of 200 bolt-action rifles, 3 Russian-made trucks, and a jeep 1 km north of the new LZ.[35] One truck in working condition would be used to haul supplies around the American camp (after having a large "Cav" patch painted on the door!). During the first night in the vicinity of the airstrip, several elements of the 1/8 Cavalry heard tracked vehicles (bulldozers or tanks?) moving to the south.

The next day, following a lead from helicopter recon, D Company uncovered a supply area containing 315 Soviet K-44 rifles, 30 flamethrowers, 600 122mm rockets (not!), ammunition, and other supplies.[‡36] The AAR confirms that 315 individual weapons were captured on this date.[37] On 28 April, D/1/8 Cavalry was placed under operational control of the 2/8 cavalry, and its activities from that date on are reported as part of the 2/8 Cav narrative.

On 27 April, the 1/12 Cavalry found an enemy truck repair facility, and elsewhere 800 gallons of fuel and more ammunition were captured.[38] Minor enemy camp and cache finds were present all around the airstrip and to the north up Route 548, with the patrols reporting a steady stream of new discoveries. Equipment and ammunition were either evacuated or destroyed in place. Many of the trails had NVA communication wire running alongside. Also found were a number of unexploded bombs.[39]

Commencing on 28 April, it was planned to leave eight helicopters (six slicks and two gunships) overnight at LZ STALLION. This would facilitate resupply of the other bases in the valley during periods when clouds and fog made flight into and out of the A Shau extremely difficult but when the ceiling within the valley was high enough for flights to originate there. After several days, a heavy enemy mortar barrage hit the helicopter pad. Thereafter, only a single helicopter to be used as a flare ship was maintained at A Luoi.[40]

The streams draining from the northern and southern ends of the A Shau Valley merge near Ta Bat, where the Rao Lao leaves the valley on

---

‡ The reported capture of 600 122mm rockets is incorrect, although repeated by many later sources. The correct information was provided in the press briefing on 2 May: "600 rounds of 122mm artillery ammunition."

the west side and flows to the southwest, winding its way toward the Mekong. Company D, 1/12 Cavalry was patrolling in the Rao Lao Valley, about 3 km south of LZ CECILE on 3 May, as the battle in the Punchbowl was reaching a climax. The Eighth Cavalry units in contact requested an airstrike to soften up the enemy positions before continuing their sweep. The NVA were dug in deep, so the ground commanders asked for heavy ordnance to be used.

During the Vietnam War, fighter-bombers waiting on station near the battlefield for targets to be identified by ground units and/or designated by FACs occasionally had ordnance that could not be used, usually because the aircraft were low on fuel and had to return to base before a target could be named. In such instances, a "dump grid" was specified, where no friendly units or civilians were located, but where a possible enemy presence was suspected. On 3 May, two F-100s were in the air over the A Shau awaiting strike orders, one carrying a load of MK-82 high-drag bombs, napalm, and CBU-2 cluster bomb units. The CBUs had to be dispensed before the "snake and nape" and were unsuitable for use in the Punchbowl (cluster bombs would not penetrate enemy bunkers and might not detonate on contact, creating a danger to allied ground troops who later entered the strike area), so a dump grid was located where the planes could unload the unwanted bomblets before hitting the NVA.

What followed was a communications snafu of the first order ("areas of overlapping confusion"). Helicopters of the 1/9 Air Cav were in the area near the dump grid, and it was known that American troops were nearby. The FACs, helicopters, ground troops, and fighters were ultimately unable to coordinate between all of the units in the Rao Lao Valley that day. After much radio traffic between the various aircraft, the F-100 pilot believed that a new target near the dump grid had been identified and cleared. "The ordnance was delivered on target as directed, with the CBU impacting directly across D Company, 1/12 Cavalry, and wounding 23 cavalrymen troopers (all later expected to recover)." While final responsibility for the incident was placed on the forward air controller, "all during the chain of events, numerous occasions presented themselves which, if properly utilized by all parties involved, could have averted the incident."[41] Such is the nature of war; stress and confusion are endemic and "friendly fire" accidents inevitable, although taken individually, they may appear avoidable. On 3 May, no one was killed. Sadly, the outcome would be much worse on two occasions in the valley in August 1968.

At 1830 on the evening of 4 May, 25–30 60mm mortar rounds were fired at LZ STALLION from the west, but there were no reported casualties.[42] However, SP4 Jerry T. Evans of C/1/12 Cavalry was KIA in the valley on this date.[43] On 5 May, elements of the 1/12 Cavalry were operating about 6 km southwest of LZ CECILE and within 3 km of Laos, when they engaged an estimated enemy platoon. The following day, both A/1/12 and C/1/12 were in contact with the NVA. C Company's engagement was a stalemate, with no blood drawn, but A Company lost two men, 1LT Bruce N. Huff and SP4 Ronald E. Baird, versus four enemy KIA.[44]

After securing the location for the ARVN regiment arriving in the valley on 29 April, the 1/12 would continue to operate up Route 547A until linking with the 3rd ARVN Airborne Task Force on 9 May.[45]

On the morning of 9 May, an OH-13 scout helicopter from the 1st Brigade Headquarters flight was shot down over the valley. WO Dayton W. Lanier and crew chief Chris W. Frankhauser were killed on impact.[46]

At 1115 on the morning of 12 May, A/1/8 Cavalry, on patrol within a stone's throw of where D/1/8 had made the first contact in the Punchbowl two weeks earlier, found an underground NVA Hospital, likely one of two under command of Binh Tram 7. Several hundred pounds of penicillin and sulfa drugs were on hand. This was the last major enemy installation to be taken during Operation DELAWARE, although significant caches continued to turn up as the end of the cavalry raid neared.[47] On 13 May, in an encore performance of its first day in the A Shau, A/1/8 Cavalry uncovered 900 rounds of 37mm AA ammunition.[48]

"On 14 and 15 May, friendly units in the vicinity of A Luoi received incoming artillery, estimated to be 122mm and 130mm. This is the first time artillery has been used by the enemy in this area. The firing positions were described as west of A SHAU Valley. Most probably, units of the 38th Artillery Regiment were involved."[49] Before the NVA could accurately register their artillery concentrations, on 15 May 1968, the troops at LZ STALLION were withdrawn as Operation DELAWARE neared an end. C/1/8 Cavalry was the last element to leave, even as a North Vietnamese army unit was reportedly closing in.[50]

12

OPERATION DELAWARE: ARVN, AIR, AND CONCLUSIONS

The ARVN 3rd Regiment

The third component of Operation DELAWARE in the valley was the 3rd Regiment, 1st ARVN Division, under COL Hoa. The ARVN would have the responsibility to carry out what was, in execution if not in title ("reconnaissance in force"), a search and destroy mission between Ta Bat in the central valley and A Shau at the southern end.

The 1/12 Cavalry, from its base at LZ STALLION (A Luoi), had been conducting platoon and company air assaults over a wide area to the southeast and was tasked with connecting with the 101st Airborne Division and the 3rd ARVN Airborne Task Force, moving south along Route 547A. The junction of Routes 547A and 548 was near Ta Bat, the designated location for one of two ARVN LZs. On the morning of 29 April, elements of the 1/12 moved overland from STALLION to secure what would be called LZ LUCY (YC424944). The 1st Battalion, 3rd ARVN Regiment was the first to arrive and went to work securing and constructing the new base. On the following day, the regimental command post and 2/3 ARVN Battalion arrived, and the force was completely assembled on 1 May, when the 1/2 Battalion and the 3rd Regiment's 105mm howitzer battery were inserted.[1]

On its first day in the valley, the 2/3 ARVN found a large ammunition cache. It would find another on each of the next three days.*[2] The ARVN haul included 1,500 hand grenades, 3,000 rounds of .50 cal ammo, 10,000 rounds of small arms ammo, and 960 rounds of 37mm AAA ammo.[3] There was little enemy contact until after dark on 3 May, when an ARVN patrol about 7 km southeast of LUCY was approached by a 10-vehicle convoy. Artillery fire was called in, destroying two trucks and causing one large secondary explosion. The rest of the enemy column vanished into the night.[4]

By 4 May, the thorough search of the Ta Bat area had been completed. LZ LUCY was reinforced with a battery of U.S. 105mm howitzers (A/6/33 Artillery), and while the 1/3 remained as security, the 2/1 ARVN on the north side of the Rao Lao and the 2/3 to the south commenced sweeping southeastward toward the old A Shau airstrip. That same day, moving south, the 3rd Regiment located two tunnel complexes containing three tons of arms and equipment.[5] By 7 May, the attacking troops had reached the maximum range of their supporting howitzers, having discovered additional small caches and met little opposition, and the final phase of the attack plan began. The U.S. 1/12 Cavalry, with an attached battery of howitzers, was moved from LZ STALLION to LUCY. At the same time, the ARVN 2/3 battalion moved to a location about 4 km northwest of the A Shau airstrip and secured LZ LILLIAN (YC496856). The next day (8 May), the regimental howitzers and A/6/33 Artillery were lifted to LIL-LIAN, providing artillery coverage to the far south end of the valley, the 1/3 battalion moved south, and the ARVN search of the valley down to its southern limits, beyond the air strip, was executed. After the discoveries on 4 May, no significant enemy caches were found.

By 9 May, all of the ARVN objectives had been met. On 10 May, First Cav helicopters lifted the 2/1 ARVN out of the valley, and all other units in the ARVN AO were extracted on 11 May.[6] If there were any veterans of

---

* This is the battalion that Colin Powell had been with in the valley, over five years previously. This is also the same unit that would fight at Hamburger Hill one year later. Unlike their U.S. counterparts, the ARVN soldiers did not get to hang up their helmets and go home after a 12-month tour of duty.

the earlier era still present, it probably came as a considerable relief to the 2/3 troops that they did not have to walk home, as they had done in 1963.

In a monograph titled *RVNAF and US Operational Cooperation and Coordination*, LTG Ngo Quang Truong later noted that LAM SON 216 (the ARVN part of DELAWARE) was the first large-scale combined operation joining the U.S. First Cavalry Division with the First ARVN Division. The Cav "at first was not enthusiastic about cooperating with ARVN forces. The combat effectiveness of the ARVN 3rd Regiment was held in serious doubt by U.S. forces."[7] As a result, the ARVN was assigned the easiest objective, producing only modest results. When the ARVN arrived in the valley 10 full days after the assault on LZ TIGER, it was most unlikely that the NVA would have left behind any valued ordnance, yet the ARVN soldiers did contribute materially to the destruction of the enemy's caches. They also moved rapidly from Ta Bat to A Shau in less than a week at a time when it was critical to complete the mission before the withdrawal schedule began. While coordination of ground maneuvers with U.S. ground troops was not required, the ARVN worked well, through their U.S. advisors, with the American artillery and aviation units.

The positive performance of the ARVN impressed the First Cavalry leadership. Over the next several months, after continued joint operations between the 3rd Brigade of the American division and the 3rd Regiment of the ARVN division, the latter "became one of the best ARVN combat units,"[8] an unexpected yet encouraging outcome from an operation that was otherwise disappointing in many ways.

Yet that silver lining, too, came with a cloud. The ARVN regiment in Operation DELAWARE was supported by a single howitzer battery of its own, relying heavily on American artillery, helicopter, and fixed-wing air support. A comparable U.S. brigade would have had one battalion of artillery in direct support, and likely one or more additional battalions in indirect support. In the A Shau, the 3rd Regiment was fully supported by the superb helicopter units of the First Cav. What would happen to the ARVN infantry, even the best ARVN infantry, when it had only its own supporting arms, albeit much enhanced by "Vietnamization," with which to face the North Vietnamese Army? Seven more years would pass before that question would be answered.

Air Support for Operation DELAWARE

There was no supply route from the allied bases along the coast to the A Shau Valley. Route 547 wanders southwestward from the paved road network near Hue to Route 548 in the valley, 40 km away. But that route was controlled by the VC at the time of Tet 1968, and would have to be retaken, cleared, and widened, one objective of the 101st Airborne Division's participation in Operation DELAWARE (not covered here). The 101st occupied Firebase BASTOGNE on Route 547 early in the operation, bringing in 175mm guns, but this was the only artillery that could reach the A Shau from outside. Thus, air operations would play a central role in the First Cavalry's plan. One hundred percent of the supplies to support DELAWARE had to come in by air (including all of the artillery ammunition), and support from marine, navy, and air force bombers (including B-52s) would provide the heavy punch to overcome enemy strongpoints or to obliterate the base areas once mapped.

Some air actions (e.g., the B-52 and tac air support during the insertions of the 3rd Brigade) have been previously discussed. Let us now look at the overall effort. This narrative relies heavily on the *Project CHECO Southeast Asia Report #91, Operation Delaware, 19 April–17 May 1968*, issued 2 September 1968.[9]

On 1 April 1968, the Seventh Air Force (7AF) initiated its southwest monsoon interdiction campaign in southern North Vietnam, Laos, and areas along the SVN/Laos border. In the A Shau Valley and the adjoining base areas 607 and 611 in Laos, this campaign was named GRAND CANYON. Along route 547, it was called BUFFALO (not to be confused with a Marine ground operation of the same name). Both operations would end on 18 April, with 375 B-52 ARC LIGHT and 468 tac air sorties having been logged.[10] During DELAWARE, air force, marine, and navy aircraft provided ground support, with the 7AF most commonly doing the pre-planned missions and the Marines providing the on-call service.[11]

ARC LIGHT strikes had been used sporadically in the A Shau prior to the run-up for DELAWARE. In February, 7 missions were targeted at the northern part of the valley, in the areas where LZs TIGER and GOODMAN would be sited. Another 22 missions hit A Shau and Route 547 during the period 24–29 March.[12] Finally, "for the eight days from 16 to 23 April, the A Shau Valley and Route 547 received 70 percent of all ARC LIGHT sorties in Southeast Asia."[13] A typical B-52 ARC LIGHT mission used three planes (sorties) and carpet bombed an area 1 km wide by 2 km long. The areas in the vicinity of LZs CECILE, STALLION, and LUCY (including the A Luoi and Ta Bat airfields) were primary targets.[14]

On D-Day, the USAF provided 41 preplanned and 10 immediate tactical air sorties, and the USMC contributed 14 preplanned and 22 immediate sorties.[15] As the operation continued, tac air employment was often limited by weather, particularly during the two days after the initial landing (20–21 April) and again from 27–30 April. Not including support for the ARVN LAM SON 216 operation, the Air Force provided 1,208 tac air sorties and the USN/USMC 1,316. Not a single fighter was lost to enemy fire. On May 6, two A-4 Skyhawks from the Marine First Air Wing (VMA 223 and VMA 311) collided at 10,000 feet while one was entering and the other leaving A Shau. Both pilots were able to eject and be rescued.[16] It was reported that the combined effects of the ARC LIGHT and ground attack strikes included 13 trucks, 24 gun positions, and over 400 bunkers destroyed, in addition to over 250 secondary explosions and 83 enemies KBA (killed by air).[17]

If there is one group of participants in the A Shau campaigns (not just DELAWARE, but continuously over the years) who may be considered among the unsung heroes of the American war in the valley, and who also certainly saw more of the A Shau than anyone else on the allied side, it would have to be the forward air controllers (FACs). For DELAWARE, the FACs were in the air along with the 1/9 Cavalry, starting on 13 April. In the beginning, they had to keep their small O-2 Cessna Skymasters at high altitude due to the heavy NVA AAA fire, but as the operation proceeded and the enemy withdrew its heaviest weapons from the valley, they could drop down and better identify targets for air strikes or artillery. In addition to dodging enemy fire from below, the FACs didn't always know when a B–52 raid was coming, and sometimes had to dodge bombs from above! Weather was a persistent problem.

> It would chase the FACs out of the valley usually by mid-afternoon. Sometimes we could work one hole [in the cloud cover] generally in the area of A Luoi, which stayed clear for some unknown reason. We could spiral our FACs and fighters into the hole, but it was always a problem. We had to watch the weather real closely and, in certain instances, we had trouble getting FACs back to the east.[18]

Most of the time, two or three FACs worked above the valley simultaneously to ensure adequate support for all of the patrols down below. Due to the remoteness of the A Shau, air-to-air and air-to-ground communications were often difficult. Yet for all the challenges, the FACs consistently and successfully managed fighters from all three American services, as well as the Republic of Vietnam Air Force, with many different types of ordnance and sometimes quirky bombing methods. One Forward Air Control O-2 was shot down during a night flight on 28 April, resulting in the loss of CPT James F. Lang.[19] No other FAC aircraft were hit by enemy fire, in spite of their continuous presence overhead.[20]

The survival of the men on the ground in the A Shau would be entirely dependent on supply by air, from C-rations to bulldozers and ammunition. It was planned from the outset that reopening the abandoned A Luoi airstrip would be the key to maintaining a viable presence in the valley. This was, of course, the reason why the original plan had been to insert the 1st Brigade into the center of the valley on D–Day.

> The importance of airlift to the 1st Cavalry's operations in the A Shau was apparent in the careful timetable written into the division's DELAWARE OPLAN 5-58. On D+2, the 1st Brigade was to seize A Luoi Airfield and be ready the next day to accept Flying Crane delivery of earth moving equipment. On D+5, the airfield would be open to C-7s and the next day to C-123s. At the same time, a three-day stock level of all classes of supplies would be put into A Luoi, at a rate of 225 tons per day from D+3 to D+8.[21]

In the event, A Luoi was not even occupied until D+6 (25 April), and bad weather delayed the heavy equipment so that the real work on the airfield did not start until 29 April. The work was completed in 48 hours, and on 2 May the first C-7 Caribou light transport landed. It was the first fixed-wing aircraft on the ground in the A Shau since Bernie Fisher's Medal of Honor landing in March 1966.[22] The larger C-123s and C-130s were able to land two days later.[23]

To compensate for the delay in opening the airstrip, C-130s dropped cargo by parachute into A Luoi commencing on 26 April. They continued to do so for nine days (the weather deteriorated again on 27 April and remained bad for several days), averaging not 225 tons, but 18 sorties and over 250 tons per day, and they did so with no ground navigation aids of any kind and in weather that, on the first two days of the drop, prevented even helicopters from flying in the valley.[24]

On the first day of the airlift, the ceiling was at 500 feet above the valley floor. The big transports had to drop down through the clouds into the valley from the northwest about 3 km from the airstrip, fly down the valley through heavy AAA and AW fire, drop their loads from a ramp in the back of the plane, and then exit by a sharp climbing right turn. Of the first 20 C-130s to make the perilous approach, 7 were hit by enemy fire. The big aircraft could take a lot of damage and keep going. Because of the low clouds, fighter aircraft were unable to engage the enemy gunners; helicopter flights in the valley were stopped mid-morning as the weather deteriorated.

In the afternoon, the ceiling dropped to 300 feet. As the afternoon air drops continued, a C-130B from the 772nd Tactical Airlift Squadron, 463rd Tactical Airlift Wing approached the valley, breaking out of the cloud cover a bit further to the north than most and offering the enemy a plum target. It received multiple hits from NVA 37mm AAA and heavy machine guns.

As the airplane came over the drop zone, the crew could be seen trying to jettison the load, but it would not release. As the C-130 passed over the drop zone, the combat control team could see holes in both wings while one engine was streaming either smoke or fuel and smoke flowed from the open cargo doors. Apparently, the load of ammunition had been set fire [sic]. Major Stow attempted to make an emergency landing, but as the airplane made a descending turn toward the runway it struck trees and exploded.[25]

Killed in the crash were the pilot MAJ Lilburn R Stow, crew members MAJ John L. McDaniel, CPT James J. McKinstry, Jr., SSG Beryl S. Blaylock, SGT Larry R. Todd, A1C Kenneth L. Johnson, and USAF combat photographers SGT Daniel J. O'Connor, and TSG Russell R. Fyan. Air drops were cancelled for the rest of the day.

The air drops resumed the next day and the supply line was kept open. General Tolson, commander of the Cav, later wrote to the 7th Air Force Commander: "On 26 and 27 April in the A Shau Valley, I witnessed your C-130 crews in one of the most magnificent displays of courage and airmanship that I have ever seen. . . ."[26] This from a general who himself had hundreds of helicopter air crews under his command!

Commencing with the decision on 7 May to pull out of the valley, the same aircraft that just days earlier had carried into the A Shau hundreds of tons of supplies were called on anew to haul some of the same gear back to whence it came. On 10 May, 575 support troops were evacuated by plane.[27] On 11 May, as daily afternoon rains grew in intensity, A Luoi airstrip and LZ STALLION were closed.[28] All remaining troops and gear were extracted by helicopter.

Operation DELAWARE – Mission Accomplished?

Operation DELAWARE has been variously described as a "spoiling attack,"[29] a "reconnaissance in force,"[30] or a "cavalry raid."[31] In response to President Johnson's concern "that the size of the operations . . . offered ground for suspicions that the allies were escalating the war just when the United States was proclaiming a search for peace,"[†] GEN Westmoreland

---

† The peace talks in Paris first convened on 13 May 1968.

had "directed his commanders to stop using the term 'search and destroy' to denote attacks on enemy units and base areas. To replace that phrase, which was 'overused and often misunderstood particularly in lay circles,' commanders were to substitute in their reports such standard military terms as 'spoiling attack' and 'reconnaissance in force.' These terms, the MACV commander hoped, would sound less violent in the public arena."[32]

In spite of MACV's initial instruction to the media, DELAWARE certainly was not a spoiling attack, as the NVA was not "in the process of forming or assembling for an attack." Also, this maneuver was "usually employed . . . on enemy assembly positions in front of a main line of resistance or battle position."[33] A Shau was not, however, a forward assembly point for NVA assault units but rather a very secure base along their main line of communication and supply for the I Corps region. I suspect that GEN Westmoreland and his staff liked to use the term *attack* wherever possible, as it implied initiative and an aggressive strategy.

Reconnaissance was not the *primary* purpose of DELAWARE, even though the mission statement in the 1st Cavalry Division's After Action Report (p. 7) specifies, "Conduct a recon in force, throughout the A Shau Valley." Certainly, "to discover and/or test the enemy's strength or to obtain other information"[34] was part of the goal, and if evaluated solely based on the letter of the mission statement, Operation DELAWARE was a success, although full identification of the enemy order of battle would not occur until all of the captured documents had been translated. Areas of the valley where the enemy offered strong resistance were not searched (e.g., west of Tiger Mountain, south of LZ GOODMAN, and west of the Punchbowl), limiting the scope of the recon. But for the company commanders who led the maneuver elements, reconnaissance was, as the preceding chapters have shown, strictly secondary to the capture and destruction of enemy supplies and equipment. So Shelby Stanton's "Cavalry Raid" nomenclature is, in traditional military terminology, most apropos: "[A]n operation conducted by a detached force of cavalry, often reinforced by other mobile troops, for the purpose of suddenly damaging the enemy at a sensitive point."[35]

In his final briefing on DELAWARE to the media on 17 May 1968, MG Tolson offered his analysis of the 1st Cavalry Division's "major

accomplishments in the valley": "Large caches of weapons and ammunition with other items of supplies were found within the valley confirming the existence of a large supply depot and way station." As MG
Tolson devoted most of his remarks about the Cav in his final briefing to a
lengthy reiteration of the most significant captured equipment, it is pretty
clear that the senior commanders concurred with their subordinates (or
vice versa?) about what really mattered during the operation.

MG Tolson concluded his remarks about the First Cavalry's role in
DELAWARE:

> The major accomplishments achieved by the 1st Air Cavalry Division in
> the valley have been cutting the main supply route through the A Shau
> Valley. The supply routes of Highways 547 and 547A into the area west of
> Hue and into southern Vietnam have been *temporarily* [emphasis added]
> cut off [this is actually a reference to the 101st Airborne Division's mis-
> sion!]. Large amounts of ammunition, weapons and equipment have been
> denied to the enemy, and *possibly* [emphasis added] the enemy's present
> and future plans have been disrupted. Also, we have detailed intelligence
> on the layout of this large depot area.[36]

So the supply line was *temporarily* cut, military plunder was plentiful,
and we might *possibly* have disrupted enemy plans. I wonder if the media
picked up on the adverbs. These remarks track very closely the three
"accomplishments" listed in the After Action Report.[37] Number one was
the disruption of the enemy supply system. Number two was the successful reconnaissance, which "continues to be of assistance in further disrupting his operations with more accurate and precise targeting for future
USAF bombing." And number three, not covered by the First Team's
Commander, was the "psychological impact on the enemy." The latter,
of course, was unlikely to lend itself to empirical proof, short of surrender
by the NVA.

How long would the supply line disruption endure? The 7th Air
Force Weekly Air Intelligence Summary issued 1 June 1968, just two
weeks post-DELAWARE, concurred that the "most immediate [result

of Operation DELAWARE] was serious disruption of the enemy's infiltration through A Shau." That report went on to note: "Most of these effects will be short-lived for there is no Allied force occupying the valley. Already the enemy has been observed using trucks in the valley."[38] CPT Donald Abbott, a FAC who flew over the valley throughout the operation and knew, as a matter of life and death, the locations of NVA AAA batteries, observed: "As they closed out the valley it [is] known as we evacuated [LZ] Tiger that the North Vietnamese immediately reestablished automatic weapons positions to the point where we put the northern part and the southern part of the valley off limits to low level flying and would only get down low in areas controlled by friendly forces. . . ."[39]

The NVA would be back very soon and perhaps with even heavier weapons. An NVA soldier who surrendered to the ARVN in November 1968 reported that in late May he had observed two 100mm anti-aircraft guns and two Soviet T-34/85 tanks near the junction of Routes 547 and 548.[40] On the other hand, during future air operations in the A Shau Valley up to the end of the American deployment in late 1971, enemy antiaircraft fire would sometimes be intense but almost exclusively limited to weapons of 12.7mm or 14.2mm or smaller, including RPGs. Never again would massed fire from 23mm, 37mm, or heavier weapons be encountered.

As for the value of the information that would enable more accurate future air strikes, many of the enemy installations were destroyed in the course of the operation, and the NVA tactic of constantly shifting bases, combined with the use of deep bunkers and caves, hidden by impenetrable camouflage, would minimize the long-term impact of that intelligence.

If we now consider the maneuvers of the companies scouring the valley and adjacent hills, and take into consideration the metrics most often cited to validate DELAWARE—the lengthy list of arms, equipment, ammunition, and food stores captured, and the ever-tabulated "body count" – a more nuanced picture of "success" emerges.

Noted author Shelby Stanton concluded:

The raid into the A Shau Valley achieved its objectives admirably. The raid determined enemy dispositions and area utilization, disrupted a principal supply area and infiltration route, and harassed NVA forces. The tangible success of this division cavalry raid was evidenced by the incredible amounts of enemy equipment captured, including 1 tank, 73 vehicles, 2 dozers, more than a dozen 37mm antiaircraft guns, 2,319 rifles and submachine guns. . . .[41]

Let us look at the line items in this summary in a bit more detail.

- 1 tank – previously destroyed by air
- 73 vehicles – a heavy majority of which (at least 50) had been "badly" damaged by air and were out of service.[42] Whether they were repairable or only usable for cannibalization is unknown.
- 2 dozers – previously damaged by air strike
- 13 37mm antiaircraft guns
- 2,319 rifles and submachine guns – This reads as though it were enough weapons to arm an entire NVA regiment. Alas, that is not the case. Of that total, 2,290 were bolt action rifles, almost obsolete by 1968, and at least 200 of these were unserviceable. Many of these were Soviet K-44s, which with scope would continue to be in service as sniper weapons.[43] However, there is no indication that any of the captured rifles had scopes. All of these rifles were probably weapons that had been withdrawn from frontline NVA infantry (replaced with the AK-47) and/or were suitable only for use by support units or trail watchers. The 29 submachine guns were likewise not weapons for infantry or sappers. In fact, a total of 33 semi-automatic and 17 automatic (assault) rifles were taken – about enough to equip one enemy platoon![44]

Another oft-noted accomplishment was the capture of 135,000 rounds of small arms ammunitions.[45] By comparison, the U.S. 227th Assault Helicopter Battalion itself expended 325,000 rounds of 7.62mm machinegun ammunition during Operation DELAWARE.[46]

Compounding the limited value of the captured weaponry on the operational scorecard, but rarely mentioned in the after action documents, is the tremendous volume of men and equipment that evaded the allied raid. One need only review the roster of NVA units assigned to the area in April – 3+ antiaircraft battalions, a bulldozer battalion, a company of tanks, and at least four truck battalions (over 500 trucks!) – to perceive the magnitude of equipment that evaded the raid and also to comprehend why the NVA was able to reoccupy the A Shau and resume business as usual so quickly.[‡47] MG Tolson, summing up why so many of the NVA were able to flee to safety in Laos despite his division's huge airmobile force, remarked: "According to old French records, April was supposed to be the best month for weather in the A Shau Valley. As it turned out, May would have been a far better month – but you don't win them all."[48] This is disingenuous. The deterioration in the weather in early May 1968 (the airstrip at A Luoi was unserviceable starting on 11 May) is evidence that May was not appreciably better than April, and the timing and locations of the initial air assaults did not close any exits from the valley except the Route 548 spur on the south side of Tiger Mountain.

MG Tolson had informed GEN Westmoreland that he "would like to see the enemy come after us for a change, particularly the forces from Laos."[49] When this wish came true, as the 5/7 Cavalry was first stopped in its tracks and then bombarded by NVA troops operating along the Route 548 connector to Route 922 in Laos, American air power was unable to win the day, and, perhaps the critical shortcoming in the operational plan for DELAWARE, there were no reinforcements available to employ in maneuver to envelop or outflank the NVA strongpoint.

In 1968, both GEN Westmoreland and GEN Abrams stated that they had enough resources to get the job done in Vietnam. Judging by what happened in the A Shau, both were egregiously incorrect. The mission to the valley was temporary, when only a permanent presence could have denied the NVA this key bastion. The forces assigned to DELAWARE were far too weak, and lacked sufficient helicopter reserves, to close key exit routes simultaneously, trapping the enemy rather than squeezing him out, or to provide ground reinforcements when they were needed to

---

‡ The U.S. Army official history states that elements of two artillery units, one with 122mm rockets and one with 85mm guns, were also in the A Shau at the time Delaware commenced – they, too, escaped unmolested. Or not: see following section (Postscript) for subsequent artillery and vehicle captures that were probably a result of DELAWARE.

overcome enemy units that may have been defending important caches. The enemy, as a matter of practice, defended strongly only when they had something that needed defending, or when they saw the opportunity to inflict heavy casualties.

What equipment did the Cav give up to "achieve its objectives?" The After Action Report lists the following helicopters as "dropped from accountability" during DELAWARE: 12 OH scouts, 31 UH-1 Hueys, 4 AH-1G Huey Cobras, 9 CH-47 Chinooks, and 1 CH-54 Flying Crane – a total of 57 helicopters.[50] Thus, the balance sheet does not look so favorable when both credits and debits are acknowledged. If anything, the helicopter losses might have served as a bright red warning of what could transpire when they were employed in a high-risk, AAA-dominated environment. This experience was a harbinger of what would happen in 1971, when the ARVN went into Laos.

Perhaps the most significant intermediate-term impact of the operation was on the enemy food supply. Over 71,000 lbs of food stores (primarily rice) were captured or destroyed, and the effects of this interruption in the NVA rations would be felt for months to come.[51] As result of "rice denial" operations throughout northern I Corps, in July, a "counter-famine move-ment" was initiated by the NVA in some areas. By August, "enemy units in MR Tri-Thien-Hue remained in their base areas and were suffering severe food shortages." These conditions persisted until at least October 1968.[52]

Looking back on the dominant role of weather during Operation DEL-AWARE, MG Tolson later concluded that timing the raid into the valley to coincide with the expected lowest rainfall period (April–May) was not a valid decision. "*Ceilings* and *visibilities*" [emphasis in original] had a much greater impact on helicopter operations than monsoon thunderstorms and rain. As a result, future allied forays into the A Shau would not be limited to the period between the monsoons.[53]

In a 1975 monograph in the Department of the Army's *Vietnam Studies* series, LTG Willard Pearson observed:

> The A Shau Valley campaign occurred after friendly forces had been absent from that area for two years. In a way, this operation [DELAWARE] signaled an end of one phase of the conflict. It marked the loss of enemy control of

a long-held fortress and also demonstrated the control which the U.S. and South Vietnamese forces were reestablishing in the wake of the enemy's Tet offensive.[54]

This conclusion is typical of the "we were winning when I left" syndrome that to this day pervades much of the literature on the Vietnam War. The search and destroy mission (for such it truly was) in the A Shau Valley in April–May 1968 did indeed mark the return of allied ground forces to the valley. Neither then, nor as the result of any of the subsequent American operations that would continue until late 1969, did the allies ever control that valley. At best, they contested it for brief interregnums between longer periods of North Vietnamese hegemony.

Operation DELAWARE – Postscript

There is a very distressing footnote to Operation DELAWARE. The allies engaged in a division-sized operation of almost a month's duration to rack up the scorecard just discussed. The Communists achieved more on 19 May 1968 in a solitary attack with a dozen rockets:

> Though the enemy had failed to prevent the 1st Cavalry Division from raiding his supply bases in the A Shau Valley, he managed to strike back at the U.S. logistical system just as DELAWARE was coming to an end. On 19 May, the 1st Brigade, 1st Cavalry Division was in the midst of refueling its helicopters and taking on supplies when twelve 122-mm rockets slammed into Camp Evans [the main base of the First Cav]. The incoming rounds hit an open-air ammunition dump, igniting 3,400 tons of ordnance and causing a fire that spread to the nearby airfield, eventually damaging or destroying 124 aircraft. The catastrophic loss of helicopters and ammunition rendered the cavalry brigade ineffective for at least a week. . . .[55]

There were also some discoveries by the 101st Airborne Division along Route 547A that indicated that enemy equipment that had been used to support the attack on Hue, and/or stockpiled for future action in that direction, had been stranded east of the A Shau Valley by Operation Delaware:

On 19 May, B/1/327 Infantry found two 85mm guns and a 1½ ton truck;

On 20 May, B/1/327 Infantry found three 1½ ton trucks;

On 22 May, B/1/327 Infantry captured two 23mm AA guns with 1500 rounds of ammunition;

On 22 May, Recon Platoon and C/1/327 Infantry found three 85mm guns

On 28 May, A/1/327 Infantry found 48 1½ ton trucks.[56]

While not credited to the First Cav in the final report on DELAWARE, these captures were undoubtedly made possible by the NVA retreat route having been cut.

13

INTERDICTION AND RECONNAISSANCE: MAY–AUGUST 1968

Ground Interdiction and Area Denial Operations

As the last allied ground troops departed the A Shau in mid-May 1968, engineer teams engaged in a number of "area denial" projects. "1 ACD initiated extensive denial measures to maximize disruption to the enemy's activities after extraction. The DS [direct support] Engr Company of the 3d Bde executed 32 separate denial targets in the Bde's AO, completely dropping the road (Route 548) south of LZ TIGER. The DS Engr Plat supporting the 3d ARVN Regt executed 16 separate denial targets." In the 1st Brigade AO, another 26 denial targets were executed.[1]

"Area denial" included a variety of options:

1) **Destruction of roads, e.g., by using explosives to cause landslides where NVA routes traversed steep hillsides ("choke points")**

2) **Destruction of bridges**

3) **Laying minefields[2]**

4) **Seeding enemy caves and tunnel complexes with long-lasting riot gas (CS-1) crystals**

5) **Placement of "electronic ambushes," i.e., booby-traps of various sorts that would be tripped by enemy movement, not requiring in-person monitoring.**

As with other scientific approaches to inhibiting the enemy supply routes (e.g., Operation POPEYE, COMMANDO LAVA), the A Shau Valley would be a test bed for the latest generation of acoustic sensor devices.

During Operation Delaware, our intelligence effort was supplemented by a small Department of Defense team of electronic experts. They came all the way from Washington to emplace the then-new acoustic sensors in the A Shau. The team supplemented our other intelligence capabilities during the operation and placed some sensors that would remain to monitor enemy activity after our forces left the area.[3]

Aerial Interdiction – May to August 1968

In the fall of 1966, Secretary of Defense Robert McNamara was the driving force behind the initiation of a plan to stop NVA infiltration by creating a strong point/obstacle system (SPOS) along the DMZ separating North from South Vietnam. As he envisioned it, this barrier would run from the South China Sea about halfway to the Laotian border. It would become known as the McNamara Line; it would never come anywhere near to completion. Less well known, but equally critical in the scheme to stop the flow of enemy manpower and supplies into the south was the second part of this program: "an air-supported anti–infiltration subsystem extending westward from the strong point/obstacle subsystem into central Laos to include what popularly was known as the Ho Chi Minh Trail."[4]

The original code name for the two-part plan was PRACTICE NINE. This was changed several times, whenever it was thought it might have been compromised, and by July 1967 it was known as DYE MARKER. On 8 September 1967, after Secretary McNamara publicly revealed the strong point subsystem, it was decided to retain the DYE MARKER name for the strong point barrier and to create a new codename for the (supposedly) still-secret air-supported sensor system that was being set up from the western part of South Vietnam into southern and central Laos. This would be known as MUSCLE SHOALS. The equipment and operational techniques that were implemented and improved in MUSCLE SHOALS would be used in the A Shau Valley beginning in mid-1968, under the final project code name IGLOO WHITE. The problems that impeded the effectiveness of the system in Laos would likewise be in play when implemented in the A Shau.

There were five basic components in the MUSCLE SHOALS / IGLOO WHITE system.

1) **Seismic (ADSID, HANDSID, HELOSID) and acoustic (ACOUBUOY) sensors, that would detect enemy movement on the ground and transmit that information to circling control aircraft**

2) **Very small camouflaged mines called Gravel, Microgravel, and Dragontooth. These devices would injure personnel and damage vehicles; detonations would also help to trigger the sensors**

3) **EC-121 aircraft that would receive the sensor signals and relay on to the ISC**

4) **An infiltration surveillance center (ISC) located at Nakhon Phanom Air base in Thailand (code name DUTCH MILL), where the raw data from the sensors would be evaluated and target information passed on to 7th Air Force attack aircraft**

5) **Forward air controllers (FACs) and strike aircraft, which might include fighter-bombers or fixed-wing gunships.**

The first action required to start the air interdiction ball rolling was to implant the sensors. These were air-dropped (also delivered by helicopter and Special Forces teams after March 1968), camouflaged, spike-shaped devices that in theory would implant nose-first in the ground, with sensory electronics in the nose and radio transmission circuitry in the tail. In the pre-GPS days, it was necessary to know, with as much precision as possible, where each spike fell to earth (acoustic devices could also be dropped by parachute into the jungle canopy), as that would be the basis for calculating the expected enemy location. The seismic sensors were dropped in "strings," initially by propeller driven OP-2E Neptune aircraft based at Nakhon Phanom (Naked Fanny) Air Base.

The effectiveness of the sensor screen was subject to several limitations:

1) **Placement problems: sensors might strike foliage or rocks while landing and fail to penetrate, or they could land in soft soil and penetrate too much;**

2) **Battery life was short (initially 30, later 45 days) and often unpredictable;**

3) **Placement was uncertain. One study found "errors in range from 262 meters to 715 meters, and errors in deflection from 143 meters to 248 meters."[5]**

4) **Because of (1) and (2), "strings" had to be regularly "reseeded," requiring additional time and aircraft, and adding to the calibration issues arising due to (3).**

The Gravel and Dragontooth mines were dropped via Cluster Bomb Units (CBUs) and had a chemical timer deactivating them after a fairly short period. These also therefore required regular reseeding.

The greatest success of the sensor-based interdiction program came immediately prior to Operation DELAWARE, when an intense target identification project was focused on the Khe Sanh area and used in coordinating the air power that ultimately won the day there (code name NIAGARA). The experience gained during the Khe Sanh battle would also be invaluable in the evolution of the equipment and methods employed in future IGLOO WHITE operations. On 9 March, with NIAGARA still underway, COMUSMACV asked that MUSCLE SHOALS sensors be seeded into the A Shau Valley.[6]

In late March 1968, GEN Westmoreland sent an inquiry to 7th Air Force Headquarters, asking about the possibility of using aerial mining to impede enemy movement in the A Shau Valley.* The response was negative. Heavy enemy AAA in the valley made low-level delivery of gravel mines too hazardous, and the 500 lb. Mk-36 Destructor Mines had a 6-month self-destruct feature (plus or minus 3 months) that would be a problem if any allied ground forces were to enter the area.[7] Another report cites alternate reasons: the high dud rate of the M-36 and the NVA capability to defuse the chemical fuses. In either of those cases, the explosives in the bombs would likely be used against the allies.[8]

In late March, 7AF asked for approval of "'designated strike areas' where strikes can be put in under FAC control without further clearance." GEN Cushman responded that all tac air strikes would be treated the same as close air support, and the MACV Tactical Air Support Element (TASE) said there were no plans to create air free strike zones inside South Vietnam.[9]

Up to this point in time, air-to-ground mission authorization required a multistep protocol:

> Close air support was requested by a ground commander and passed to MACV and its Tactical Air Support element (TASE), which passed the approved requests to 7AF for accomplishment. In short, the *ground commanders* [emphasis added] designated the in-country targets. Since, as a practical matter, the ground units concentrated on ground operations, they rarely organized or implemented concentrated air interdiction campaigns. Consequently, what air interdiction occurred was done by scattered FACs flying border reconnaissance and back-country surveillance.

The predictable end result: "No significant in-country air interdiction took place."[10] In the A Shau, the control of air operations was further

---

* This would have been after Westy's message to LBJ concurring with GEN Cushman that ground operations were not feasible until at least late May, but before the success of Operation PEGASUS in early April reinvigorated the "recon in force" option.

complicated because the area along the Laotian border, which generally lay outside of the designated Area of Operations for any U.S. unit (except when temporarily assigned, as in the 1st Cavalry and 101st Airborne Division operations in 1968–1969) fell under the responsibility of the ARVN, who also seldom ventured there.[11]

The exception that was granted and that opened the door for future *direct* Air Force control over target designation was the NIGARA bombing that commenced on 22 January and ended on 31 March 1968, effectively breaking the Siege of Khe Sanh.

On 1 April as part of its southwest monsoon campaign in Laos and southern North Vietnam (Route Package I), 7AF implemented operations BUFFALO, covering Route 547, and GRAND CANYON, encompassing the A Shau Valley and contiguous Base Areas 607 and 611 in Laos. The establishment of these in-country interdiction missions was in response to a mandate from GEN Westmoreland on 30 March, instructing 7AF to "impede and harass enemy and equipment entering South Vietnam." Oddly, these operations never became "official," as III MAF never gave the necessary clearance. Cleared or not, 300 B-52 sorties and 227 tac air sorties were put in during GRAND CANYON. These operations ended on 18 April, as Operation DELAWARE began.[12] Results of Grand Canyon included 76 secondary explosions, 3 enemy KBA (killed by air), 22 road cuts, and 3 trucks destroyed.[13]

On 6 April, unable to reconcile the III MAF refusal to allow the Air Force to unilaterally direct air strikes with COMUSMACV's order to interdict enemy movements inside South Vietnam, the 7AF Commander sent a message to MACV stating that absent the establishment of designated strike zones, an in-country interdiction campaign could not be carried out. GEN Westmoreland, in turn, sent a message to his boss, CINCPAC, stating his intention to authorize 7AF control of air operations in Specified Strike Zone(s), in a sustained air campaign targeted especially against enemy truck traffic entering RVN.

On 26 April, 7AF requested III MAF to designate four SSZs in I Corps, including Route 547A (the new NVA road connecting Route 547 with Route 548), Route 548 in the A Shau Valley, Route 922 connecting the northern end of the valley with Laos, and "Route 548 Extended," connecting the southern end of the valley with the NVA road network in the direction of Da Nang.[14] As a result, after the conclusion of Operation DELAWARE, on 31 May 1968 Specified Strike Zone Victor was created, covering the A Shau Valley and surrounding mountains.[15] The Air Force at last had a free hand in the valley.†

For the remainder of 1968, except during Operation SOMERSET PLAIN, 7th Air Force would be responsible for interdicting the enemy movement through the valley and destroying his installations. There were two air strategies for reducing the NVA "throughput" of supplies (applicable in the A Shau, as well as in Laos and North Vietnam).

"Truck killing" relied on the sensor network, with FAC on-site evaluation of sensor-reported activity to designate "live" targets. Truck parks and maintenance facilities were also targets.‡

"Line of communication (LOC) interdiction" involved the identification of "choke points" on the enemy road network that could be closed with bombing and kept closed by continued attacks on engineers attempting to rebuild. The most common choke points were roads running along steep mountain sides or through narrow passes, and at river crossing points. To prevent NVA road repairs once a choke point was closed, the target area would be seeded with antipersonnel trip-wire mines (CBU–42), Gravel, Dragontooth, Mk–36 500-lb Destructor Mines, and CS Riot Control gas (using 55-gal drums or cluster bombs [BLU-52]).[16]

A United States Senate committee report in 1971 describes a choke point:

Three types of munitions were involved: First, highly accurate guided weapons are used to cut the road at a point difficult to bypass. Next, anti-material landmines are emplaced in the areas adjacent to the road. These mines will destroy a truck if one leaves the road and enters the mined

---

† It is noteworthy that *all* of the SSZs in I Corps were in or adjacent to the A Shau Valley.

‡ The Project CHECO Report (#124) on Interdiction in SEASIA November 1966– October 1968 contains a very interesting analysis of "LOC Interdiction Versus Truck Killing" as Appendix II. It concludes, "Interdiction of choke points is more effective and costs less than truck-killing alone."

area. Third, antipersonnel landmines are emplaced over the antimaterial mines to deter the enemy's mine clearing operations. Fourth, sensors on both sides of the munitions package determine the amount of activity and whether truck traffic is getting through the package.[17]

It is interesting that although the preceding paragraph was included in a formal report to the U.S. Senate in 1971, a mid-1969 Project CHECO report was a bit more cautionary re: the enduring result of cutting roads:

Once the road cut was made, 7AF did not have adequate long-term area denial weapons to prevent rapid enemy repairs. Mention has already been made of DIAs [Defense Intelligence Agency] credit of three hours' closure for a road cut and how 7AF calculations suggest a closure time of under one hour would be more realistic. Area denial weapons such as gravel (XM-41), Dragon tooth, delay-fused bombs, and MK-36s were used often and in several combinations but with generally unsatisfactory or unknown results.[18]

Another problem that haunted the interdiction effort in 1968 was the lack of a centralized intelligence exploitation center. From June through November 1968, only 10 sorties per day were assigned for SSZ Victor,[19] and timely identification of the most important targets was key to making the best use of limited aircraft. This was often not accomplished. Another important shortcoming of the SSZ system was the inadequacy of reconnaissance assets. Photo reconnaissance was wanting, and bomb damage assessment (BDA) never strong enough to provide feedback to planners about actual target composition or the level of destruction achieved.[20] Such reconnaissance teams as were assigned to BDA endeavored at the greatest risk to complete their missions, but the NVA anti-recon teams were extremely effective in denying them access to the valley.

Of course, it was not only the recon teams on the ground that suffered casualties. On 9 June, a Marine A-4 Skyhawk from VMA-121 was one of several planes in direct support of allied troops in contact (unit unknown) about 4 km southeast of Tiger Mountain. His aircraft fatally damaged by enemy ground fire, the pilot, 1LT Walter R. Schmidt, Jr., flew to the north, ejected, and was observed by other pilots to have caught his parachute in trees. Immediate radio contact was established, and 1LT Schmidt reported that he had a broken leg and could not move. An SAR (search and rescue) mission was initiated and an Air Force HH-3 Jolly Green Giant arrived at the crash scene, about 2 km northwest of Tiger Mountain, as enemy troops were observed approaching. As the helicopter hovered over 1LT Schmidt, it was hit by heavy fire and crashed to the ground, disintegrating on contact. The four-man crew, LT Jack C. Rittichier, United States Coast Guard, CPT Richard C. Yeend, Jr., USAF, SSG Elmer L. Holden, USAF, and SGT James D. Locker, USAF, were killed on impact.[21] The remains of all four were repatriated in 2003 and buried with honors. LT Jack Rittichier is one of only seven U.S. Coastguardsmen lost in the Vietnam War, and the only one to die in the A Shau Valley.

Early the next morning, a ground search team was inserted into the area but could find no trace of 1LT Schmidt. He had been captured and was later seen by other POWs, but did not return when the prisoners were released in 1973 and is believed to have been executed by his captors in 1971.[22]

The effectiveness of the air campaign in Strike Zone Victor is very difficult to determine. Although the airmen believed "day and night operations closed the valley," interdiction on the flat valley floor was problematic, as the NVA excelled in constructing alternate well-concealed routes. Summer and fall are the wet season, so a significant reduction in enemy traffic was in any case expected from May to October. In November, NVA engineers constructed a bypass around one of the interdicted areas in the northern valley, and the tit-for-tat continued as air strikes interdicted the new road in five places. The most notable single result of a strike came on 30 July when an ARC LIGHT mission produced 300 secondary explosions.[23]

A May 1968 MACV report to CINCPAC included the problem area "ROUTE/AREA DENIAL": "Existing means for route and area denial have not effectively denied either route or areas from the enemy. Additionally, most of the available area and route denial means require excessive time for installation and are hazardous to friendly troops and civilians."[24]

Eighteen months later, the next report in the "Problem Area" series still included "ROUTE AND AREA DENIAL": "There is no effective means of denying routes and large remote areas to the enemy."[25] Within the A Shau, a December 1968 briefing paper for the 7AF commander concluded that "it is extremely difficult to keep the Route [548] interdicted."[26]

Ground Reconnaissance – MACVSOG May to August 1968

On 3 May, while Operation DELAWARE was underway, MACVSOG was tasked with sending a recon team "across the fence" into Laos to look for the NVA units that had fled from the A Shau during the allied incursion. While some sources claim that Recon Team (RT) Alabama was looking for "an entire NVA division that had been pushed into Laos when the 1st Air Cav had swept the adjacent A Shau,"[27] that is likely hyperbole, as (1) there are no contemporaneous intelligence documents supporting the presence of large enemy combat units in the area, and (2) if there had been a division of NVA regulars present, there would assuredly have been stronger attacks launched on the First Cav's firebases and patrols. In any case, RT Alabama, consisting of SSG John Allen, SP5 Kenneth Cryan, PFC Paul C. King, Jr., and six Nung tribesmen, was inserted into Base Area 607 in Laos about 13 km southeast of LZ LILLIAN.§

The insertion went off without opposition. For its first hour on the ground, the team moved warily. They then surreptitiously approached what was soon observed to be a major NVA headquarters facility. Enemy soldiers suddenly began to be seen in increasing numbers and all around. The patrol was spotted, and a running gun battle ensued, until the team climbed a knoll and took shelter in a large bomb crater. From that strong defensive position, they fought off repeated enemy attacks until dark, supported by air strikes "danger close" to their position. PFC King and one Nung were killed in this day's actions, and SP5 Cryan was severely wounded. An attempted extraction by Huey just before dark was aborted by enemy fire. Throughout the night, the surrounded team repulsed

---

§ This team is in some sources mistakenly referred to as ST Alaska.

multiple enemy probes, throwing hand grenades so as not to expose their positions.

At dawn, the NVA launched an all-out assault, supported by multiple machine guns and rocket-propelled grenades. When the enemy was stopped, four more Nungs lay dead. A FAC then arrived overhead, and soon, F-4 Phantoms were back on station, pummeling the enemy with bombs and cannon fire. The pilots reported six 12.7mm machine guns encircling the crater and 37mm AAA farther out. One Phantom was shot down, but both crew members ejected and were picked up.[28] When the Phantoms ran low on fuel, an A-1 Skyraider arrived to continue the air support. More A-1s and an Air Force HH-3 Jolly Green search-and-rescue helicopter arrived. The Skyraiders were to keep the enemy heads down, while the helicopter dropped a jungle penetrator, which could only lift two men at a time due to the high-density altitude. SSG Allen secured SP Cryan and the surviving Nung in the penetrator, and the helicopter started to rise. Instead of firing at the aircraft, the enemy on the ground focused all of his weapons at the two men dangling below the helicopter, striking and mortally wounding both. SSG Allen was suddenly alone in Laos, the sole survivor of RT Alabama.

Allen asked that the pilots dump all remaining bombs on top of his position, while he sheltered at the bottom of the crater. Once that was done, he bolted through the still-stunned enemy's line, killing several NVA soldiers he encountered, including the five-man team manning a 12.7mm machine gun. Reaching another shelter, he once again asked the FAC to blast the area around him. That done, an RVNAF H-34 Kingbee from the 219th Special Operations Squadron rolled in to attempt a rescue. It was shot down and the entire crew killed. SSG Allen continued to evade his pursuers while fighters with 20mm cannon, cluster bombs, rockets, and machine guns laid down a wall of fire around him. Reaching another possible extraction point, a Huey tried to lower a McGuire rig. It too was hit and crashed, although the crew survived and was rescued. Finally, after two more hours of running and directing air strikes against his pursuers, SSG Allen reached a large open field where another RVNAF helicopter picked him up, receiving no fire. His ordeal was over; for it, he was awarded the Silver Star.[29]

After the loss of RT Alabama, a 130-man Hatchet Force was inserted into Base Area 607 to continue the reconnaissance mission in much greater force. On 20 May, the company was operating about 4 km due south of where SGT Allen had been surrounded. In the early morning, the unit became engaged in a bitter battle with an enemy unit of unknown size. The American leader of one of the platoons of indigenous troops was Special Forces MSG Robert D. Plato. MSG Plato sacrificed his own life to save his men and was posthumously awarded the Distinguished Service Cross:

Master Sergeant Plato distinguished himself by exceptionally valorous actions on 20 May 1968, as the leader of a Vietnamese platoon on a reconnaissance-in-force patrol. His unit had become surrounded by a numerically superior enemy force which began closing on one squad that was isolated from the rest of the platoon. Exposing himself to enemy fire, Sergeant Plato joined the squad, rallied its members and directed their counterfire against the insurgents. As the fight grew more intense, the enemy received reinforcements, and Sergeant Plato realized that his men would not be able to hold their ground much longer. Ordering the squad to join the main force's perimeter, he remained behind and placed devastating fire on the insurgents to cover the withdrawal. As he began his own maneuver toward the perimeter, he discovered that not everyone had returned. Disregarding his own safety, he ran back to his former position, saw three missing men, and made his way to them. Only one was still alive. Sergeant Plato held off the assaulting enemy long enough for the man to reach the safety of the perimeter. Realizing it was too late to return there himself, he chose to hold his ground as long as possible. He fought the enemy with deadly fury until his position was overrun and he was killed. His gallant stand diverted the insurgents' attention from the main perimeter and totally disrupted their assault.[30]

The Hatchet Force was able to break contact and move away from the battle area, calling for a helicopter with ammunition resupply and for the evacuation of several wounded soldiers. The designated LZ was atop Hill 1095, about 6 km south of the southern end of the A Shau Valley and near the junction of two major enemy trails, Routes 923 and 614. The force reached that location and established a perimeter, but they became aware that as the morning wore on, the enemy once again had them surrounded. Around noon, an H-34 Kingbee of the RVNAF arrived and hovered over the hilltop, preparing to land. Ambushed as it descended, the Kingbee was struck by intense enemy fire, including an RPG. As it tried to climb and escape to the east, it caught fire and crashed, exploding either in the air or upon contact. Aboard was a Vietnamese crew and one American, Green Beret SFC John H. Robertson. Just days earlier, SFC Robertson had served as the aerial observer in a FAC aircraft during RT Alabama's ordeal. Now he, too, joined the list of those missing in action in Laos. A search-and-rescue attempt was aborted under heavy fire, and MSG Robertson's remains were never found.[31]

The next day, MACVSOG CCN (Command and Control North), FOB-1, made another attempt to put a recon team on the ground in Laos, this time about 7 km due south of Tiger Mountain and almost directly on the Laos-Vietnam border. RT Idaho was led by SFC Glen O. Lane, with SSG Robert D. Owen and four Vietnamese troops. After insertion on the morning of 20 May 1968, the team made radio contact at 1024, indicating that they were surrounded. They were never heard from again.[**][32]

A short time later (sources list the date variously as 22 May, 23 May, or 24 May), RT Oregon with two Americans and 10 Nungs was inserted into the same LZ from which Team Idaho had disappeared. About fifty meters from the LZ, they came upon an area where concussion grenades (employed to stun an enemy for capture rather than for killing) had been used. While investigating this site, RT Oregon was attacked by a company of NVA. One Nung was KIA and seven others were wounded. The survivors were able to evade the enemy and be extracted under fire.[33]

According to author Robert Gillespie, "[t]he decimation of the FOB's Nung element in the A Shau led to a mutiny of the Cambodian members of the exploitation force in June. The men refused en masse to return

---

** The missing-in-action date listed on the Vietnam Wall is 23 May, but it does appear that no word was heard subsequent to insertion on 20 May.

to the A Shau and threatened their officers. The Cambodians were then forcibly removed from the FOB at gunpoint."³⁴ The Cambodians, like the Montagnard Nungs, were paid mercenaries. Unlike the Green Berets, they could "opt out" of the war if the risk/benefit ratio became too unfavorable. Special Forces men could only soldier on.

FOB-1 lost one more member on 15 June. MSG Francis E. Manuel was leading a platoon of indigenous troops on a patrol down a mountain in the A Shau. The point man tripped a mine and was severely wounded, and a strong enemy force opened fire from ambush positions. Although he himself was hit with shrapnel from a rifle grenade, MSG Manuel managed to reach the injured trooper and drag him to safety. In spite of weakness due to blood loss, MSG Manuel continued to lead his men in a counterattack, until he was fatally wounded by small arms. The Silver Star awarded posthumously to this courageous Green Beret was one of many earned by SOG soldiers who risked their lives to save their indigenous fighters.³⁵

Ground Reconnaissance – Company F, 58th Infantry (LRP) May – July 1968

With the departure of the First Cavalry Division from the valley, and while the "Sneaky Petes" of MACVSOG were patrolling just across the border in Laos to the west, the 101st Airborne Division took over the reconnaissance zone in the valley proper. The assignment fell to the Ranger company attached to the division, Company F, 58th Infantry (Long-Range Patrol), which was attached to the divisional reconnaissance squadron, the 2/17 Air Cavalry.

In late May, a team from the ranger company was trained in the emplacement of seismic intrusion devices and sent into the valley to do just that. "The devices . . . were black plastic boxes about the size of a shoe box, with thin little antennas disguised to look like twigs. Once in place along infiltration routes, they were supposed to pick up the vibrations of passing traffic and send off radio signals that would bring on artillery or air strikes. . . ." The team was led by SGT "Teddy Bear" Gaskell. The Lurps joined a company of 1/327 Infantry already in the valley, split off from the grunts in the morning, and positioned the first of two sets of three boxes at one end of a stream valley draining into the A Shau. They then linked up with another element of the same battalion. After a couple of days with the second company, the Ranger team again went its own way, this time escorted by a platoon of infantry. The combined unit was moving on a road (probably 547A) "terraced into a hillside, with the high ground on the right and a clear ravine to the left" when it was ambushed. There were several dead and wounded, including "Teddy Bear," who was shot in the buttocks and received multiple grenade shrapnel wounds. When the firefight was over, medevac was called in and the emplacement of the second set of sensors was aborted.[††36]

On about 10 June 1968, another recon team from F Company, 58th Infantry, was inserted into the A Shau Valley to keep an eye on the NVA's return. Team 10, led by SGT Ray Martinez, was about 5 km west–northwest of abandoned LZ PEPPER, between the valley and Laos. The six-man team was scheduled to be in that area for 5–6 days. Almost immediately, they were spotted by NVA scouts and soon began to receive mortar fire. The Rangers moved off the LZ, finally occupying a night position in an abandoned NVA bunker complex. Meanwhile, more NVA soldiers arrived and continued to search for the LRPs well into the early morning hours.

After a sleepless night, the Rangers soon located enemy soldiers close by and decided that the time had come to call for extraction. At 0900, word came that the pickup flight of two slicks and two gunships would be arriving shortly. As they were waiting, the recon team observed eight 122mm rockets launched from about 300–400 meters away, apparently firing at Signal Hill, which continued to operate as a radio relay station for recon patrols in the A Shau. SGT Martinez directed the gunships to the area where the rockets had originated. As the first gunship began its run at the target, two or more .51 cal weapons opened fire from the ridge below

---

†† The 1/327 Infantry in late May was operating near LZ VEGHEL along Route 547A as part of Operation NEVADA EAGLE. Although the author places this action in the A Shau Valley, as no grid coordinates are known for the F/58 Infantry patrols in 1968, I cannot confirm that this was the case. It is probable from the description of the ambush site and the presence of the 1/327 Infantry that this patrol was actually northeast of the A Shau Valley.

the pickup zone. Soon, two Marine F-4 Phantoms arrived and attacked the rocket site; they, too, received strong AAA fire. One of the helicopter gunships and one of the Phantoms were hit, but neither went down. Six more air sorties were directed against the enemy, and the LRP team took advantage of the enemy's preoccupation with the air activity to escape on one of the slicks that was able to land without opposition.

It was the NVA's reception welcoming the 101st Airborne to the A Shau. A Blue team (air cavalry infantry platoon) from the 2/17 Air Cavalry, the divisional reconnaissance squadron of the 101st, tried shortly thereafter to land in the same area but was beaten off before debarking.[37]

On 22 July, three teams from F/58 Infantry were inserted simultaneously to emplace sensor devices. The LZs on the eastern side of the valley were all "green," and the mission, unlike most in the A Shau, went off without a hitch. Each team left behind a dozen well-hidden devices along trails in the valley floor, and all 18 men were picked up five hours later.[38] The unexpected always took place in the A Shau.

14

OPERATION SOMERSET PLAIN

The 101st Airborne Division's First Operation in the Valley

The First Cavalry Division wasn't long gone from the A Shau Valley before the former tenants started moving back in. In early June, a new NVA road under construction had been identified. It went about four kilometers to the northeast from near the abandoned LZ VICKI around the western side of Co Pung Mountain (YD300088 to YD335136).[1] Unknown to the allies, this track, later known as "T-7," would connect Route 548 with the eventual 1970 battlefields at FSBs RIPCORD and O'REILLY. This area had not been explored during Operation DELAWARE. By early July,

Indications are that the enemy has reoccupied northern A Shau Valley and is reestablishing truck parks, POL [petroleum, oil and lubricants] and general storage depots, as well as AA and ground defense positions. Recent aerial photography and visual reconnaissance flown over this area has revealed the presence of four new AA positions, two of which are occupied, and the other two having a possible control center. In addition there are numerous newly-constructed bunkers, occupied trench and fighting positions, supply areas, several large truck parks and indications of recent vehicle activity.[2]

An ARC LIGHT strike near Ta Bat on 30 July caused over 300 secondary explosions and fires, indicating that a major ammunition storage dump had been hit.[3]

In light of this intelligence, LTG Rosson, CG, XXIV Corps (formerly Provisional Corps, Vietnam) was contemplating a new ground operation in the A Shau Valley. MG Melvin Zais, CG, 101st Airborne Division, volunteered his unit. The mission: "The 101st Airborne Division will conduct offensive operations in the A SHAU Valley with one U.S. brigade, plus a two-battalion task force from the 1st Division, Republic of Vietnam, to interdict enemy lines of communication, to destroy North Vietnamese Army/Viet Cong forces, caches and installations, and on order, to establish designated barriers."[4]

As was not uncommon for allied initiatives in the A Shau Valley, there would be some differences of opinion among the U.S. commanders after the fact about the success or accomplishments of Operation SOMERSET PLAIN, the initial incursion of the 101st Airborne Division into the valley. The III MAF Command Chronology lists SOMERSET PLAIN as one of the operations that "did not have any significant contact" with the enemy and "met with little success, as contacts were very light and sporadic."[5]

The After Action Report of the 101st Airborne Division reached a different conclusion:

Operation's Effectiveness—Operation SOMERSET PLAIN successfully blocked for 15 days all enemy use of the major line of communication to DA NANG, at a time when DA NANG felt the pressure of an enemy build-up around the city. Even after the Division's departure from the A SHAU Valley floor, crossing the valley was not safe for the enemy because of our extensive minefields and sensor devices implanted in the valley. . . . Operation SOMERSET PLAIN was a highly successful operation in which all assigned missions were accomplished.[6]

Viewed through a narrow lens, each of the specific tasks assigned was, as noted in the After Action Report, successfully completed. It is equally true, however, that the duration of the impact on the NVA line of communication was transitory, while the casualties inflicted were minor, especially in comparison to allied losses, and even more so when the extraordinarily high level of "friendly fire" (a.k.a. "short round") losses is considered.

B/1/9 Cavalry in the Reconnaissance Phase

The 101st Airborne Division in early August 1968 was completing a transition to become the second airmobile division in the United States Army. One of the last steps in that evolution was the conversion of the division's ground reconnaissance squadron, 2/17 Cavalry, to an aerial reconnaissance role. As that process would not be completed until well into 1969, Troop B of the 1st Cavalry Division's 1/9 Cavalry was OPCON to the 101st Airborne for SOMERSET PLAIN.

B Troop commenced reconnaissance in the valley at the beginning of August. As in OPERATION DELAWARE, the primary mission was to identify, and where possible destroy with either its own gunships or with fixed-wing airstrikes, the enemy anti-aircraft artillery positions in the operational area, in this case the vicinity of A Luoi / Ta Bat. The Air Cav was also to select landing zones for the infantry and identify the safe(st) approaches to those sites for the helicopter-borne assault force on D-Day (4 August). The level of NVA ground fire was not as intense as it had been four months earlier but was still a factor to be reckoned with. "The [1/9 Air Cavalry] troop averaged seven sightings or engagements daily beginning 1 August, located numerous trails and bunker complexes, and confirmed proposed landing zones for assault forces."[7]

When the scout helicopters on 2 August probed the area initially selected for the southernmost landing zone (LZ #1), about 2 km southeast of Ta Bat, they encountered AK-47 fire from an estimated enemy company, combined with heavier caliber anti-aircraft weapons. One UH-1C gunship was shot down, resulting in the death of aircraft commander WO1 James W. Arvidson and crew chief SP5 Richard I. Bornheimer.[8] As a result, the LZ for the 2/327 Infantry was moved to the abandoned Ta Bat airfield.

For the entire operation, B/1/9 Cavalry had two helicopters destroyed and five shot up too badly to fly, while five troopers were wounded in addition to the two aircrew KIA. The Cavalry claimed only seven enemy KIA.[9]

Hitting targets likely identified by the air cavalry, B-52 ARC LIGHT strikes achieved some notable success. On 3 August, a six-plane strike resulted in 11 secondary explosions. The next day, another six-plane mission connected with an enemy ammunition cache, causing 8 second-ary explosions.[*][10] For SOMERSET PLAIN, a total of 12 strikes using 68 aircraft were employed. As there were no reported bomb damage assess-ments (BDAs) by troops on the ground, evaluation of the impact of ARC LIGHT was limited to the observation of secondary explosions.

Preparations Outside the Valley and Along Enemy Routes into Laos

Several lessons from Operation DELAWARE were taken to heart by the staff of the 101st Airborne Division in planning for SOMERSET PLAIN.

First, given the unpredictable but often poor flying conditions and the limited hours of flying time that could be expected over the valley on any given day, it was important to establish the major logistical base for the next operation as close to the A Shau as possible. For that purpose, the existing Fire Support Base BIRMINGHAM, about 35 km northeast of A Luoi on Route 547, was developed in late July into a major facility. Between 26 and 30 July, helicopter pads and expanded refueling points were added (flight time would be about 10 minutes shorter per trip), a five-day level of most classes of supply for all units assigned to the oper-ation was stockpiled, and other enhancements to support an airmobile brigade in the A Shau Valley were completed.[11]

Second, the artillery support for SOMERSET PLAIN was an immense improvement compared to the earlier First Cavalry Division operation. When the 1st Cav went into the valley the previous April, heavy helicopter losses on D-Day precluded the insertion of all but a single battery of 105mm howitzers at LZ TIGER. The only artillery outfit outside the valley with the range to provide fire support was a 175mm gun unit at FSB BASTOGNE, but that battery was firing

---

[*] It is not entirely clear that these strikes were in the A Shau Valley, but it is considered likely as that was the focal point for B–52 activity in Thua Thien Province at the beginning of August.

at maximum range, limiting accuracy. The howitzers at TIGER also turned out to be of little value during the first several days, as almost their entire ammunition supply was destroyed when fire spread to the ammo dump after a CH-47 was shot down on the LZ. The lack of tube artillery impeded the response to enemy attacks on First Cav Landing Zones, precluded effective engagement of enemy vehicular movement out of the valley, limited options to patrolling infantry units in contact with enemy forces, and provided no alternative during hours when air support was unavailable.

The 101st Airborne Division artillery plan provided for eight howitzer batteries (three 155mm, four 105mm, one ARVN 105/155mm) stationed at four firebases in the mountains just east of the A Shau. There were also two batteries of 175mm guns at FB BASTOGNE.[12]

FB ZON (aka SON) (YD473011), 10 km due east of the A Luoi LZ, began on 19 July as a one-ship landing zone. By 24 July, bulldozer-equipped engineers had prepared positions for 18 howitzers, as well as command bunkers for the 1/327 Infantry and the 1st Regiment, 1st ARVN Division.[13]

At the site known to the First Cav as Signal Hill (Hill 1487), 2,000-lb. "daisy cutter" bombs were dropped to clear away more jungle (along with any enemy mines or booby traps), making room for the much larger FB EAGLE'S NEST (YD406036). This site, too, was quickly transformed into a three-gun artillery base. Two additional firebases, 1 and 2 km southeast of the EAGLE'S NEST, completed the roster of mutually supporting artillery positions. These LZs were called GEORGIA (YD413025) and BERCHTESGADEN (YD423012). GEORGIA could accommodate 15 howitzers and BERCHTESGADEN 12, as well as the HQs for the 1st Brigade, 2/327 Infantry, and 2/502 Infantry. All guns were in place by 2 August. The three firebases in the west could each support units operating on the valley floor well to the west and southwest of A Luoi and Ta Bat, while the guns at ZON could cover the ARVN operational area to the south (see Map 10).

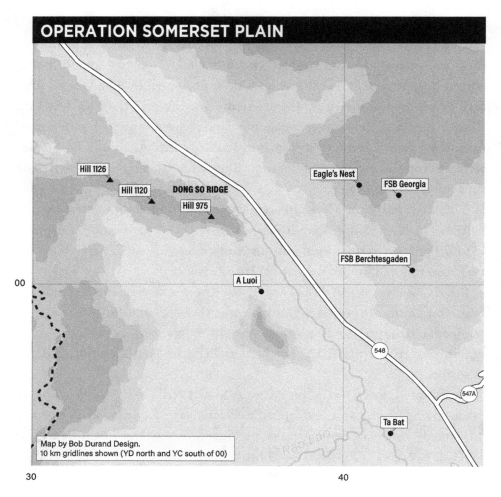

OPERATION SOMERSET PLAIN

Hill 1126

Hill 1120   DONG SO RIDGE

Hill 975

Eagle's Nest

FSB Georgia

FSB Berchtesgaden

A Luoi

00

548

547A

Ta Bat

Map by Bob Durand Design.
10 km gridlines shown (YD north and YC south of 00)

30   40

MAP 10

Source: Bob Durand Design

All of the artillery that would be in support of SOMERSET PLAIN was in the mountains east of the valley and hence would not be subject to the morning fog that often blanketed the A Shau. The downside of the firebase construction, not to be ignored, was that the enemy knew very well at least a full week before D–Day that the allies were about to venture once again into the floor of the valley. This provided ample time for Binh Tram 7 to withdraw troops and equipment into Laos, and to even better conceal its remote camps and caches.

The third element of SOMERSET PLAIN that differed markedly from DELAWARE was the use of chemical warfare. Earlier employment of persistent CS-1 riot control powder to contaminate enemy caves, bunkers, and tunnels would continue. In the open, the old method of using 55-gal. drums dropped by helicopter with limited accuracy to hamper enemy efforts to repair and use road interdiction sites was replaced by the new Air Force BLU-52 CS gas munitions. Delivered by F-4 Phantom jets, the new ordnance was more accurate and more effective, and 120 of the gas bombs were dropped at 15 sites during SOMERSET PLAIN.[14] On D-Day, eight BLU-52 strikes hit three choke points on routes leading away from the central valley: (1) 3 km north of A Shau at a key Route 548 junction and stream crossing; (2) southwest of Ta Bat along the trail roughly paralleling the Lao River valley into Laos; and (3) just east of the Laotian border along a main route running west from the A Luoi / Punchbowl area. Six days later, on 10 August, six additional persistent CS gas strikes were made on three routes between A Luoi and Laos, one target in the Rao Lao Valley, and along Route 5481 near Tiger Mountain.[15]

Phase I (Air Assault): 4–5 August 1968

While its two sister battalions (2/327 Infantry [LTC Charles A. Beckwith] and 2/502 Infantry [LTC Daniel G. Sharp]) in the 1st Brigade, 101st Airborne Division (COL Harold I. Haywood) were in standdown at the 101st Division's CAMP EAGLE base camp at the end of July, preparing for the air assault into the valley, the 1/327 Infantry (MAJ Duane G. Cameron) was at work in what the enemy would call "preparing the battlefield."[16] A Company secured FB ZON while B Company was at BERCHTESGADEN, after earlier guarding the construction of EAGLE'S NEST (26–30 July). C and D Companies were working their way along Route 547A.[17] On D-Day, the 1/327 was responsible for screening along Route 547A to the east of the assault force on the valley floor. A and B Companies would guard the firebases throughout the operation.

On 4 August, 50 UH-1 troop-carrying helicopters ("slicks") conducted multiple trips to ferry the 2/327 and 2/502 Infantry into the A Shau

Valley. With the initial lift departing BIRMINGHAM at 0913, 2/502 went in first to LZ #2 at A Luoi. Flight time to the valley was under 20 minutes. By around 1100, the third and final lift was completed and the entire battalion was on the valley floor. Enemy fire had been light, and no slicks were lost, although one gunship was shot down with no crew losses. The airborne infantry immediately secured the A Luoi airstrip and commenced reconnaissance in force (RIF) to the east.

The first lift to LZ #1 was in the air at 1130 and ran into "moderate to heavy ground fire" on approaching the Ta Bat airfield. Fortunately, as earlier in the morning, none of the slicks was shot down. Two gunships were destroyed. One crashed about 200 meters north of the LZ and the crew was rescued. The second, a UH-1C from C/101 Aviation Battalion, crashed in flames, killing crew chief SP4 Harry Kim.[18] The other crew members were all wounded but were able to be rescued by troops on the ground. Once all three lifts had been delivered, the 2/327 Infantry established security at the LZ and began its RIF to the southwest.[19]

The following day, the helicopters were in the air again after an hour delay due to morning fog, inserting the ARVN 2/1 and 3/1 Battalions at Ta Bat in Operation LAM SON 246. The much smaller ARVN formations were delivered in just two lifts between 0830 and 0945. On the valley floor, the American paratroopers continued their reconnaissance. The 2/327 had an uneventful day, but the 2/502 ran into trouble when two companies approached Hill 696, 2 km southwest of the A Luoi airfield. A Company made contact first with an enemy platoon at around noon about 400 meters east of the hilltop. Artillery, helicopter gunships, and airstrikes were all called against the dug-in enemy, but contact continued until the evening. PFC Dennis M. Hyland was KIA and five men were wounded.†
On the other side of the mountain, 400 meters west of the summit, C Company ran into two NVA squads about two hours later. Supporting arms were employed and the NVA finally withdrew at 1740 hours, leaving behind one body, a large abandoned bunker, trench, and multiple blood

---

† The operational records do not record a KIA in A Company on this date; the information is taken from the Virtual Wall.

trails. C Company also lost one man in this battle, CPL Gary M. Mabrey, and three others were wounded in action.[20] On 6 August, A Company swept through the enemy position of the previous day, finding 16 medium to large bunkers and 25 spider holes and capturing one M-60 machine gun and a 60mm mortar.

In an event that is not well reported in the historical record, a "friendly fire" incident happened to the ARVN component on 5 August. The 1st Brigade, 101st Airborne Division Daily Staff Journal reports: "1320–1325 hours: Short rd fall into ARVN Loc, possible short rd, also possible 155 rd results 6 ARVN KHA, 3 US KHA, 16 ARVN WHA." The Case Study summarized: "Tragedy also stuck in their [ARVN] area when a 155mm shell fired from SON fell short and killed ten and wounded eighteen. Two of those killed and two wounded were U.S. advisers; the remainder were ARVN soldiers."[21] It is unknown whether the firing battery was US or ARVN, as both had 155mm howitzers at FB ZON (C/2/11 Artillery and the ARVN A Bty, 11th Artillery), or where responsibility was assigned. The number and nationality of the casualties are also mysteries. The final casualty report for the operation lists 11 ARVN KNHA (killed, nonhostile action) and 20 ARVN WNHA, most or all likely attributable to the short round.[22] But there are no American advisors reported as hostile or nonhostile losses, and the Virtual Wall lists no advisor losses on 5 August or for the next several days, so it is possible that all of the Americans involved were only wounded.

Phases II & III (Recon to West / Search and Destroy): 6–10 August 1968

For the next 10 days, the units in the A Shau Valley were assigned to conduct reconnaissance and search and destroy operations, supplemented by additional platoon-sized helicopter assaults (eagle flights), in the direction of several designated objectives. Two of the objectives, BEAK (YD3502) and HOPTOWN (YC3599), were due north and south of the Punchbowl area, 3 km apart. A third was located in the Rao Lao Valley, southwest of Ta Bat (PERRY at YC3793), and the last was about 2 km south of FB BERCHTESGADEN (ESSEX at YC4299).

On 6–7 August, the Hac Bao (Black Panther) Company of the 1st ARVN Division arrived, completing the roster of units assigned to SOMERSET PLAIN. American and South Vietnamese units spread out across the AO, finding many small, abandoned camps and caches and a few abandoned weapons, but no significant contact with enemy forces. The NVA responded to the allied presence with an increasing number of standoff attacks, with A/2/502, D/2/502, and the Recon Platoon 2/502 each receiving 10 rounds of 82mm mortar fire, the latter 2 while occupying night defensive positions on the night of 6–7 August. There were no casualties. At 0300 on 7 August, a probe of the A/2/502 perimeter resulted in six NVA KIA and one prisoner of war, with no U.S. casualties. On the afternoon of 7 August, about 900 meters east of Hill 696, 2/502 Infantry troops found an enemy complex, including a large cave and four 2½-ton trucks. Three were damaged, but the other was an operating refrigeration model.[23] This would be the only important vehicle discovery during the operation. Also on that day, ARVN troops sweeping several km to the southeast of Ta Bat on the east side of the valley found four enemy base camps and a cache containing over seven tons of rice, the largest discovery of enemy rations reported for SOMERSET PLAIN.[24]

By 8 August, the NVA had begun to be increasingly active in opposing the allied infantry in the valley. At 0935, two Huey gunships were escorting a slick sent to extract the NVA lieutenant who had been captured during the probe of A/2/502 early the previous morning. One of the UH-1Cs from D/101 Aviation Battalion was hit in the fuel cell by an RPG, crashing inverted, with the loss of the entire crew: aircraft commander CPT Gary L. Higbee, pilot WO1 Richard A. Ovaitt, crew chief SP5 Charles D. Bartlett, and door gunner SP4 Dickie L. Leach.[25] The captured lieutenant was a political officer from the 9th Regiment of the PAVN 304th Infantry Division. He reported that two enemy divisions were in the vicinity of the A Shau Valley.[26] For the first time since the enemy shelled LZ STALLION in May, NVA artillery was employed. Firing from positions estimated to be along the south side of Dong So Ridge, 25–30 rounds of unknown caliber impacted the area occupied by D/1/327 Infantry, killing PFC Ronald L. Noldner and wounding nine other soldiers. In the afternoon, FB GEORGIA received three rounds from enemy artillery. Two kilometers south of Hill 696, D/2/502 encountered an NVA platoon in bunkers supported by snipers in trees and mortars. Fourteen soldiers were wounded and seven NVA were KIA in this firefight. The ARVN made contact with an estimated enemy battalion late in the evening, with minimal casualties on either side.[27]

On 9 August, both ARVN battalions continued to engage squad and platoon size enemy units, but the NVA evaded and casualties were light for both antagonists.[28] D/2/327, patrolling in the Rao Lao Valley, had one paratrooper KIA by sniper fire, SGT Victor B. Meyers. For most of the Americans operating in the A Shau, this was a day of relative calm.

For the ARVN, 10 August brought more contacts and caches. The 2/1 Battalion and the 3/1 Battalion were each in contact with platoon–sized enemy units. Five kilometers south of Ta Bat, the 3/1 swept through the contact area after an airstrike broke the enemy resistance, finding another 5 tons of rice as well as 4,000 meters of commo wire and 10 kg of medicine.

In the American AO, enemy mortar barrages fell on FB GEORGIA and B/2/502 Infantry, neither causing any KIA or WIA.[29] The heaviest losses this day came not from enemy fire, but once again from a misdirected air-strike. D/2/327 Infantry was patrolling west of the A Shau, on the south side of the Rao Lao as it flows westward into Laos. The company was just 3 km south of where a short round incident in May 1968 had wounded 23 men in D/1/12 Cavalry. Late the previous afternoon, D Company had been in contact with a small enemy force occupying a hill to the west. On the morning of 10 August, the company commander requested air strikes (a FAC was on station) to soften up the enemy position before renew-ing the attack. A first flight of Marine F-4B Phantoms placed its bombs on target, and a pair of F-100s from the 37th Tactical Fighter Wing at Phu Cat, armed with bombs and 2.75" rockets was overhead to provide a follow-up punch. The FAC, meanwhile, was confused about the actual location of D Company. When the first F-100 made a run firing rock-ets, the impact was directly on the center of the paratroops. Seven men were killed[30] (PFC Daniel L. Ault, SGT Antonio V. Garcia, PFC Roger Hulsey, PFC Samuel R. James, II, PFC John J. Matarazzi, Jr., PFC Steven M. Schlosser, and SP4 Edward Stewart, a medic from HHC/2/327) and

54 wounded, devastating the company's ranks.[31] A subsequent review of the incident identified a number of contributory factors to this tragedy:

- Rocket attack heading was off.
- Rocket range error was long.
- Strike pilot was not aware of the extent of the friendly troop location.
- Lack of prominent terrain features made orientation more difficult.
- Low fuel added urgency.
- Ground commanders should consider marking two or more points along company boundaries.
- FAC might orbit over friendlies to give vertical dimension to their location.
- Communication among all three parties must be complete.
- FAC must complete all parts of the briefing. Insure understanding, and then mark the target.[32]

As the United States ground troops had been operating in Vietnam for over three years by the time of this incident, it would be naive to attribute this event solely to the "friction" of war and combat. Frequent and repeated turnover of U.S. personnel, especially those at the platoon and company command levels on the ground, contributed inevitably to mistakes, sometimes fatal, that experienced commanders could have prevented.‡ One soldier from D/2/327 was killed by enemy action this day, PFC Kurt P. Stephenson.[33]

Phases II & III (Recon to West / Search and Destroy): 11–15 August 1968

The divisional long-range patrol company (F/58 Infantry) was called on 10 August to provide two six-man recon teams in support of SOMERSET PLAIN: "Two teams were inserted, one just north of the A Shau Valley, with the mission of looking for an NVA radar-guided .51-caliber

---

‡ The Virtual Wall cites an Army Command and General Staff College report from 1982 describing the aircraft involved as Navy A-7D bombers. I have relied on the contemporaneous account in the Project CHECO series, identifying it as an F-100 from Phu Cat.

antiaircraft gun, and the other team six kilometers to the southeast, with the mission of performing a general area reconnaissance." As it turned out, it was the second team that ran into the heavy machine gun. SP4 Glen "Marty" Martinez surprised the gun crew, killing two and sending the other two running, and became the proud possessor of the .51-cal weapon on a tripod that was evacuated to Camp EAGLE. Both teams were withdrawn without casualties on 11 August.[34]

For the three days commencing on 11 August, the U.S. and ARVN patrols continued to find numerous small enemy trails, camp sites, bunkers, ammunition caches, and occasionally a weapon or two. On 11 August, 22 rounds of 82mm mortar fire were aimed at B/2/502 Infantry, but no hits were scored.[35] For the next two days, contact continued to be very light and no NVA indirect fire attacks were noted. In the ARVN Area of Operations, 3/1 "maintained contact with an estimated battalion" on the west wall south of Ta Bat on 11 August, killing 20 enemy in the initial engagement with air and artillery support. Shortly thereafter, the ARVN found a gravesite with 22 NVA bodies buried about two days earlier, and then killed 12 more NVA in a second firefight 1 km to the north (YC434896 & YC435908).[36] The enemy all but disappeared on 12 August, but on the 13th, the 2/1 ARVN, patrolling the hills on the east wall between the A Shau and Rao Nai watersheds (YC477924), killed 10 NVA and captured 11 weapons.

On 14 August the pace of action increased across much of the operational area. For the actions of D/1/327 Infantry commencing on this date, please see the next section. At midday, B/2/502 spotted 5–6 NVA moving down a trail. In the ensuing firefight, PFC Michael W. Sinibaldi was KIA and another solider wounded and evacuated. There were no known enemy casualties. A/2/327 Infantry engaged an enemy force in late afternoon; PFC Robert A. Bulmer was KIA and one soldier was wounded and evacuated. One NVA soldier was KIA in this action and an airstrike caused a secondary explosion.[37] The major contact of the day was between the 2/1 ARVN Battalion and an enemy company, about one kilometer east of the previous day's action. The ARVN lost just one man KIA and two WIA, but accounted for 10 NVA bodies and captured a 12.7mm HMG, an 82mm mortar, 7 SKS rifles, and 25 B-40 rockets.[38]

15 August was another "light contact" day. A/2/502 received some very inaccurate mortar fire around 0930, and then had five men wounded by another barrage of 60mm fire at 1100 hours. B/2/502 was operating about 1 km west of the A Luoi airstrip, on the east side of the Punchbowl, when it was bombarded by 25–30 rounds of mortar fire. Paratrooper SP4 Phillip J. Essig was killed and nine men wounded in this barrage.[39]

Dong So Ridge: The Ordeal of D Company, 1/327 Infantry (Airborne)

On the night of 8–9 August, D/1/327 crossed the valley floor on foot, preparing to climb Dong So Ridge and search for the North Vietnamese artillery reported to be located on the reverse slope.

The Dong So Ridge is the dominant terrain feature in the north-central A Shau Valley, and the eastern end had been designated as Objective BEAK. It extends for about 5 km from northwest to southeast, with peaks of 1126 meters in the west, just over 1100 meters in the center, and 975 meters in the south. During Operation DELAWARE, LZ PEPPER was located just west of the central peak, and the 1/8 Cavalry had heavy contact along the southern slopes of the eastern end of Dong So, adjacent to the Punchbowl. One kilometer northeast of Hill 975, the valley floor is at its narrowest point, just a few hundred meters wide. Earlier attempts to create a choke point on Route 548 here were unsuccessful, however, as the terrain offered many possibilities to rapidly develop alternate trails. By August 1968, the eastern end of Dong So had largely been stripped of vegetation by repeated applications of Agent Orange.[40]

The move up the ridge began inauspiciously. Point man Roy Jenkins, armed with an M-79 grenade launcher, believed he spotted enemy movement to his front. Conducting reconnaissance by fire, he fired a 40mm high explosive projectile. The round exploded just a short distance away and a small piece of shrapnel hit Jenkins, who died of shock just a short time later.[§][41] No enemy was present.

---

§ The Virtual Wall lists Jenkins as KIA by hostile fire, but this is incorrect.

The next day, 10 August, D/1/327 received about 40 rounds of artillery fire and 20 rounds of 82mm mortar fire at 1320 hours and an additional smaller mortar barrage at 1700. Three men were wounded and had to be evacuated. The NVA artillery was believed to be located 6–8 km to the west and within 2 km of Laos.[42] It was apparent that the enemy was disturbed by the U.S. troops moving up the eastern finger of Dong So toward Hill 975.

For the next three days, D Company continued to move up the eastern end of Dong So Ridge, slowly advancing to and beyond Hill 975. On 13 August, they left the defoliated area and moved back into the jungle to the northwest, following the ridgeline.[43] On 14 August, nearing the central peak of the ridge at just over 1100 meters (550 meters above the valley floor), they began to find increasing signs of recent enemy activity. First, a recently used trail. Then a single NVA soldier, who left behind a blood trail as he fled. Just after noon, the company received five rounds of artillery fire. In midafternoon, abandoned enemy fighting positions and two fresh trails were found, and at the same time 60mm mortar fire began to come in. The scene was described as follows by the leader of the 3rd Platoon on point: "Dong So's summit stands before me, a great stone staircase of upright boulders 30 feet high with a path twisting through the boulders. . . . On our right flank is a steep incline covered with trees and underbrush. The top of the Dong So seems to be a rock fortress."[44]

D/1/327 had begun an engagement with a dug-in enemy force holding a virtually impregnable position. The NVA had two-level bunkers and would retreat to the lower level when airstrikes were inbound, then pop up again to oppose the infantry. Trying first to advance through fire and maneuver, and leapfrogging fire teams up the "staircase," the troops soon found that the enemy held the high ground and the high cards. The platoon leader estimated the enemy force at 40–50 men, reinforced with 60mm mortars and heavy weapons. While withdrawing his point element and sighting machine guns to cover likely NVA counterattack positions, LT Charles W. Newhall, III, requested his artillery and air forward observer (FO), SP4 Lural L. Blevins, III, to call in an airstrike on the entrenched enemy. Blevins had just two days left on his tour of duty. The first bombs overshot the enemy occupying the narrow ridge. The second plane dived into the attack, to be struck by fire from three .50 cal machine guns and crash into the valley floor.** More aircraft arrived and were directed against the enemy bunkers, forcing the NVA to at least keep their heads down so that the rest of the exposed point element of D company could withdraw. Secondary explosions were also reported. In spite of the presence of the F-4 Phantoms, NVA 60mm and 82mm mortar teams persisted in firing at the 3rd Platoon. By 1800, almost four hours after the battle began, the American troops were able to complete their withdrawal. Miraculously, they suffered only five wounded, including platoon leader Newhall.[45] D/1/327 Infantry claimed seven NVA KIA in this first day of the battle for the ridge.

On 15 August, D Company's 3rd Platoon spent most of the day hacking out a landing zone about 1 km back down the ridge, so that the wounded could be evacuated and resupply brought forward. At 1400 hours, the 2nd Platoon under SGT Herman advanced into the area now known to the company as the "Devil's Staircase." SP4 Blevins, the FO, again accompanied the assault platoon. Following a wall of fire laid down by AH-1G Huey Cobra gunships, the point was able to reach the area at the top of the boulders marking the high point of the 3rd Platoon's advance, but when the gunships ceased firing, the enemy quickly reappeared all around. The rest of the platoon was unable to ascend the "Staircase," as an NVA heavy machine gun was now cited to fire directly down the passage, and the enemy strength was now estimated as a full company. Under heavy fire, SGT Herman and his men were able to withdraw by 1800, although Herman himself was wounded and would require evacuation, and SP4 Blevins, who should have been on his way back to "the World," received shrapnel in the head but remained with his comrades.[46]

Third Platoon had tried. Second Platoon had tried. On 16 August, it was the turn of the First Platoon of D/1/327 Infantry to assault the enemy bastion at the top of Dong So Ridge. LT Copeland, the newly arrived

---

** Although Newhall's narrative describes this crash and states that the pilot did not eject, *Vietnam Air Losses* records no aircraft downed with crew losses in the A Shau Valley during the period of 1–20 August 1968. The AAR lists the loss of only a single F-4 during SOMERSET PLAIN, the aircraft that made it to the Gulf of Tonkin before crashing on 16 August.

platoon leader, decided to try to take his men up the sides of the "Staircase" rather than up the middle. SP4 Blevins, back for a second encore performance, directed five hours of heavy airstrikes ahead of the attack, pounding the enemy with 36 tanks of napalm, plus 250-, 500- and 750-lb bombs. At 1509, an F-4 Phantom supporting the attack on Dong So was hit by ground fire. The pilots were able to reach the Gulf of Tonkin and eject; they were picked up by SAR.[47] When the preliminaries were completed and the paratroops moved out, it was into an even stronger hail of fire than had been seen on the previous days. PFC Jose A. Graniela, Jr., was one of the first to fall, his leg cut off by an explosion. He expired soon after. PFC David Chisum, moving to try to save Graniela, was hit by two rocket-propelled grenades and killed instantly. The enemy employed all the weapons of the prior battles, but now had added recoilless rifles, buried satchel charges, and snipers in trees to his arsenal. Although at least seven snipers were killed, the enemy never ceded the upper hand. "As usual, Blevins is in the middle of the action. He has an M-16 and is taking up slack for the machine gun on the top of the stairway. . . . At the same time he brings in Cobra gunships, directing white phosphorous rounds to mark the bunkers for the gunships. The gunships start to fire, but snipers and an NVA machine gun target Blevins. He is shot in the forehead. He dies instantly."[48] Sp4 Lural L. Blevins, III, of A Bty, 2/320 Artillery, received no medal for his actions of 14–16 August 1968. He should have been on his way home alive, one of the many tragedies of the A Shau Valley.

The battle for Dong So Ridge was over. Once again, as in the draw west of Tiger Mountain in April, the enemy had shown that it would and could defend terrain that was key to its defensive plan, even in the face of heavy applications of air power. In this case, the Americans would never know the extent of the enemy artillery and other deployments west of the ridge.

Phase IV (Denial): 16 August 1968

While D/1/327 was making a final attempt to surmount the central peak of Dong So, the 2/502 Infantry was searching about 500 meters west of where B/2/502 was mortared the previous day. The Currahees found an enemy cache containing four RPD machine guns, four 60mm mortars, and a case of grenades. Several hours later and 1500 meters farther west, C company was checking out a tunnel and bunkers on the south side of the ridge, close to where the enemy hospital was found during the final phase of Operation DELAWARE. They were hit with small arms and mortar fire, losing one man, PFC Crawford Jackson, Jr., and evacuating four wounded; two enemy were reported KIA. In the ARVN sector, both the Black Panther Company and the 2/1 Battalion were in contact with enemy forces southeast of Ta Bat and lost four men KIA and six WIA in firefights.[49]

To execute the denial portion of the Division's mission, Division units implanted three minefields at choke points in the A SHAU Valley. On 16 Aug two six-man Long Range Patrol (LRP) teams and elements of C/326 Engr were lifted from FB VEGHEL (YD550035) to the minefield sites. During daylight hours, the sites were reconnoitered, lanes marked, and mines brought in. During the hours of darkness, two minefields were installed during the night of 16-17 Aug and one the following night. Each mine was booby trapped and fitted with a delay fuse to destroy the mine after a set period of time. Sensors were placed within each minefield to detect enemy activity near the minefield. A monitoring station was established on FB EAGLE'S NEST with a direct telephone line to the nearby artillery units so that artillery fires could be placed quickly on the areas of enemy activity.[50]

The locations of the minefields were (1) the junction of Routes 547A and 548, (2) Route 548 at the Dong So Ridge choke point, and (3) 2 km south of Ta Bat airfield, where Route 548 crosses a stream.[51]

Phase V (Extraction): 17–20 August 1968

The extraction of all units from the valley commenced on 17 August, when the Hac Bao (Black Panther) Company was first to depart. The next morning, under cover of smoke screens around the pickup zones to blind enemy mortar and artillery forward observers, both ARVN infantry battalions were extracted, ending LAM SON 246. On 19 August, the American infantry battalions were picked up a company at a time from their many positions. The enemy had not opposed any of the withdrawals.[52]

Although the NVA did not attempt to use artillery or anti-aircraft artillery against the helicopters swarming over the valley during the extraction, the enemy did make known his intention to rapidly reestablish his hegemony over the A Shau. PFC Charlie Perkins, Jr. of C/1/321 Artillery based on FB GEORGIA was KIA on 19 August. While unconfirmed, he was probably the last KIA of OPERATION SOMERSET. PFC Perkins was posthumously awarded the Bronze Star for Valor for an earlier action on 25 May 1968.[53] On 20 August, EAGLE'S NEST was attacked by enemy artillery; there were no casualties. FB GEORGIA was closed that same day, and as the allied artillery sites adjacent to the A Shau were decommissioned at later dates, 30 barrels of persistent CS gas were dropped on each location by helicopter.[54]

SOMERSET PLAIN / LAM SON 246 *Scorecard*

The intelligence on the NVA units "possibly present" in the A Shau Valley at the beginning of August 1968 included two regiments, one independent infantry battalion, one artillery battalion, one engineer battalion, Binh Tram 7, one signal battalion, elements of an anti-aircraft battalion, and possibly a tank battalion.[55] In fact, only two battalions (816th and 818th) of the 9th NVA Regiment were identified during the course of the operation, and allied intelligence discredited the NVA lieutenant's report of two divisions in the area.[††56] While enemy artillery fire originating in Laos was considered possible, the allies do not appear to have expected NVA artillery units to be based inside South Vietnam, and the failure of the assault on Dong So Ridge prevented the allied forces from entering the hills between there and Laos, where NVA guns were believed to be located, or from conducting bomb damage assessment (BDA) of airstrikes against suspected batteries.

The missions assigned to the 101st Airborne for SOMERSET PLAIN were, if narrowly defined, all completed. Quantitatively, however, the results were less than impressive. Only 45 individual and 13 crew served

---

†† The known NVA battalions opposed the 3/1 ARVN Battalion; the enemy infantry that fought on Dong So Ridge was not identified.

weapons were captured, along with 12 tons of rice, 7 trucks, and a relatively small amount of munitions, according to the AAR. The III MAF Periodic Intelligence Report provides different numbers:

> There were 198 tactical airstrikes and 78 B-52 sorties flown in support of the operation. Cumulative total enemy casualties for the operation were 80 enemy KIA and one detainee. There were 24 individual and 11 crew served weapons captured during the operation. Operation LAM SON 246, which ran in conjunction with Operation SOMERSET PLAIN, terminated on 181200h August. Cumulative total enemy casualties for the operation were 53 enemy KIA and three detainees. There were 241 individual and one crew served weapon captured during the operation.[57]

The allies lost one F-4 Phantom, four helicopter gunships, and one light observation helicopter. Eleven additional helicopters were damaged.

The body count, although appearing at first blush to almost achieve the 10 to 1 "kill ratio" that was the COMUSMACV benchmark, was in reality quite disappointing. In the AAR for the combined operations, 185 NVA soldiers were listed as KIA or captured against 21 U.S. soldiers KIA or MIA. However, the ARVN also lost 17 KIA and MIA. When the casualties of the two major "short round" incidents are added (7 U.S. KIA and 11 ARVN KIA), the total allied deaths are 56, with an additional 161 WIA and 84 wounded by friendly fire, for a total allied casualty count of 301![58]

While the 1st Brigade's After Action Report referred to "extensive minefields and sensor devices implanted in the valley" as a future hindrance to enemy movement, those denial areas were only established at three points in the valley. Following de rigueur U.S. Army practice, the concluding remarks in the Case Study of Somerset Plain display the requisite After Action Report hubris: "The most significant lesson learned was the enemy does not own the A SHAU Valley. The 101st Airborne Division can effectively operate for extended periods of time in the valley. The division can make it extremely costly to the enemy, if he should decide to defend his installations against the superior firepower, mobility, and aggressiveness of the Screaming Eagles."[59] Not including the insertion and extraction days, the paratroops were in the valley for exactly two weeks.

The brief recapitulation of SOMERSET PLAIN in the U.S. Army official history contains several errors, e.g. "All told, fifty helicopter flights were needed to bring the allied force into the A Shau," and "U.S. deaths came to seven, all from Company D of the 2d Battalion, 327th Infantry. . . ."[60] The USMC history for 1968 (III MAF was the overall HQs for the northernmost provinces of RVN) does not mention SOMERSET PLAIN.[61]

15

AIR INTERDICTION: SEPTEMBER 1968–FEBRUARY 1969

Air Interdiction from the End of SOMERSET PLAIN to 9 Dec 1968

With the conclusion of Operation SOMERSET PLAIN and the withdrawal of the American and South Vietnamese "grunts," the A Shau Valley reverted to its status as air interdiction Specified Strike Zone Victor. The southwest monsoon season was in full force, so air strike activity was severely limited, as was the enemy's movement of supplies. That changed significantly around the end of October, not only with the advent of better weather, but also with the bombing halt over North Vietnam ordered by President Johnson on 1 November 1968. With the end of ROLLING THUNDER, a major reallocation of air resources took place. The Ho Chi Minh Trail in Laos, from the mountain passes between North Vietnam and Laos south to the Cambodian border, was much more heavily bombed. Additional attention was also directed to the A Shau Valley.

MACVSOG from SOMERSET PLAIN to the End of 1968

Base Area 607, due west of Da Nang and lying in an area where Laos protrudes between the South Vietnamese provinces of Thua Then and Quang Nam, was where a number of Special Forces insertions had turned out badly the previous May. One of the teams mauled in May, RT Alabama, was rebuilt under SSG James D. Stride, Jr., and inserted into the area south of the southern end of the A Shau on 5 October. Almost as soon as they were on the ground, the three Americans and six Nungs came under attack by 50 or more enemy soldiers. SSG Stride and one of the Nungs were the first to be killed. The radio operator, SGT Lynne Black, Jr., took charge and formed a perimeter while calling for immediate extraction. Wounded by an enemy grenade, SGT Black briefly lost consciousness but resumed leading the team as soon as he came to. The team split up, and Black and two Nungs fought their way through the enemy line and evaded pursuit, twice fighting through more NVA.

Reaching a viable pickup site, the men awaited the arrival of two HH-3 Jolly Green Giants from the 37th Aerospace Rescue and Recovery Squadron (ARRS). When the first helicopter tried to lower a recovery hoist, it was quickly shot up by the NVA, who had again surrounded the SOG warriors. The first two choppers had to abort the mission and return to Da Nang. Two more HH-3s came on station and, while A-1 Skyraiders pounded the enemy, tried again to drop a jungle penetrator to the men below. As one of the giant machines came to a hover over the team, it too was riddled by enemy fire. The pilot tried to gain altitude but crashed in flames about a quarter mile away. The copilot, MAJ Albert D. Wester, and the flight engineer, SGT Gregory P. Lawrence, died in the crash, but the other two crew members were able to escape the wreck and evade the enemy. SGT Black, now surrounded by an estimated 100 or more enemy troops, decided to make a run for it and hook up with the downed crewmen.

For a third time, an HH-3 hovered overhead and cranked the three-man hoist to the ground as the Skyraiders continued to keep the NVA's heads down. Just at this point, the other American member of RT Alabama arrived at the pickup zone. As there was room for only three men on the hoist, the sole surviving Nung and the two surviving helicopter crewmen (the pilot was badly wounded) were sent up in the first load. While the penetrator was completing its ascent, SGT Black suddenly found himself face to face with two AK-47 armed soldiers. Raising his hands in surrender, as they approached, he grabbed their weapons, butt-stroking one in the face and clubbing the other over the head. He then raced to the relowered hoist, where he joined the other surviving Green Beret.

They ascended, but as they clambered aboard the rescue ship, it was hit on the underside by an RPG and badly damaged. The pilot managed to fly the helicopter as far as a clearing on the other side of the ridgeline and set it down there. The now-six members of the two downed helicopter crews and the surviving Nung were soon picked up by a fourth HH-3, which took over 30 hits during the extraction. As the final chapter to this incredible story, an AH-1G Cobra helicopter, which has no passenger cabin, landed, and the last two members of RT Alabama rode to safety on the ammo bay doors. SGT Lynne Black received the Silver Star. It was his first SOG mission.[1]

One final SOG KIA ended a dismal year for the special ops teams in the A Shau Valley and the adjacent NVA base areas in Laos. On 18 December, an SF team was inserted into Base Area 611, several kilometers into Laos and due west of Tiger Mountain. At dusk, they were attacked by 15 NVA soldiers; SGT Norman Payne (codename Bison) was split off from his teammates and disappeared. His team leader searched along Payne's likely evasion route, but he could not be found. The rest of the team, with several wounded, was pulled out the next morning. A new team was placed on the ground but departed after 24 hours, having also found no sign of SGT Payne. He remains missing in action.[2]

Air Interdiction 9–31 December 1968

On 7 December 1968, a meeting of the key allied players took place at Horn Direct Air Support Center (DASC), the control facility for U.S. air strikes in I Corps. The focus area was a new, more intense air interdiction campaign in the A Shau Valley; the group's recommendations were submitted to the 7th Air Force commander for approval. The report proposed that the initial focus of the campaign be on three choke points on Route 548:

1) At the northern end of the valley, where it runs along the south side of Tiger Mountain (precisely where the engineers with the 5/7 Cavalry had "destroyed" it at the conclusion of Operation DELAWARE nine months earlier)

2) Just north of A Luoi in the center of the valley, where the valley narrows to a width of about 150 meters at the foot of Dong So Ridge (about 3 km east of the former LZ Pepper location)

3) The very southern end of the valley, where the road traverses the side of a narrow ridge.*

These points were to be closed with heavy daylight bombing attacks and then to be kept closed by a combination of

1) Daylight visual and radar-directed bombing, with FACs on station continuously to direct. The FACs would also seek out enemy truck parks or other installations.

2) Radar bombing (COMBAT SKYSPOT) would be used as needed.

3) Use of riot gas bomblets (CS gas in BLU–52 cluster bombs) seeded at choke points to restrict enemy engineer repair operations

4) Night–time coverage by OV-1 Mohawks using side-looking radar, with on-call air strikes by Marine A-6 Intruder all-weather bombers equipped with "moving target indicators"

5) An additional aerial night surveillance program using star-light scopes (night vision devices) and flares.

The results of the first week of this new effort (9–15 December) were less than stellar. The Tiger Mountain road was successfully cut, but the NVA had already built a bypass through the jungle around the north side

---

* Two of these are also two of the three locations "denied" to the enemy at the conclusion of Operation SOMERSET PLAIN the previous August.

of the mountain. In the center of the valley, even just 150 meters proved too wide a channel to shut down with bomb craters; the enemy drove around them or down the bed of the A Shau stream. As for the third target, low ceilings precluded accurate visual bombing, and the road at the southern end of the valley remained open most of the time. There was little enemy AAA fire during this period.[3] B-52 strikes continued to be used sporadically in the valley. Of four ARC LIGHT missions in I Corps during the week of 9–15 December, one was targeted at enemy installations along Route 922 in western Thua Thien. The initial BDA reported destruction of one truck, one bunker, and one installation.[4]

During the last two weeks of the year, NVA road repair work and anti-aircraft fire were on the increase. One RF-101 reconnaissance plane was hit but returned safely to base. Route 548 on the south side of Tiger Mountain remained blocked most of the time, and new targets were identified along the recently detected bypass route, which was hit with three ARC LIGHT strikes, exposing areas of the well-camouflaged thoroughfare. Five trucks were also destroyed.

On 22 December, LTG Cushman sent a message to GEN Abrams expounding the

> immediate need to concentrate an aggressive interdiction effort in A Shau Valley, SSZ Victor. . . . Efforts to date have been worthwhile but insufficient to prevent the enemy from resupplying his forces by way of A Shau or to cure the deficiencies . . . , specifically to stop enemy vehicular movement, prevent road reconstruction, maintain visual observation and decrease the flow of enemy supplies and manpower into I Corps. . . . In A Shau this headquarters estimates that at this time a minimum of 16 night/all weather and 24 daylight attack/visual reconnaissance sorties per day are required. Emphasis should be placed on closing routes at specified choke points, interdicting vehicular movement, and destruction of priority targets.[5]

As a result of this request, a second meeting was held at Horn DASC on 27 December with GEN Abrams and a number of senior officers to review the status of "Air Interdiction in the A Shau Valley." BG McLaughlin of 7AF reported a current daily level of 20 daylight and

12–18 night sorties, and the Air Force officers present stated that this level was adequate. In response to GEN Cushman's request that the air sorties for SSZ Victor be "in addition to sorties presently allocated in direct support of I Corps," i.e., that they be an add-on to the sorties designated for Horn DASC, rather than drawn from the air assets already assigned, the Air Force opposed any reduction in sorties for Laos (from where, presumably, any increase in sorties for I Corps would have to be taken).

GEN Abrams began his response to the briefers by stating:

Nothing should be dedicated to the A Shau but the A Shau is critical to us. The daily effort should be adjusted every day based on weather and aircraft availability. General CUSHMAN is correct in calling attention to the A Shau and the ground commanders should continue to watch A Shau. A Shau is a part of our major effort—both for now and for future operations. . . . He directed MG Townsend (MACV J-3) to reply to III MAF's message—saying the subject has been reviewed and *the effort will continue as it has* [emphasis added]."[6] (Words are those of the meeting Memorandum for the Record and not verbatim.[7])

Air Interdiction January – May 1969

From 1–15 January 1969, the number of sorties increased, with a maximum of 57 in a single day. One bulldozer and four trucks were put out of action, but the main focus was still on the choke points. At least one of those was kept closed every day, and there was evidence that the enemy had begun to unload trucks on one side of an interdiction target, portage the supplies by hand to the other side, and then load another vehicle. To combat this in the north, yet another interdiction point farther to the west was added. Additional ordnance was brought into service. Delay-fused bombs (FMU-72) set to explode at night were dropped extensively, and a new area-denial form of the Gravel munition (XM-42) was seeded. The enemy countered with even more antiaircraft weapons, including 37mm AAA. The fire was particularly strong at the north end of the valley and at night.[8]

On 21 January 1969, the same day the USMC commenced Operation DEWEY CANYON, President Nixon requested that his National Security Advisor, Henry Kissinger, poll the State Department, Secretary of Defense, Chairman of the Joint Chiefs of Staff (JCS), and U.S. Embassy in Saigon on the "Situation in Vietnam." The result, issued in its final form in March 1969, was National Security Study Memorandum (NSSM) One. Among the focus areas were the PAVN Channels of Supply (Tab 10), the Effects of B–52 Raids (Tab 27), and air interdiction in Laos (Tab 28). The study found "that during 1968 enemy forces required a daily average of about 250 short tons of supplies."[9] Most of that was food, and most of the food came from sources inside South Vietnam. About 80 tons per day was needed from external sources (i.e., Laos, Cambodia, or directly across the DMZ from North Vietnam), of which 43%, or 34 short tons per day, flowed through Laos. This amounted to 10–15 truckloads per day. Truck sightings in Laos during November and December 1968 were almost 10,000, with 981 claimed as destroyed and 363 damaged, containing some 1,767 tons of supplies, and another 3,000 tons destroyed in fires and explosions. This amounts to about 80 tons per day. In spite of the massive air power employed in Laos after the end of bombing in North Vietnam (1 November 1968), "the enemy has pushed through sufficient tonnages to provide the bulk of his external supply requirements . . . experience during the first months of intensified operations in Laos indicates that the current campaign may not significantly limit enemy supply flows into South Vietnam. The external supply requirements of VC/NVA forces in South Vietnam are so small relative to enemy logistic capacity that it is unlikely *any* air interdiction campaign can reduce it below the required levels" (emphasis added).[10] Once again, air interdiction could not, under any circumstances, turn off the North Vietnamese supply tap. Only by physically occupying the enemy line of communication could it be cut. Not only had the VC and the NVA been kept supplied, but also, there were now thousands of tons of arms and ammunitions stockpiled in base areas throughout South Vietnam as witnessed by the caches in and around the A Shau Valley.

The B-52s also came under scrutiny in NSSM-1. There was substantial disagreement between the JCS and the DOD on the effectiveness of the big bombers: JCS, relying on ground follow-up and POW interrogations, estimated 2.5 enemy KIA per sortie, while the DOD figure, using ground and air BDA, was about one-sixth of that, 0.43 per sortie. It was "generally agreed that a feasible method for analyzing ARC LIGHT effectiveness has not yet been devised. It is not possible to make any definitive statements regarding the effectiveness of ARC LIGHTS." While the study concluded that "there is general agreement that ARC LIGHT strikes have been effective on numerous occasions in preempting enemy attacks, blocking lines of communication, and supporting troops in combat with sizeable enemy forces,"[11] it was clear that anecdotal evidence for occasional highly effective employment of the B-52s did not equate with a great deal of "bang for the buck" on the average mission.

There was a reduction in the level of air strikes in SSZ Victor during the period 16–31 January. Partly, this was due to the start of Operation DEWEY CANYON by the Marines to the north, in the Da Krong Valley. On 21 January, the 9th Marines launched Phase I, establishing several fire support bases in the north of the operational area. In Phase II, commencing on 22 January, they began to move south, and two days later assumed responsibility for the northern end of the A Shau Valley, including the area around Tiger Mountain. Visual recon of the area on the north end of the valley continued, and Marine air coordinated air strikes along the road, with an emphasis still on choke points.

Overall, there were fewer daylight strikes because there were fewer targets in SSZ Victor. On 19 January, a FAC operating along the Laotian border near where Route 922 and 548 merge spotted a truck and some supplies, and the subsequent airstrikes uncovered a large enemy supply center, resulting in seven trucks being destroyed, as well as 33 secondary explosions.[12] Other targets were identified and attacked along Route 922, on both sides of the border. Night activity continued at the previous level, and this period saw the introduction of C-119 fixed-wing gunships into the area. They engaged vehicles moving on the road and also sampans on the Rao Lao River during several night sorties.

Ground fire continued to increase in intensity. On 17 January, an A-6A Intruder from VMA(AW)-242 flown by CPT Edwin J. Fickler (pilot) and 1LT Robert J Kuhlman, Jr., (bombardier/navigator) failed to return from a low-level night mission over the A Shau.[13] The aircraft was never found. Two F-4s were shot down, one on 24 January and the other the following day; all crewmen were recovered. Other pilots reported dodging intense AAA and AW fire.[14]

During the first half of February, weather was poor over the A Shau, and on several days, only radar-guided bombing took place. There was some success early in the month, when a bulldozer and four trucks were knocked out on Route 922 in Laos. Enemy anti-aircraft fire was also less intense. Toward the end of the month, air strikes were curtailed even more and were mainly in the southern end of the valley. On 1 March, SSZ Victor ceased to operate, as Operation MASSACHUSETTS STRIKER was beginning.[15]

Was the air campaign from 9 December 1968 to the end of February 1969 a success? As was the case with air-only campaigns in the A Shau Valley throughout the years 1966–1968, measurement of the impact of allied air strikes on enemy activity and installations was hampered by the very limited intelligence assets available to the air managers. Forward air controllers and reconnaissance aircraft were the primary source of information, but the enemy's superb camouflage and heavy anti-aircraft fire made effective evaluation, like effective interdiction, more miss than hit. Ground reconnaissance continued to be challenged by the highly trained NVA anti-recon teams. In his final report on the campaign, USAF COL Frederick Webster, Director of the Horn DASC, noted: "The most apparent conclusion derived from this program was the value which the enemy placed on this supply route. . . . To cut off this route would be to deny the enemy easy access to major population centers, Hue and Da Nang. . . ."[16]

As was also the case each time the allies resorted to air power alone to try to interdict Route 548, there was a constant cycle of tactic and countertactic between the U.S. air and NVA ground commanders. The U.S. modified day and night operations, developed and used alternative munitions and aircraft (e.g. fixed-wing gunships), designated new and deleted unproductive target areas, mixed high- and low-level strikes, and so on. The NVA engineers swarmed over critical choke points, conducted repairs and truck movement at night, rapidly developed alternate routes, constantly relocated AAA and AW positions, created dummy AAA emplacements, moved supply caches, and so on. In this fluid environment, there was no ultimate winner or loser. Each side was compelled to commit extensive resources to the valley or to cede it to the enemy.

COL Webster's final conclusion was as follows:

> The degree of success and the total impact that this operation has had on the enemy is difficult to assess accurately at this time.... [NVA] attempts to keep the road open became more and more futile. Whether this is because the enemy chose other routes or because the massive bombardments surpassed their repair capability is unclear. They never ceased their road work, though, until ground units approached the road.... [T]he number of choke points that were closed for long periods of time indicate that the interdiction program may have been responsible for the location of the supplies. The road to the west remained open most of the time.[17]

These final remarks are prescient. The backup of enemy supplies along routes 548 and 922 in the area between Tiger Mountain and the Laotian border would lead to the single largest enemy supply cache capture of the Vietnam War, when it came time to "send in the Marines" to the A Shau.

16

OPERATION DEWEY CANYON

New Year, New Strategy

On 26 December 1968, the U.S., South Vietnamese, and South Korean commanders in I Corps promulgated their Combined Campaign Plan (CCP) for 1969.[1] This document would be the master plan for LTG Robert E. Cushman and the III MAF during the ensuing year.

The basis for the I Corps CCP was two documents issued by the RVN Joint General Staff (JGS) in coordination with MACV in late 1968. The first was a Pacification and Development Plan (PDP), which prioritized objectives across a range of security, antiterrorism, local government, rural economic development, and other civilian and paramilitary issues. The second document was the national Combined Campaign Plan for 1969, addressing military operations. Only the second document had relevance to the A Shau Valley.

The underlying assumptions framing the 1969 CCP were similar to those of a year earlier, when General Westmoreland's headquarters had been the issuing authority.

1) **U.S. and ARVN troops to continue to operate against enemy bases, while screening populated areas from infiltration and attack;**

2) **Extension of areas controlled by government forces and denying enemy access to rice-producing areas and government centers (district and provincial capitals);**

3) **Population security and other actions in conjunction with local self-defense forces and civic action teams (i.e., coordination with the PDP).**

There were also, however, several quite significant changes from 1968:

1) **Recognition of the ongoing peace talks in Paris;**

2) **An assurance that allied force levels would remain stable [not to last];**

3) **A much greater emphasis on the pacification aspect of the war than previously (i.e., same as (2) and (3) above, but much more so).**

Among the 12 major objectives specified for 1969 were "destroy or neutralize enemy base areas" and "reduce the enemy's ability to conduct ground and fire attacks against military and civilian targets."[2]

The acknowledgment of the ongoing peace talks and the perhaps-not-so-assured status of forces were simply recognition of the changing nature of the war. No longer was the enemy's military capitulation considered to be the only conceivable outcome. The war might end by negotiation, rather than force of arms (in which case it might be assumed that current occupation of territory would be a factor in any agreement), and public pressure might have an impact on the continuing American commitment to the war. The third change may appear to be just a subtle realignment of priorities; it was in fact a vital and powerful reflection of the new "One War" policy of General Creighton Abrams, who had assumed command from Westmoreland in July 1968. "One War" mandated profound changes in the armament, training, leadership, and operational level of the ARVN.[3] This would be visible in the A Shau in the continuing commitment of ARVN forces alongside the U.S. units operating there.

Of equal importance to the change in overall emphasis in the 1969 Combined Campaign Plan for I Corps was a change in tactics. This also came about because of the new face of MACV leadership. GEN West-moreland's primary goal "was to grind down the enemy."[4] In execution, this meant using American "grunts" to go after the NVA and VC main force troops wherever they could be found, and then pounding them with artillery and air to keep friendly losses to a minimum; notching up the body count as he sought to reach the "crossover point" where the NVA losses would exceed the North Vietnamese birthrate, or something like that.[*5] Base areas, unless they held the possibility of bringing the enemy to combat, were not a priority. Westmoreland did not, in practice, pay much attention to the development of the ARVN or local self-defense forces. As of early 1968, only about 20,000 M–16s had been issued to the entire RVN armed forces.[6] Most fought with WWII era M–1 rifles, BARs, and carbines against a much better-armed enemy, and the artillery and Air Force of the RVN were more than seriously inadequate compared to those of the American forces.

These things changed materially under GEN Abrams. In the A Shau valley, GEN Westmoreland launched exactly one ground operation (DEL-AWARE) during his four years in command. Under Abrams, SOMER-SET PLAIN took place in August 1968. The day before that operation started, Abe observed, "Now, on the A Shau thing—I've been attracted by the idea of going back in there. There's some evidence that when we went in there before [i.e., Operation DELAWARE] it really shook them up a little bit. . . . It <u>looks</u> like it <u>disturbed</u> them." [7] He had moved quickly, having taken command only one month earlier.

This was followed by the Marines' DEWEY CANYON and then four operations by the 101st Airborne Division in 1969. The enemy's sup-ply line and bases in the A Shau Valley, built up in the period March 1966 through April 1968, would be plugged and plundered by the allies throughout 1969.

---

* See unattributed document in CIA files from Acting Chief [redacted] to Chief [redacted]: "In general, we believe that the crossover point has probably been reached, but the sharpness of the break and the rate of the trend are much less pronounced than the MACV materials indicate."

Operation DEWEY CANYON – One of a Kind

Operation DEWEY CANYON was unique in several remarkable ways. In the A Shau, it would be the only exclusive United States Marine Corps operation to penetrate the valley, and then only the very northern end, as most of DEWEY CANYON took place in the Da Krong watershed. It was also the first time American regular, nonmechanized infantry swept into the valley on foot. The U.S. Army infantry before DEWEY CAN-YON flew into the valley, or at least as far as landing zones on the valley rim. One Army battalion did eventually drive down Route 547 to the A Shau with its armored personnel carriers and tanks, the 3rd Battalion, 5th Cavalry, a ground reconnaissance squadron that arrived later in 1969. And elements of the 101st Airborne Division would also approach the A Shau on foot down Route 547A at various times that year. The Marines, on the other hand, fought their way to Tiger Mountain.

DEWEY CANYON also was notable for two incredible landmarks in the Vietnam War. It would be the only time United States infantry crossed the border and engaged the enemy inside Laos. It would also result in the capture/destruction of the single largest enemy weapons cache of the entire war, over 500 tons of material. This latter event is even more incredible in that the cache was captured, not by the highly vaunted "vertical envelopment" tactics of the Army's two airmobile divisions (1st Cavalry and 101st Airborne), but by Marine infantry who conducted a ground sweep operation. It is probable that the USMC approach, using a traditional conventional infantry tactic, was sufficiently unconventional in the A Shau setting to catch the enemy by surprise! As the flying weather in January–March in western Quang Tri and Thua Thien almost cer-tainly ruled out an airmobile approach to the area, the NVA command thought its encampments and caches, well protected by infantry, artillery, and anti–aircraft battalions, were secure. When the Marines struck, there would not be time to withdraw into Laos enormous stockpiles cached along Routes 548 and 922 in far western Thua Thien. Nor would the Marine infantry be stopped, as was the 1st Cavalry west of Tiger Moun-tain in May or the 101st Airborne along the Dong So Ridge in August, by rings of staunchly defended bunkers protecting the NVA rear areas. But the Marines would pay a high price for their victory.

The USMC operation was carried out in three phases, and only in the last phase would the A Shau Valley be entered. Even then, two of the 9th Marines battalions were operating against Base Area 611, to the west of the A Shau, and only the 3rd Battalion, 9th Marines had objectives in the vicinity of Tiger Mountain (Hill 1228) and the roads between there and the Laotian border. Our narrative will therefore focus primarily on the 3/9 Marines in Phase III, covering the earlier phases and the Phase III actions of 1/9 and 2/9 Marines only in summary. The following narrative does not include two battalions of the ARVN 2nd Regiment, operating on the east side of the DEWEY CANYON AO north of the A Shau Valley.

Operation DAWSON RIVER SOUTH / DEWEY CANYON
Phases I and II

Although there had been a concentrated effort to maintain choke points and destroy enemy truck traffic in Specified Strike Zone Victor (the A Shau Valley) commencing in mid-December, by January 1969, enemy engineers were busy rebuilding roads, including Route 922 in Laos, and at times over 1,000 truck sightings per day were reported. Enemy combat units and replacements were infiltrating into I Corps from Base Area 611, establishing caches and camps, and then moving to the east.[8] A CIA report prepared for National Security Advisor Henry Kissinger in early February reported:

> The North Vietnamese 6th Regiment and elements of the NVA 9th Regiment have returned to western Thua Thien where they appear to be preoccupied with the restoration of logistics facilities to support possible future military campaigns against allied positions and urban centers along the coastal lowlands. The Communists have also expanded their antiaircraft capabilities in the A Shau Valley, suggesting a strong enemy determination to protect their base area restoration efforts.[9]

The 6th NVA Regiment would be the primary opponent of the 9th Marines in February and March.

GEN Westmoreland had belatedly sent heavy reinforcements into I Corps in late 1967 and early 1968, yet a year later the ratio of allied to enemy maneuver battalions was still weaker there than anywhere else in South Vietnam. In mid-February 1969 there were just 85 allied battalions opposing 80 VC/NVA units, compared to the nationwide ratio of 5:4 (308 to 247).[10] It was therefore mandatory that the allies aggressively seek out the enemy, destroying his camps and caches and disrupting his attempts to concentrate for attack, else the initiative would belong to the NLF and it could concentrate at will to replicate attacks such as had occurred at Hue. It was known that a "Fourth Offensive" was coming. When that started, on 23 February 1969, its impact on I Corps was stunted: "aggressive Allied operations, both air and ground, complemented each other with the desired effect—the enemy had trouble moving men and supplies into forward positions and could not gain a sustained offensive posture."[11] One of those operations was DEWEY CANYON.

MG Ray Davis, the commander of the 3rd Marine Division, observed with concern developments in the far western reaches of Quang Tri Province, especially in the area of the Da Krong Valley. The upper Da Krong runs from northeast to southwest, at its nearest point less than 6 km from Hill 1228 (Co A Nong or Tiger Mountain) at the northern end of the A Shau Valley. MGEN Davis requested authorization from III MAF to carry out an attack in the upper Da Krong to disrupt the enemy forces. On 14 January, Davis ordered BG Frank E. Garretson, commanding TF Hotel—the Ninth Marines, then conducting Operation DAWSON RIVER WEST to the west of Khe Sanh—to plan for the Da Krong operation, with a start date of 22 January.

The weather was not predicted to be favorable:

During the final months of the northwest winter monsoon, January to March, temperatures in the valley were generally chilly compared to the 100 degree temperatures of the lowlands, rarely rising above 71 degrees or falling below 51 degrees. Skies were overcast with light drizzle or occasional thunderstorms, but no significant rainfall was recorded. As a result, the mountains were continually shrouded in clouds while the valleys and numerous ravines were blanketed with heavy fog.[12]

Because of the weather and due to the limited helicopter resources available to the Marines, the plan, initially called DAWSON RIVER

SOUTH and drawn up as a sequel to DAWSON RIVER WEST, minimized reliance on helicopter transport and tac air. This would also reduce exposure of the Marine air assets to enemy anti-aircraft fire, as intelligence about the level and location of AAA was almost nonexistent. As finally approved, the operation was to be conducted in three phases. In Phase I, a series of firebases would be established in sequence, with the first located south of the major Marine base VANDEGRIFT. Each successive location was within 8 km of the previous one, ensuring that 105mm howitzer and other artillery supporting fires would always be available.

The first three firebases, all utilizing sites previously fortified and then closed by the Marines, HENDERSON, TUN TAVERN, and SHILOH, were reoccupied on 17–21 January.[13] Phase I continued on 22 January, when the 2/9 Marines flew by helicopter to establish Firebases DALLAS and RAZOR, 8 and 12 kilometers farther south. A 105mm battery, F/2/12 Marines, was in place at RAZOR on 24 January. On 25 January, 3/9 Marines air assaulted onto the top of a razorback ridgeline, Ca Ka Va, 5 km south-southeast of RAZOR, where the most important firebase supporting the operation was soon established. This was CUNNINGHAM, where between 25 and 29 January the 9th Marines regimental headquarters (COL Robert H. Barrow) and the rest of the 12th Marines (COL Peter J. Mulroney) would set up, with twelve 105mm howitzers, eight 155mm howitzers, and six 107mm heavy mortars.[14] The big guns at CUNNINGHAM could provide indirect fire support throughout the area where the Marines would be operating until mid-March. With the exception of occasional small-arms fire, the occupation of the key firebases and the early patrols nearby elicited little enemy contact.

On 24–25 January the operation name was changed to DEWEY CANYON as Phase II began. The mission during this part of the operation was to conduct intense patrolling around the firebases and gradually maneuver south to the upper Da Krong valley, designated as Phase Line Red, from which position Phase III would be launched. One additional firebase, ERSKINE, was built during Phase II, 4 km southwest of CUNNINGHAM, to which F/2/12 Marines would be moved, placing all of the 12th Marine batteries within range of the ground troops.

For the most part, patrols during Phase II encountered small enemy units, engaging in brief firefights and causing relatively few casualties on either side. On 1 February, F Company occupied FB ERSKINE. Also, early in Phase II, F Company found the very recently abandoned NVA 88th Field Hospital near the Da Krong River. It soon became clear, however, that as the operation continued, the NVA would contest the Marine advance across the entire front of over 10 kilometers. On 2 February, the North Vietnamese fired 30–40 rounds from 122mm field guns at Firebase CUNNINGHAM, killing four Marines and wounding five others.[15] The enemy artillery, it was later determined, was located in Laos, just beyond the range of the 155mm Marine howitzers.[16]

The first significant battle of DEWEY CANYON occurred on the western end of the front on 5 February. Company G (CPT Daniel A. Hitzelberger) was assigned to sweep the Co Ka Leuye, a high ridge just east of the Laotian border on the right flank of the Marine advance, and had begun to ascend the steep, rocky ridge on 31 January. On the morning of the next day, the weather closed in on the Da Krong watershed and would remain bad for the next nine days, delaying the artillery move to ERSKINE until February 10 and halting helicopter operations throughout the area. On 4 February COL Barrow decided to pull his companies back to firebases or nearby locations where they could be supplied regardless of the overcast. All were able to comply quickly, except Company G, which had continued to climb Co Ka Leuye and was now stuck on the top of that remote ridge to the west. On 5 February, CPT Hitzelberger's command had begun to move back down the mountain when it was ambushed by an estimated 30 well-hidden NVA soldiers in bunkers. In the battle that followed, 5 Marines were KIA and 18 WIA. LCPL Thomas P. Noonan, Jr., was posthumously awarded the Medal of Honor for his heroic actions on 5 February 1969.[17] When the weather finally cleared on 10 February, the third battalion of the regiment (1/9) was brought forward and preparations undertaken for the final phase, with the lead companies closing on the Da Krong – Phase Line Red. After a weather delay of nine days, during which the NVA 812th Regiment was able to strengthen its defenses to the south, the 9th Marines were at last able to begin the final chapter of DEWEY CANYON.[18]

Operation DEWEY CANYON: Phase III – the Move South Begins

In spite of the Marine presence just a short howitzer shot north of Route 922, it had been hoped that the element of surprise would still apply to the advance on Base Area 611 and the northern end of the A Shau Valley, by giving the appearance that the Marine operation was a search-and-destroy effort limited to Quang Tri Province and culminating in the Da Krong Valley. For this reason, it was critical that no one jump the gun on Phase III. The Laotian border and Route 922 were 7 km or less to the south (see Map 11).

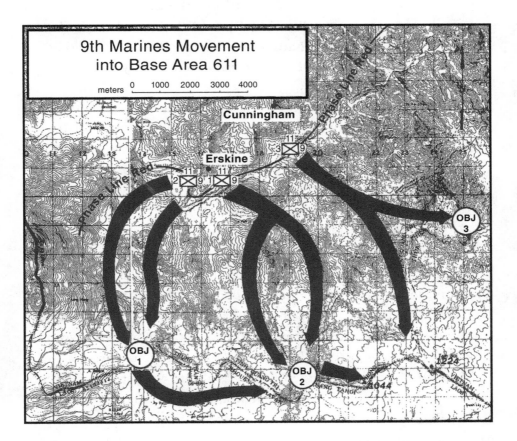

MAP 11

Source: Smith, Charles R. *U.S. Marines in Vietnam: High Mobility and Standdown, 1969* (Washington, DC: History and Museums Division, Headquarters, U.S. Marine Corps, 1988) p. 40.

Objectives for Phase III were as follows:

1) **3/9 Marines (LTC Elliott R. Laine, Jr.) on the left (east) was to send one company to Hill 1228 (Tiger Mountain) at the north end of the A Shau Valley and two companies to Hill 1224 (Tam Boi), where Route 922 from Laos crosses the border and joins Route 548. L Company initially defended FB CUNNINGHAM.**

2) **1/9 Marines (LTC George W. Smith) in the middle was to advance down two parallel ridgelines to the abandoned village of Lang Ha on the Laotian border on Route 922 and at an important junction of several well-used enemy trails.**

3) **2/9 Marines (LTC George C. Fox until 4 March) on the right (west) would protect that flank while following a wide valley and mountain ridge to a point about 5.5 kilometers west of Lang Ha. G Company initially defended FB ERSKINE.[19]**

Each battalion was to advance with two companies in the lead and the other one or two following in trace. This provided for flexible maneuver to either flank in the event that a major NVA strongpoint was encountered.

When the troops marched forward on 11 February, the enemy immediately made manifest its intention to oppose any advance south of the Da Krong. M/3/9 Marines to the east had a tough day, with contact at 0800 and 1145 in the morning, and then two 60mm mortar barrages after dark. The company lost 3 KIA and 16 WIA, reporting only 1 NVA KIA and 1 POW. I/3/9 on the far left flank was also in contact, and a 122mm artillery barrage on the 3rd Battalion wounded five. At 2200 hours, the 1st Battalion ran into a large enemy unit, apparently in the process of preparing an assault against FB ERSKINE. It was a tough fight, costing the Marines 2 killed and 9 wounded. The Marines counted 25 enemy bodies when the battle ended. (Note: The 9th Marines AAR appears to overstate the threat level of this enemy concentration when it says, "On 11 February he attempted to overrun FSB Erskine."[20]) F Company in the 2nd Battalion area also had two men wounded this day.[21]

On 12 February, the left flank of the advance, moving in the direction of Tiger Mountain, was again heavily contested. I, K, and M Companies all came under fire, and 30 of the 31 Marine casualties this day were in 3/9 Marines, although only 1 man was KIA. Few enemy soldiers were accounted for. After two days of Phase II, the 9th Marines had incurred over 80 casualties, more than in both earlier phases of DEWEY CANYON combined.[22] The last few miles to the south were going to be a tough trek.

On the 13th, the combat moved to the center, where C Company fought to overcome a heavily defended enemy hilltop position, and then to repulse a counterattack after taking that objective. For the day, the company had 2 KIA and 21 WIA, while accounting for 24 enemy bodies. There was also light contact elsewhere as the front moved south. For the next three days, enemy resistance was light. The heaviest contact was by K Company on the left flank. On 14 February, K lost 2 KIA and 11 WIA, and two days later had 1 KIA and 10 WIA while repulsing an enemy attack. On 15 February, an aerial observer reported one NVA 122mm gun destroyed by air beyond the western end of the advance.[23]

At about 0400 hours on 17 February, the last and most successful enemy ground attack on a U.S. firebase during DEWEY CANYON hit FB CUNNINGHAM, defended by L Company and the 2/12 Marines. Enemy sappers, supported by a reinforced company, penetrated the base and caused widespread casualties before being driven off by L Company and the artillerymen. At daybreak, 37 enemy bodies were found, 13 within the perimeter. Four Marines were killed and 46 wounded. Later that day, the hapless G Company became engaged in an all-day running gunfight with an enemy force of comparable size. In this engagement, the Marines suffered 9 killed and 14 wounded, eliminating 41 NVA infantrymen.[24]

Operation DEWEY CANYON: Phase III – The 1/9 and 2/9 Marines Reach the Border

In the center and on the right, Route 922 and the border with Laos would soon be in sight.

During the five days from 18–22 February, the 1/9 Marines would suffer heavy casualties: A Company: 15 KIA and 79 WIA; B Company: 4 KIA and 15 WIA; C Company: 7 KIA and 65 WIA, and D Company: 3 KIA and 19 WIA. It began on the 18th, when A Company ran into a strongly fortified enemy position. For three days, A and C companies alternated attack roles, overrunning successive enemy lines of defense and killing, in conjunction with air strikes and artillery fire, over 150 enemy troops.

On 20 February, on the slopes of Hill 781 about 2½ km north of Lang Ha, C Company captured two 122mm guns and their half-track prime movers, the largest enemy artillery pieces taken during the war. A Company passed through C Company and captured the battalion objective, Lang Ha. That morning, a destroyed 122mm gun was sighted by air about 1.5 km to the southeast. The reason for the enemy's tenacious defense began to become clear. On the 21st, west of Lang Ha, another NVA hospital complex was taken and destroyed.

The battles of the 1st Battalion culminated on 22 February, when A Company, under 1LT Wesley L. Fox, was tangled in an all-day fight near Lang Ha, finally routing the enemy but losing 11 Marines. 1LT Fox received the Medal of Honor for his heroism on this day.[25] B Company, about 1 km to the south, also took heavy casualties, when a mortar barrage killed 4 and wounded 15, while C Company killed 9 NVA, but had 17 Marines WIA. For 22 February, the 1/9 Marines had 111 casualties, the heaviest single-day losses of DEWEY CANYON.[26] After the battles near Lang Ha, the 1/9 was redirected to the east along Route 922, in the direction of Tam Boi and the A Shau Valley.

On 26 February, D/1/9 took Hill 1044. After defeating the enemy covering force at a cost of 2 killed and 9 wounded, the Marines uncovered a previously unknown enemy technique for concealing caches. The North Vietnamese would dig out the bottom of a large bomb crater on or near the road to create a storage area and then cover it with a bamboo mat and a thin layer of dirt.[27] From the air, these caches would appear no different from any other crater, but could be loaded and unloaded at night. The cache revealed by D/1/9 included 629 rifles, 108 machine guns and mortars, and over 100 tons of ammunition.[28] This was the largest, but only one of many enemy caches that were discovered daily throughout the area near Route 922 during Phase III of DEWEY CANYON.

The rest of the 1/9 Marines moved to Lang Ha and then on to Hill 1044. As the battalion had taken its objectives, the next several days were spent in continuing to search for and evacuate enemy equipment, while engaging in several small firefights and being subject to frequent mortar barrages. Battalion HQs and C and D Companies moved onto Tam Boi, from where they would be evacuated. A and B Companies remained at Hill 1044, waiting for the arrival of a Special Forces unit inserted on 7 March into Base Area 611.

While LTC Smith's troops were battling their way down the center of the AO, Companies E and H of the 2/9 reached the border to the west on 20 February against only light opposition, occupying positions on a ridge inside Vietnam from which enemy convoys could be seen moving on Route 922, 1700 meters away.

Operation DEWEY CANYON: Phase III – The 3/9 Marines Advance to Hill 1228 and Tam Boi

By 18 February, the left flank (3/9) was nearing Hill 1228, site of LZ Tiger and the battles of the 5/7 Cavalry during Operation DELAWARE. The enemy opposition in this sector became lighter as the Route 548 bypass north of Hill 1228 was reached. At a location just northwest of the saddle between Hill 1228 and the next ridgeline to the west, where 5/7 had been brought to a standstill in May, a large enemy burial area was found, yielding 185 bodies buried in June. These were casualties either from the 5/7 Cavalry attacks or from the air strikes, including ARC LIGHT, which were directed against the enemy during and after the DELAWARE engagement.

On 19 February, I Company reached Hill 1228,[29] where FSB TUR-NAGE was established the following day.[30] The howitzer battery from FSB Erskine was moved to this location on 28 February.

On the 21st, M Company found a damaged bulldozer, front-end loader, three 5-ton trucks, and over 300 empty 50-gallon fuel drums along the Route 548 bypass on the west side of Hill 1228.

24 February marked the day the L/3/9 Marines reached the Laotian border at Hill 1224 (YD240039). For the next three days, K/3/9 engaged in a number of small firefights on the south side of the Tam Boi area, as the NVA made a last-ditch effort to prevent the exposure of the huge supply complex scattered over the ridges and valleys adjacent to the border. The company had 6 KIA and 11 WIA in these skirmishes. Company I searched the area around Tiger Mountain, while the other three companies and battalion headquarters operated near the border with Laos. L Company patrolled to the west down Route 922 almost 2 km, while M Company, after helping to search the Tam Boi complex, went almost 3 km to the east.[31] A sampling of some of the discoveries, taken from the 3/9 After Action Report:

2/25   I Co at YD250093: "a 30 foot tunnel leading to storage room 25'x10'x5'"
M Co at YD243045: "2 lodging type tunnels in side of mountain"
L Co at YD230044: "a drive-in type garage 6' high, 10' wide, and 12' deep"
K Co at YD242027: "3 shelters and a reinforced tunnel 125 meters long, six feet high, and six feet wide"
M Co at YD243043: "3 storage shelters and a tunnel 40 feet long"

2/26   K Co at YD244039: "a cave in side of hill, 50' long, 5' wide, and 6' high" and then "a mess hall and an auto repair shop"
M Co: "a destroyed full tracked, 10 ton vehicle"
K Co at YD244040: "another 10 ton tracked vehicle. Appeared to have been hit by air strike."

2/27   I Co at YD256095: "a 26 bunker complex"
L Co at YD233043: "a tunnel containing 1400 gallons of oil and gasoline in 50 gallon drums"

2/28 I Co at YD255096: "a 20 bunker complex"

3/2 Bn CP at YD244045: "2 chests with 75 lbs of automotive tools"
L Co at YD249042: "Located 11 tunnels averaging 150–250 feet in length and heavily reinforced. Complex consisted of a hospital area, living, and messing areas. Ceiling in hospital section was made of air strip matting." This complex had "caves cut in solid rock and fronted with heavy iron doors . . . successfully withstood an ARC LIGHT strike."[32]

On 27 February, west of Tam Boi in the vicinity of the Route 922/548 junction, L/3/9 found two 85mm artillery pieces.

In each of the enemy installations, vehicles, weapons, munitions, documents, rice, and other materials were found in abundance. It took days to evacuate or destroy in place the captured caches. On 3 March, while operating at the eastern extreme of the AO about 3 km east of Tam Boi, M company was ambushed. In this battle, PFC Alfred M. Wilson threw himself on an enemy grenade, sacrificing his life for his fellow Marines.[33]

The enemy continued to harass the Marines of 3/9, who fought their last battle of DEWEY CANYON on 5 March, when K Company repulsed an enemy squad, with 11 Marines WIA but no fatalities.

Operation DEWEY CANYON: Phase III – Into Laos (Part 1)

As early as 3 February, in response to the NVA use of Laos-based 122mm artillery to shell FB CUNNNGHAM, MG Davis had requested permission for the 9th Marines to enter Laos and eliminate the enemy guns: "From the present position of the 9th Marines, a raid by a force of two battalions could be launched quickly and effectively to cut road No. 922 and moving rapidly back to the east, destroy art[iller]y forces and other forces and installations which threaten us." The commander of the 3rd Marine Division had earlier (30 January) asked that MACSOG redirect its PRAIRIE FIRE operation inside Laos toward Base Area 611 to increase reconnaissance and exploitation of that NVA stronghold. This was approved on 31 January.[34] The request for a USMC raid into Laos was shelved for the moment, as that did not appear to meet the Rules of Engagement (ROE) "hot pursuit" guidelines regarding cross-border actions.

Route 922 crossed from Laos into Vietnam about 1 km southwest of Lang Ha and less than 10 km from Tiger Mountain. In Laos, it continued westward, parallel to and about 1–2 km south of the border. When H/2/9 Marines reached the border on 20 February and could see parts of Route 922, it was obvious that the enemy was moving large numbers of trucks and tracked vehicles to the west. The distance was too great for accurate direction of air strikes or artillery fire.

On the night of 21–22 February, while a renewed request to cross the border was winding its way up the chain of command to GEN Abrams, COL Barrow had determined that there was sufficient justification, without higher authority, for the 9th Marines to take "necessary counteractions against VC/NVA forces in the exercise of self-defense and to defend . . . against armed attacks with all means at their disposal."[35] He ordered CPT David F. Winecoff to take Company H into Laos, conduct an ambush on Route 922, and return to Vietnam not later than 0630 hours the following morning. Although CPT Winecoff asked for a 24-hour delay to rest his worn-out men, that was denied, and two platoons of H/2/9 set off for the NVA highway inside Laos, almost 1 km away, just after dark. By 0100, the ambush was set. At 0230, after allowing several smaller enemy units to pass unmolested, the ambush was triggered when eight trucks approached from the east (direction of South Vietnam). The result was the destruction of three enemy trucks, eight NVA KIA, and not a single scratch on the Marines. After sweeping the ambush zone, H Company marched back into South Vietnam.[36]

Around noon on the 22nd, General Abrams's response to the request to cross the border was received, specifying that only MACSOG teams were authorized to conduct operations in Laos. The Marine command, taking the stance that "self-defense" required that enemy units just inside Laos be attacked, did not concede. The Marine request, endorsed by Army LTG Stillwell, XXIV Corps Commander,† was refiled with COMUSMACV, who finally relented on 24 February, with the proviso that any actions be within two kilometers of the border and restricting public discussion of the incursion.[37]

Within hours of General Abrams's approval, Company H, 2d Battalion, 9th Marines was again instructed to move 'back down onto the bloomin' Route 922.' According to Captain Winecoff, the men's 'morale zoomed way down because the company was extremely tired, [and] we were afraid that we were going to have to go off and leave our supplies.' . . . The plan was for Company H, followed by Companies E and F, to move into Laos, and then drive eastward along Route 922, forcing enemy troops into the waiting sights of the 1st and 3rd Battalions. In addition, intelligence indicated that the NVA were desperately trying to evacuate their remaining artillery pieces in the face of the other two battalions' push southward. In essence, the direction of the operation was now toward removing the enemy threat to the regiment's right flank.[38]

Operation DEWEY CANYON: The First Radio Battalion Direct Support Team

Whence came the intelligence about the enemy's planned artillery movements? On 25 January, a USMC First Radio Battalion Direct Support Team was sent to FSB RAZOR to provide top secret communications intercepts in real time in support of DEWEY CANYON. This four-man team was led by a gunnery sergeant and consisted of three Marines and a Vietnamese soldier-translator. On the day the team arrived at RAZOR, M/3/9 discovered a four-strand, tree-mounted NVA telecommunications line connecting Laos with Base Area 101 in the direction of Quang

---

† General Stillwell replaced LTG Rosson in command of Provisional Corps, Vietnam on 31 July 1968, and PCV was redesignated XXIV Corps on 15 August 1968.

Tri.[‡39] The Marine Corps official history reports that the team "tapped the wire and eventually broke the NVA code, . . . no information was provided the 3d battalion as it was 'presumed to be of strategic rather than tactical value,'" according to the 3/9 commander, LTC Laine.[40] According to the NSA version, this was not the case: "No wiretap attempt was made on this line which, if sampled, might have provided lucrative intelligence. The line was destroyed."[41] Whatever happened with regard to the nonexploitation of the enemy's land lines would soon be overshadowed by the information gleaned through interception of its radio communications.

On 11 February when Phase III kicked off, the team moved to FSB CUNNINGHAM, where it would remain until the completion of the operation. "By this time, the Comint team was exploiting the enemy transportation group net and the artillery net, which consisted of 122mm gun batteries and their forward observers scattered throughout the area of operations."[42] On 17 February, a Marine officer arrived to take command of the team, and on the 23rd, additional personnel were assigned.

The wide front of the 9th Marines' advance coupled with pressure on Route 922 was causing problems for enemy artillery movements. NVA artillery radio net intercepts were quickly translated, decrypted, and provided to the 9th Marines HQs. "As units of 1/9 advanced toward A Bum [Ap Boum], enemy communications revealed the location of unidentified artillery pieces in relation to the advancing Marines. . . . units swept the area and captured two Soviet-made 122mm field guns and four 85mm field guns."[43] The USMC history validates the NSA story with respect to the attempted withdrawal of the NVA guns: "With the information provided by Winecoff's company, and the intelligence gathered by SOG teams and 1st Radio Battalion intercepts, indicating that the enemy was

---

‡ By early 1969, the North Vietnamese hard-wired telecommunications network into southern Laos and northern South Vietnam was expanding along with its road network. A CIA report in March 1969 titled "Open-Wire Telecommunications Network Laotian Panhandle" descries 500 NM of lines, including "Branch Line 2a (which also includes two wires) extends east from LA/VS Line Two and generally parallels route 922 and LA/VS Line One, *crossing the Laos/South Vietnam border into the Ashau Valley, through a salient of Laos [Base Area 607], and then back into South Vietnam [Quang Nam Province]*" (emphasis added).

evacuating its heavy artillery westward out of the reach of the 9th Marines, Lieutenant General Stillwell revived Davis's initial request."[44] This reference also recognizes the role of the signals intelligence in support of the USMC's second request for cross-border operations authority.§[45]

The 3/9 Marines' advance into the Tam Boi area south of Tiger Mountain also benefitted from the Radio Team's work. Intercepted 559th Transportation Group messages revealed NVA concern that the loss of the Tam Boi area would be a serious problem and that a radio relay site was situated there. As a result, a B–52 ARC LIGHT strike was requested and took place on 20 February, just before the 3/9 Marines reached the area.[46] Two more 122mm guns were captured when Tam Boi was overrun, and the entire area was honeycombed with tunnels and caves.

In addition to providing timely information about the enemy dispositions and movements to and from the front, the communications intelligence proved its value at the fire support bases. After the team became operational, FSBs CUNNINGHAM and ERSKINE were able to receive immediate advance warning of an impending artillery barrage. "When the enemy transmitted his preparatory firing order to this forward observer located somewhere in the vicinity of the FSB, the incoming alert was sounded."[47] As a result, there were no KIAs from indirect fire at CUNNINGHAM after 25 January.

The radio intercepts proved their value one final time when the Marines were departing from the operation.

The most significant contribution during this final phase affected the lift-out of the regimental CP and other units on FSB CUNNINGHAM. This FSB

---

§ While there is no doubt that the work of the radio intercept team was important during DEWEY CANYON, I am unable to confirm that all of the artillery captures credited by Rayburn to the team are accurate. There were a total of four 85mm guns taken during DEWEY CANYON according to all reports. The commander of the 2/9 Marines reported capturing "an entire battery of 85mm guns with almost 600 rounds of ammo. . . . There were four guns in this battery. They had blown the breach blocks and that was all." The 9th Marines AAR Execution daily report, on the other hand, records "2 85mm arty pieces" taken in Laos by L/3/9 on 27 February (YD233043) and "found 2 85mm guns" on 28 February (YD167032) by H/2/9. The 1/9 AAR does not claim any 85mm guns.

had been under daily fire from enemy 122mm guns located high up on the ridges of Co Ka Leuye. Enemy artillery shooters had the primary landing zone bracketed. The decision was made to clear a new landing zone in the jungles approximately 300 meters outside the northern perimeter. This new landing zone would be used for the lift-out. An intercept on 16 March disclosed that enemy artillery fire was going to be readjusted to this northern landing zone. Several rounds were fired for adjustment, and it was evident that the enemy was, in fact, shifting his fire to the new landing zone. The enemy continued his fire until he had the new landing zone zeroed in. Activities on the FSB continued to give the appearance that this landing zone was going to be used for the removal. At the last moment, the regimental commander directed that the helicopters land at the primary landing zone to execute the flyaway of Marines off the FSB. While enemy mortar and other ground fire in the area was rapidly shifted back to the primary landing zone, the 122mm guns were unable to reorient their fire in time, and no casualties were suffered from 122mm artillery fire during the operation.[48]

The NSA article notes several problems with the COMINT operation, including the lack of a VHF direction-finding capability. Nonetheless, the 9th Marines AAR for DEWEY CANYON has the final word: "A special Radio Detachment . . . was deployed. . . . This is the first time a unit of this type was deployed in a manner which offered a minimum of facilities and required rapid displacement. *Intelligence obtained utilizing this capability was invaluable*"[49] (emphasis added).

Operation DEWEY CANYON: Phase III – Into Laos (Part 2)

On 23 February, H/2/9 was headed back into Laos. After another ambush on Route 922 on the morning of the 24th, the company led a battalion sweep to the east. On 25 February, H Company made contact with the North Vietnamese in the morning, killing 8 but losing 2 Marines KIA and 6 wounded. Moving through the area abandoned by the enemy, a 122mm field gun and two 40mm antiaircraft guns were captured. In the evening, CPT Winecoff's men got into another tough fight with enemy troops in well-fortified bunkers. This time around, only 2 enemy were killed, but Marine losses were 3 KIA and 5 WIA. During this action, CPL William D. Morgan gave his own life to save several injured Marines. He was posthumously awarded the Medal of Honor.[50] Sweeping through the battle area, a second 122mm gun was captured.

Over the next several days, Companies E, F, and H continued to sweep along both sides of Route 922 toward the border near Lang Ha. The battalion moved eastward (5 km in five days), engaging in only minor firefights and capturing an entire battery (four) of 85mm guns, several antiaircraft guns, and a number of ammunition and rice caches. The Marines in Laos had a number of close calls with "friendly" air strikes, but fortunately came away unscathed by allied bombs and rockets. Some if not most of the air attacks were due to not everyone being "in the loop" that the Marines were operating "over the fence." One very large final rice cache was discovered right on the border: "anywhere you sat down you came up with rice on your fanny."[51] On 3 March, the 2/9 Marines were extracted to FSB VANDEGRIFT, the first unit to depart from DEWEY CANYON.[52]

SOG CCN operations in support of DEWEY CANYON

While the USMC sweep along Route 9 was taking place against light opposition, other allied forces inside Laos were overmatched and endangered. When MG Davis requested MACSOG support on 30 January, it set in motion a chain of events leading to not one but two SOG missions inside Base Area 611.

The Special Operations Group at this time was conducting operations in Laos under the code name PRAIRIE FIRE. As six-man recon teams had not fared well against the NVA counter-recon tactics in Base Area 611, SOG was employing platoon- or company-sized units of Nung mercenaries led by Special Forces cadre. In conjunction with DEWEY CANYON, it was decided to insert a company, reinforced with a recon team, south of the Marine Area of Operations, but within range of the larger 155mm howitzers at FSB CUNNNINGHAM.

On 25 February, Company A of the SOG CCN (Command and Control North) Hatchet Force was scheduled for insertion about 6 km into Laos.[53] The apparent mission was threefold: (1) to provide reconnaissance on the enemy deeper inside Base Area 611; (2) to serve as a diversionary element, drawing away enemy forces that might otherwise be committed against the Marines then on Route 922 in Laos and along the Lang Ha-Tam Boi corridor; and (3) to exploit any identified targets by engaging with air or artillery. There were only enough helicopters (H-34 Kingbees) to bring in a single platoon on the first lift; the landing soon turned into a nightmare. When the first lift came in, the troops had to jump into a 10–15 foot tall bamboo forest, and one Nung broke a leg, while one Kingbee was turned back by enemy fire. Only 28 men were on the ground, led by young SF 1LT Robert Burns on his first mission "over the fence," when the weather closed in and left the understrength platoon alone in the enemy stronghold.

Because the patrol was south of the mountains south of Route 922, radio contact with the artillery at FSB CUNNINGHAM could not initially be established. The platoon moved north for about an hour before being ambushed for the first time. The Covey FAC who had been on station during the insertion was no longer overhead, but fortunately by this time the platoon was within 300–400 meters of the ridgeline and able to call on the howitzers for support. By the time the artillery fire was adjusted, however, the enemy ambushers had already withdrawn. 1LT Burns consulted with his platoon sergeant, SFC Ralf Hawkins, whose experience he relied on heavily, and they decided to move to the ridgetop before settling in for the night.

On the morning of the 26th, the weather still prevented the arrival of the rest of the Hatchet Force, so the 1st Platoon of A Company waited in place. The enemy was present and tried to lure the platoon into an ambush by making an appearance and then moving off, feigning ignorance of the allied force. The ruse (used hundreds of times by the enemy throughout the war, and this is almost the only time I have heard of the Americans *not* falling for it) did not work, and the platoon remained in place until ordered to proceed with the mission by headquarters. When they moved out in the afternoon, the direction was to the west, the opposite of the expected enemy ambush. Just after crossing a stream in a deep gorge, and refilling their empty canteens in the process, the Hatchet force was again ambushed. One Nung was killed, and four Nungs and SGT Lee, the SF medic, had been wounded by the time the patrol reached higher ground and found a place to laager for the night.

The following day passed without enemy contact, and the wounded were able to be lifted out by McGuire Rig. On 28 February, with just 4 Americans and 17 Nungs still manning the perimeter about 3500 meters from the SVN border, a Covey FAC spotted a force estimated at 300 NVA soldiers to the northwest, massing to attack the beleaguered platoon. The aerial observer called in air strikes, helicopter gunships, and artillery, shattering the enemy force. The remnants retreated to the north, and the platoon received not a single round of enemy fire. That night, the B-52s came over and rained more hell on the NVA, just 1500 meters away.

On 1 March, with the 2/9 Marine sweep of Route 922 completed, the weather cleared, and the rest of the Hatchet Force, known as Task Force (TF) Moore (commanded by MAJ Moore, the CCN Intelligence Officer), including the balance of A Company, as well as B Company and two recon teams, was flown into the same LZ used by LT Burns on 25 February. The initial scheme of maneuver was for the reinforcements to march to join LT Burns, and TF Moore marched through the afternoon, resting for the night about 1,000–1,500 meters to the southwest. That night, the plan changed and orders went out for the 1st Platoon of A Company to move to join the battalion the following morning. Although dehydrated and worn down, they executed this move successfully, and the entire Hatchet Force was reunited at last.

On 3 March, medevac had to be called in twice to extract Nung scouts who had triggered enemy "toe popper" mines, and the Hatchet Force ended up where it started. On 4 March, the unit moved first east and then southeast. On 5 March they reversed direction and headed up the back-trail. The lead recon team was ambushed. Two SF men were killed: SGT Earl W Himes and SP4 Sanderfield A. "Sandy" Jones. Both received the Silver Star for their actions that probably saved the lives of the Nungs on their recon team during the opening moments of the battle. The battalion again formed a perimeter, and medevac came in the late afternoon to evacuate the wounded and the dead. By this time, the Hatchet Force was low on food, out of water, low on ammunition, and low on morale.

At around 0730 the next day, as the task force prepared to move out, the NVA attacked with automatic rifles, machine guns, and rocket-propelled grenades. The 155mm howitzers at CUNNNINGHAM spoke with a louder voice, and the enemy attack was broken. More casualties to be evacuated, more socked-in weather; more delay. The medevac helicopter was able to land in the afternoon. In the meantime, an intelligence report indicated the presence of a battalion-size bunker complex to the north. The Hatchet Force was to attack. Then it happened again: the NVA opened up with automatic weapons and RPGs, and the Marine artillery responded. At this point, the resupply need was critical and there were still more wounded to be sent out, while the number of combat effective troops was rapidly dwindling.

The 7th of March brought more of the same. Attacks in the early morning and evening were repulsed, and resupply by helicopter was not possible due to the overcast. An attempt to drop supply by parachute resulted in the enemy on the ridgeline receiving the goods intended for the CCN troops.

On 8 March, LTG Stillwell at XXIV Corps had had enough. After the defeat of one final enemy ground attack, the Hatchet Force was lifted out, a platoon at a time (due to a shortage of helicopters and the bad weather). The extraction was successful. The total SOG CCN casualties were 59.[54]

When the SOG Hatchet Force was surrounded and became paralyzed by an NVA force estimated at a regiment or more, the diversionary mission relative to the Marines maneuvering to the north was realized, but in a way that was not anticipated. Although one account implies that the original Hatchet Force mission was necessary because the Marines were in serious danger ("A Marine force was in deep trouble in the Dewey Canyon area"[55]), that is incorrect. A force on the ridgeline south of Route 922 would certainly guard the flank of the Marines and perhaps serve to divert enemy infantry from protecting the caches along that corridor, but the 2/9 Marines were never in "deep trouble." The SOG troops were the ones whose existence was threatened.

With TF Moore growing weaker by the minute, Captain Richard J. Meadows of SOG offered to take a small, elite force into Base Area 611 to create yet another diversion (a diversion from the diversion if you will) to draw the enemy away. Launched on 7 March, TF Meadows was led by one of the true legends of the Special Forces in Vietnam, a man who would later lead the abortive raid on the Son Tay POW camp in North Vietnam. Dick Meadows's handpicked SF team included, among others, SFC Jerry "Mad Dog" Shriver, another SOG legend who later disappeared on a BDA mission in Cambodia in the vicinity of the expected enemy COSVN headquarters. With two Nung platoons and two recon teams, CPT Meadows's unit arrived to the south of TF Moore in two lifts on 7 and 8 March, encountering small arms fire. The force on the ground numbered 77, including 10 Americans, 2 Vietnamese, and 65 Nungs. On 9 March, local patrols were sent out, finding the surrounding area in heavy use by the NVA.[56]

During the period 10–12 March, the platoons under CPT Meadows moved slowly to the north, hindered by the very difficult terrain and enemy delaying actions, headed for Hill 1044, on the border about midway between Lang Ha and Tam Boi.**

A and B Companies, 1/9 Marines, linked up with TF Meadows sometime on 12 March, and on 13–14 March they moved together from Hill 1044 to Tam Boi, where they were scheduled to be extracted.[57]

Operation DEWEY CANYON: Results

The 9th Marines After Action Report for the operation, dated 8 April 1969, listed total friendly personnel losses as 130 KIA, 1 MIA, and 920 WIA. For the enemy, 1,617 KIA and 2 POW were claimed. "Despite the large number of casualties inflicted on the enemy, only the 675th NVA

** According to Fox, *Marine Rifleman*, p. 261, Hill 1044 is where the SOG unit rendezvoused with the Marines. Meadows's biographer, Hoe, refers to Meadows's USMC contact, call sign "Delmar" as "LTC Smith", "CO of the 3/9 Marines," but Delmar Six was the call sign of George Smith, who commanded the 1/9, to which Fox's A Company belonged, and it was 1/9, not 3/9, that met the SOG warriors.

Artillery Regiment and the 4th Engineer Regiment/559th Transportation Group were identified. . . ."[58] Enemy ordnance and supply losses (captured or destroyed) included (partial list only)

12 122 mm guns and 4 85 mm guns

257 140mm and 770 122mm rockets

Antiaircraft guns: 21 20mm, 4 23mm, 1 37mm, 3 40mm, 5 50-cal, and 39 12.7mm

24 recoilless rifles and almost 15,000 rocket-propelled grenades (RPGs)

26 mortars (60mm, 82mm, and 120mm) and over 60,000 rounds of mortar ammunition

104 7.62mm and 30-cal machine guns

1,223 individual weapons

110 tons of rice and 2 tons of salt

66 trucks, 15 tractors, 6 prime movers, 1 front end loader and 3 armored personnel carriers

Throughout DEWEY CANYON, weather permitting, air support had been substantial. Fifty-four B–52 sorties and 1,410 tac air sorties accounted for large numbers of enemy casualties, the destruction of numerous vehicles and anti–aircraft sites, and dozens of secondary explosions.[59]

Losses in the Marine infantry companies were high; the highest were in A Company, 9th Marines. Company Commander and Medal of Honor recipient Wesley Fox reported total casualties of 24 KIA, 89 WIA (evacuated) and 129 WIA (not evacuated).[60] This is a rate of well over 100%. Fox reported his "foxhole strength" after the battle of 23 February at 2 officers and 63 enlisted men,[61] and according to the AAR, suffered at least 5 KIA and 40 WIA subsequently, so his company was truly decimated.[††62] For the regiment, casualties, including nonevacuated wounded, may have been greater than 50%.

---

†† The 9th Marines also received a number of replacements during the operation. Just a few days after listing his strength at 65, LT Fox reported "about ninety" effectives.

In spite of the heavy USMC casualties, DEWEY CANYON was a very successful operation. Were the effects transitory? Certainly, as shown by the increasing pressure that the recovering NVA brought against FSB CUNNINGHAM and the allied pickup zones throughout the AO during the last two weeks of the operation. But the enemy material losses were immense, and the destruction of its long-developed base in the Tiger Mountain–Tam Boi–Lang Ha region was a serious setback.

The next act in the A Shau saga was staged at the other end of the valley. The opening scene was already underway.

OPERATION MASSACHUSETTS STRIKER

Operation MASSACHUSETTS STRIKER – Overview

The designated mission of the 2nd Brigade of the 101st Airborne Division (COL John A. Hoefling) for the period 1 March–8 May 1969 was to "conduct combined airmobile operations [in the] vicinity of the southern A SHAU Valley to locate and destroy enemy forces and LOC [lines of communication], on order, continu[ing] operations southeast astride Hwy 614 in QUANG NAM Province." Ergo, the Screaming Eagles were to carry out a search-and-destroy mission at the southern end of the valley similar to the one the Marines were concluding in early March at the northern end. The "concept" of the operation included "emphasis on a rapid thrust to the border to block enemy withdrawal into Laos followed by a detailed systematic search in assigned battalion AOs to destroy caches and disrupt enemy logistical system."[1] Four American airborne infantry battalions and four battalions of the ARVN 3rd Regiment were provided, supported by 5 batteries of 105mm howitzers, 3 batteries of 155mm howitzers, and 1 battery of 8-inch howitzers.

Although the mission statement prioritized the "vicinity" of the valley, in execution most of the ground operations during MASSACHUSETTS STRIKER took place outside the A Shau watershed. Of the six firebases established for this operation, only FURY, sited at Hill 825 on the east wall of the A Shau (YC534846), 4 km east northeast of the abandoned airstrip at A Shau, was adjacent to the valley.[2] None of the other firebases was within 10 km of Route 548, and two (LASH and SHIELD) were in Quang Nam Province to the southeast (see Map 12). Because weather and the unanticipated enemy concentration on Dong A Tay delayed the original allied timetable, the prescribed "rapid thrust to the border" was never more than a dream of the operation planners, and the enemy for the most part vanished into his Laotian refuge.

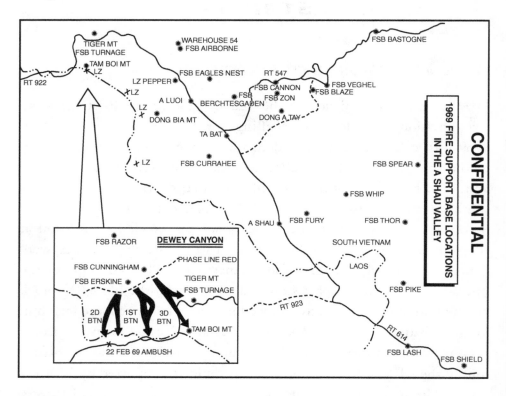

MAP 12

Source: *Project CHECO Southeast Asia Report #2 - Special Report: A Shau Valley Campaign, December 1968 – May 1969. 15 October 1969.*

The narrative in this chapter follows, as closely as possible, only actions in the valley or along the eastern and western boundaries; it is not a comprehensive account of the entire operation. There were three major focal points for actions at locations other than in the A Shau:

1) At the start of the operation, the 1/502 Infantry (LTC Donald Davis) unexpectedly ran into a buzz saw of resistance from the 9th NVA Regiment in the Dong A Tay hill mass southwest of LZ VEGHEL at the juncture of Routes 547 and 547A. The battle in that area resulted in a majority of both friendly and enemy casualties during MASSACHU-SETTS STRIKER and lasted for 33 days, from 12 March till 15 April. On 16 April, the battalion was flown south to join the concluding phase of the operation in Quang Nam.

2) On 20 March, the 2/501 Infantry (LTC Joseph C. Wilson) was inserted in the southern Rao Nai Valley. The headwaters of that valley are less than 2 km from those of the Rao Lao. There was only light action in this area, as the paratroops swept south to the Laotian border and then retraced their steps in search of something to destroy.

3) As assigned, in the final phase of MASSACHUSETTS STRIKER the focus moved south to where Route 614 from Laos enters Quang Nam Province, providing a direct path from the enemy sanctuaries in Base Area 607 to the allied sanctuary in Da Nang. This was the "Yellow Brick Road." Firebase LASH was constructed directly on this path, and the lion's share of the cache discoveries took place in this area, including the largest communications cache taken during the war, as well as a heavy machine repair shop and hospital complex.[3]

The weather in western Thua Thien was the first enemy, delaying the initial deployment of 1/502 Infantry to LZ VEGHEL to 12 March 1969 and cutting some two weeks off the time available before the monsoon rains were expected to commence. This was the same weather system that prolonged the extraction of the 9th Marines from the northern valley.

MASSACHUSETTS STRIKER in the Southern A Shau Valley[4]

The air cavalry of the 101st Airborne Division, consisting of the 2/17 Cavalry and D Troop, 1/1 Cavalry (OPCON from the Americal Division), played a leading role in MASSACHUSETTS STRIKER, using "Pink teams" of one light observation helicopter and one gunship, either a UH-1C Huey or a Huey Cobra, to scout the ridgelines, ravines, streambeds, trails, and hills of the middle and southern A Shau Valley and its tributaries for enemy activity. The cavalry would provide intelligence to optimize the employment of paratroops combing the area for enemy camps and caches and could direct air strikes as well as use its organic firepower to attack NVA troops. On any given day, there were usually from two to four pink teams in low-level flight over the valley. In addition to scouts and gunships, each air cavalry troop (company) had a "Blue team," an aero-rifle platoon (ARP) with its own slicks that could be inserted for missions such as reinforcing ranger teams or to rescue downed helicopter crews. The divisional long-range reconnaissance company, L/75th Rangers,[*] was attached to 2/17 Cav, which provided the Lurps with free rides to and from their patrol areas.

Two battalions of the ARVN 3rd Regiment air assaulted into an area east of the Rao Nai on 15 March. The 3rd Battalion moved south, making light contact, and then turned west toward the A Shau Valley. Once in the A Shau, "when the 3rd Bn began driving NE, the 2nd Bn was extracted and conducted a combat assault into the A SHAU Valley near TA BAT airfield from which they moved on two axis toward [the north]."[5]

The 2/327 Infantry (LTC Charles W. Dyke), OPCON from the 1st Brigade, air assaulted into the airstrip at A Shau on 22 March. This assignment had been planned for the 1/502, but that unit was

---

[*] Formerly F/58th Infantry – all Ranger companies in Vietnam were reflagged as elements of the 75th Ranger Regiment. Company F, 58th Infantry (LRP) became Company L, 75th Infantry (Ranger) on 1 February 1969.

still caught up in the battles at Dong A Tay, by this time being called "Bloody Ridge." The "No Slack" troopers, in three columns, would sweep to the very southern end of the valley, then retrace their steps when they reached the border with Laos. Just a few kilometers to the east, the 2/501 was simultaneously performing a similar maneuver in the Rao Nai basin. By 16 April, the 2/327 had completed its search of the southern valley, encountering only scattered enemy troops and capturing "assorted enemy equipment." The battalion was returned to its parent brigade.[6]

The air cavalry screened ahead of the grunts on the valley floor, identifying targets for exploitation. On 23 March, scout helicopters from the 2/17 Air Cav spotted truck tracks, 15 oil drums, and a four-wheel trailer along several streams 1–2 km south and southwest of A Shau. An air strike was called in and one truck was reported destroyed. More oil drums and abandoned damaged vehicles would be found in this area over the next several days, confirming the location as an NVA motor pool and refueling station. The following day in the same area, 25–30 huts and bunkers were spotted, and another air strike was called. On 25 March, 1.5 km south of the A Shau airstrip, a 37mm gun was found. Throughout the afternoon of 28 March, light observation helicopters of D/1/1 Air Cavalry made multiple sightings of bunkers, huts, abandoned automatic weapons positions, and four-strand commo wire in the vicinity of the headwaters of the Rao Lao, where the southern end of Route 548 crosses the border into Laos. Only two NVA soldiers were seen.

On 30 March, the 3/3 ARVN operating 6 km north of the old A Shau camp (YC473890) captured a Russian truck. On 3 April, near the southern end of the valley (YC503814), 2/327 Infantry found two Russian 2½-ton trucks.[7] There were other caches of varying size, most containing ammunition and explosives.

On 21 April, the aero-rifle platoon of A/2/17 Cav accompanied combat engineers to Ta Bat, where an LZ was established at 1121 in the morning. The engineers conducted soil tests near the abandoned airfield, information that would be used during the next operation in the valley the following month, APACHE SNOW. The mission was completed and the units were extracted at 1655 hours.

Also on the 21st, three Ranger teams from L/75 Infantry (Teams 11, 12, and 26) were inserted to the west of Firebase FURY, on the east side of the valley. One team (11) had communications problems and was extracted that evening. The remaining Rangers observed some signs of enemy presence and the usual signaling shots from trail watchers but saw no enemy. Teams 12 and 26 continued to patrol the hills and ravines east of the valley and just north of the abandoned A Shau airstrip on the 22nd.

On 23 April, the radioman of Team 26 was struck by lightning, causing the team's claymores to explode and wounding all six team members, one seriously. The Ranger team was safely extracted by medevac at 1730 hours. In the meantime, Team 12 carried on with its mission, making contact with a force of six or seven VC and receiving small arms fire at 1000. In the early afternoon, the team again observed two enemy soldiers and a trench line. When Team 26 was put out of action, Team 12 lost its radio relay link to base, and after successfully evading the enemy, was extracted without incident at 1800.

On 26 April, there were no American casualties as things became quiet for the most part throughout the AO, with the exception of ineffectual enemy fire directed at a LOH just southeast of the troops on the ground. Two Ranger teams (12 and 22) were inserted in the morning. They would find only minimal signs of the enemy and make no contact while patrolling a couple of kilometers to the east of FB FURY until being extracted three days later (29 April).

Warehouse 54

"An enemy prisoner captured in February 1969 in BA [Base Area] 101, northeast of the [A Shau] valley described a major supply station located in the north end of the valley. The prisoner said porters from BA 101 walked two days west and three days south to reach 'Warehouse 54.'"[8] Captain

Albert W. Estes, a FAC working with the 3rd Brigade, 101st Airborne Division (COL Joseph B. Conmy), picks up the story:

> About 1 April the brigade intelligence officer, Captain Robert Fredricks [sic], called several FACs to the brigade headquarters and outlined his idea of a lucrative storage area at YD 355075 and asked the FACs to look in a 2,000 meter radius of the point. He had developed this point from several sources including POWs and agent reports. . . . The FACs were very skeptical. . . . [t]hey looked and found a heavily canopied area with a few trails and no excessive activity. 'We didn't pay much attention to it and we'd report back every day that we couldn't see anything.'[9]

Captain Friedrich remained adamant, and on 17 April a FAC-directed airstrike on the area resulted in nine secondary explosions and a fire. The hunt was on. The name, it was later determined, derived from the proximity to the kilometer 54 marker on Route 548, at YC323063.[10]

Although initial contact came during MASSACHUSETTS STRIKER (2nd Brigade), Warehouse 54 was in the AO of the 3rd Brigade, and that HQs would control the actions there.

On 18 April, D/1/1 Cavalry began to take a closer look. A number of bunkers and huts were observed and the aero-rifle platoon was to be inserted for BDA (bomb damage assessment), but the weather deteriorated in midafternoon and the insertion was postponed. On 19 April, the air cavalry continued to investigate the area just east of Route 548, to the north northeast of Dong So, finding another large enemy complex of over 60 huts, with wet clothing hanging out to dry. Thirteen huts and several bunkers were destroyed by helicopter gunships, and another air strike produced more secondaries. A third air strike was called in on 20 April and destroyed an additional 25 hooches and 10 bunkers.[11] On 22 or 23 April, an ARC LIGHT strike was targeted at the enemy base, but the target grid was slightly incorrect and the heavy bombers missed by about 150 meters.

3rd Brigade Headquarters had decided that a landing zone needed to be established in the vicinity of what was evidently a major concentration of enemy camps in the rugged terrain northeast of Dong So, and 10 km east of Tiger Mountain. On the morning of 23 April, a mixed Ranger/ Air Cavalry squad from the 2/17 Cavalry was to go in first by rappelling down through the jungle canopy to secure a site for clearing a landing zone (YD350063). As the first helicopter neared the LZ, it was hit in the fuel cells by either a rocket-propelled grenade or tracers and exploded violently, breaking into two pieces. Killed in the crash were two crew members, crew chief SP4 Otto P. Barnhart and door gunner SP4 Henry W. Cardwell of C/2/17 Cavalry. Five passengers were also KIA: SFC William F. Rocco of Headquarters Troop, 2/17 Cavalry, SSG Julian D. Dedman of L/75 Infantry (Ranger), SP4 James P. Heim of D/2/17 Cavalry, SP4 John F. Koehler of B/2/17 Cavalry, and SP4 John W. Tiderencel of D/2/17 Cavalry.[12] The pilots' cabin came to rest within a few hundred meters of Route 548, and a rescue mission was soon underway. The aero-rifle platoon (ARP) from D/2/17 Cavalry was inserted several hundred meters from the crash site (YD344056) and spent most of the day moving toward the wreck, bypassing extensive bunker and hut complexes. Two members of the rescue party were wounded by a command-detonated claymore mine. By 1600, the ARP and the two surviving pilots had linked up, and the wounded airmen were on their way to the 22nd Surgical Hospital.

Early on the 24th (0630 and 0830), the air cavalry unit on the ground came into contact with the enemy. In the ensuing firefights, 11 U.S. soldiers were wounded and an additional 12 were wounded by "friendly fire" from helicopter gunships. The unit would spend the day constructing and defending a one-ship LZ and evacuating casualties. In the afternoon, the rifle platoon of A/2/17 Cavalry was brought in to reinforce, along with a dozen engineers to clear the landing zone. Although they had little time to conduct a search of the vicinity, 400 lbs of rice and two large bunkers containing grenades and rockets were discovered. In the morning, another Huey went down around 0930 just to the north. This time, it was a UH-1C gunship from C/4/77 Artillery (Aerial Rocket). The pilot, WO1 James A. Brown II, was KIA.[13] By 1015, Ranger Team 12 had rappelled into the crash site and rescued two surviving crewmen. The fourth crew member was alive but pinned under the wreckage; he was later freed.

The air cavalry had stirred up a hornet's nest. At 0630 on 25 April, D/2/17 Cavalry was hit with a barrage of rocket-propelled grenades, wounding two men. The medevac was completed by 0800 and the troop began to patrol around the landing zone. Soon there was another battle against NVA in bunkers, and six to eight more cavalrymen were wounded. After a second medevac was completed, three air strikes were called in on the stubborn enemy. A/2/17 Cavalry's aero-rifle platoon now attempted to move off the LZ, but at 1300 reported that booby traps were numerous and napalm from the air strikes had started the ammunition in enemy storage bunkers to cook off. Contact continued sporadically throughout the day until dusk, with D/2/17 engaged with a force of unknown size at 1940 hours. At 1700 hours, the air cavalry in the middle of the enemy bastion was finally reinforced, when B/3/187 Infantry arrived. As recon around the LZ continued intermittently, a huge store of ammunition was captured:

300 cases (600 rounds) 82mm mortar ammo
100 cases (600 rounds) 60mm mortar ammo
100 cases .51-cal ammo (10,000 rounds)
25 cases 7.62mm small arms ammo (20,000 rounds)
7 cases TNT
Plus numerous recoilless rifle and RPG rounds

The remaining soldiers of the D/2/17 Cavalry aero-rifle platoon finally walked down to the valley floor and were extracted.[14]

The type of terrain in which the storage area was situated is a mountainous area rising over 1,100 meters and covered with double and triple canopy. The contents of the cache were stored in large hut and bunker complexes with most of the bunkers being well constructed. There were numerous complexes throughout the area which comprises approximately four square kilometers. These cache sites were well concealed.[15]

The complex, estimated to be "four to sixty months old," was actually much vaster than a mere four square kilometers, as the 3rd Brigade would discover during Operation APACHE SNOW.

A 1,000-lb. daisy cutter bomb was used to clear the top of the hill for what would become Fire Support Base AIRBORNE (YD354071) on 26 April. A and D Companies of the 3/187 Infantry (LTC W. F. Honeycutt) were then inserted into both tiers of a two-tier LZ. Although CPT Estes reported "14 men killed in the first ten hours of the assault into the Upper LZ," the records do not bear this out. On the 25th, one slick pilot was KIA in the air (CPT David W. Watson), another was killed by shrapnel on the second night after being shot down at the LZ (WO2 Elton L. Searcy), and one infantryman lost his life when hit by fire while riding in a helicopter (SP4 Thomas G. St. Onge of B/3/187). The next day, one UH-1H was hit and crashed at the upper LZ, killing crew chief SSG James R Dorsey, Jr.; one infantryman, SP4 Seth E. Randolph of A/3/187, was KIA by multiple fragmentation wounds on the ground.[16] On the afternoon of 27 April, an aircrewman from A/2/17 Cavalry was mortally wounded about 1.5 km north of the unnamed LZ at the enemy cache site. SP5 Albert D. Austin, Jr., the gunner on an OH-6A Loach, was wounded in the head by small arms fire but did not survive the flight to the 22nd Surgical Hospital.[17] An air strike was directed against the area with unknown results. The weather became bad the following afternoon, halting air operations over the A Shau.

While both LZs on 26 April were hot, CPT Estes's recollection that the "troops on the two LZs were under heavy attack and the threat of being overrun" may be a slight exaggeration, given the low American casualties. In any case, the FACs were active through the night, directing air strikes and "Spooky" fixed-wing gunship fire to break up possible enemy concentrations. According to an article in the 101st Airborne Division in-country publication,[18] the "Rakkasans" killed 65 NVA soldiers during the first six days of their stay in the valley and "captured numerous ordnance, food, and equipment caches." B Company, 2/327 Infantry, made contact with the enemy, killing 10 in a brief firefight on 28 April (YD347066) and then finding 12 bodies at a gravesite nearby. On 30 April, B Company found 12 more corpses about two days old.[19] The B Company Commander, CPT Barry Robinson, said, "The enemy hasn't stood and fought for any length of time. They are fighting a delaying type action, trying to stall our drive and prevent us from exploiting lucrative caches." The NVA was leaving for now but would return.

On 2 May, an LOH from one of the pink teams was shot down by automatic rifle fire 1.5 km south of the abandoned village of La Dut, crashing in the center of the valley. Another LOH was able to land and evacuate the crew, one of whom was seriously injured. The aero-rifle platoon of C/2/17 was inserted without receiving fire at the crash site. A UH-1H Huey arrived, and the LOH was prepared for a sling-load extraction. As the UH-1H lifted with the LOH dangling below, the small helicopter began to gyrate uncontrollably. It had to be cut loose and crashed back to the valley floor, now a total loss. As another UH-1H Huey was approaching to extract the riflemen, it suffered a blade strike on a tree and it, too, settled to the valley floor a few hundred meters to the southeast. It was later successfully extracted. The rifle platoon spent the night in the valley and was extracted without incident before 0800 the next morning.

As contact with the NVA continued around the LZ, SP4 Rolland K. Fry of A/3/187 Infantry gave his life at FSB AIRBORNE on 2 May in an action for which he was posthumously awarded the Silver Star. Patrolling near the firebase, Specialist Fry's platoon was ambushed by a well-entrenched and heavily armed enemy force. From his position at the head of the column and despite serious wounds from a rocket-propelled grenade, Fry provided covering fire, enabling other members of his unit to withdraw to safety, until a second grenade mortally wounded him.[20]

In the afternoon of 4 May, another LOH was hit by fire and crashed several kilometers north of the previous day's actions. The Blue (aero-rifle) platoon of A/2/17 Cav was inserted and destroyed the badly damaged bird. A second LOH operating in the area was also hit by fire and flew south, crash-landing on the old A Luoi airstrip. Before sundown, this helicopter was able to be extracted by sling. The only injury this day was the pilot of the first LOH, who received leg wounds. Over the Warehouse 54 area, multiple new sightings of huts and bunkers were added to the list, and the complex was now known to cover several square miles. "This general area proved so rich in enemy presence that once Hill 937 [Hamburger Hill] had been captured, the 101st Airborne continued operations in the locale in May and June 1969."[21]

MASSACHUSETTS STRIKER – *After Action Report*

As per SOP for the controlling headquarters of an operation, the 2nd Brigade, 101st Airborne Division (Airmobile), gave itself a hearty pat on the back in its After Action Report.

> The 2nd Brigade in 2½ months has successfully interdicted all supply and infiltration routes used to supply NVA forces operating against allied installations from Hue City to Da Nang. All supply lines into Base area 114 from the A Shau Valley north were cut. . . . Also, there are now no supplies flowing from BA 607 to Laos and the southern A Shau Valley. Finally, the 2nd Bde built a fire base [LASH] astride route 614, the "yellow brick road," which the 559th Trans group was using to transport supplies into Da Nang, NW Quang Nam Province.[22]

Unfortunately, of the six firebases constructed, none appear to have outlived the operation. While the "Chemical" Inclosure (5) of the AAR reports "The Bde CP at Fire Base Whip was the only permanent fire base set up during the operation," the "Engineer" Inclosure (9) reports: "Charlie Company was given the mission of closing FSB Whip . . . and was able to close FSB Whip by 9 May 1969." Whoops – somebody didn't get the memo!

As it was not a permanent facility, the impact of FB LASH on the enemy supply line to Da Nang would at best be fleeting. Compared to the other operations in the A Shau Valley before and after MASSACHUSETTS STRIKER, equipment captures were modest: 30 trucks, 2 armored personnel carriers, 2 bulldozers, 3 anti-aircraft guns, 40 crew-served, and 857 individual weapons (mainly SKS rifles).[23] Most of the equipment taken was in northwest Quang Nam Province along the "Yellow Brick Road," not in the A Shau.

The Combined Intelligence Center, Vietnam, concluded in an Order of Battle study on 5 May: "It is apparent that BT [Binh Tram] 7 now known as BT 42 evacuated its area during Operations Delaware/Lam

Son 216 in April 1968, Dewey Canyon and Massachusetts Striker."[24] In its Summer 1969 issue, the in-country magazine of the 101st Airborne Division reviewed the operation and concluded: "As the days went on it became obvious that the entire area of the extreme southern A Shau had been abandoned by the NVA, who left the bulk of their equipment behind in their hasty retreat. The Yellow Brick Road was interdicted and destroyed by the 2nd Brigade, leaving enemy plans for future offensives against Hue and Da Nang extremely hampered if not impossible."[25] The parts about "abandoned," the "bulk" of the enemy's gear, "hasty retreat," and "extremely hampered" were inaccurate. Its trucks, artillery, anti-aircraft guns, and almost all of the engineering equipment had been temporarily evacuated, and the PAVN *never* abandoned the A Shau Valley. Given the operation at the other end of the valley and the battle at Dong A Tay in mid-March, Binh Tram 42 had abundant advance notice of the impending allied attack and ample time to reposition its forces and many caches. The North Vietnamese modus operandi was to fight only when the odds were stacked in their favor or when they had something important to defend. They did not fight in the A Shau Valley during MASSACHUSETTS STRIKER. APACHE SNOW would be another story.

18

OPERATION APACHE SNOW – PART I

Operation APACHE SNOW – Into the Valley

"**H**amburger Hill," the GIs' new name for Hill 937 (YC327982), the highest peak on Ap Bia Mountain (Dong Ap Bia), was located a scant 4 km southwest of the heart of the "Punchbowl," from where the Binh Tram 7 headquarters had controlled the PAVN supply line to Hue during the 1968 Tet Offensive (see Map 13). The NVA base area in the valley had been extensively "searched and destroyed" during Operation DELAWARE just over a year earlier. One of a handful of highly publicized "named" battles of the Vietnam War, Hamburger Hill is memorialized in both books and film and associated in collective memory, correctly or incorrectly, with the needless waste of American lives. Samuel Zaffiri's work provides a vivid narrative of the battle itself and the hell that the troops of the 3rd Battalion, 187th Infantry went through on Dong Ap Bia.[1] Yet there is much more than this one engagement to the history of the 3rd Brigade (reinf) of the 101st Airborne Division, the 9th Marines, and the 1st and 3rd ARVN Infantry Regiments during the four weeks from 10 May through 7 June 1969. APACHE SNOW was the second in a series of operations by the 101st Airborne Division and ARVN 1st Division that took place in 1969 in the A Shau Valley and its environs, with the objectives of disrupting the NVA's main line of communications in I Corps and destroying vast quantities of equipment, ammunition, and food supplies that were stockpiled throughout the area.

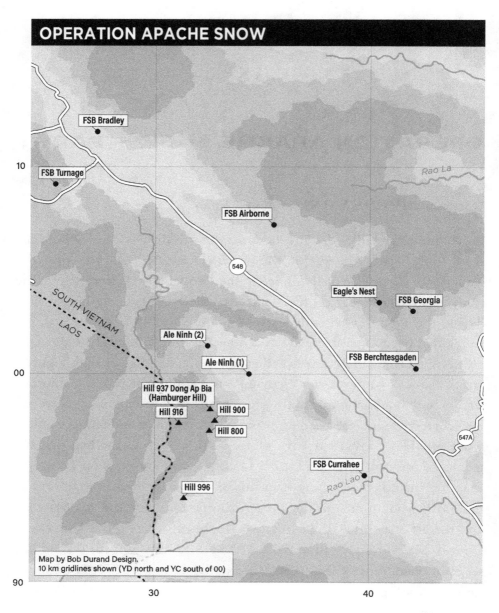

MAP 13

Source: Bob Durand Design

Whereas previous operational plans for the A Shau had paid lip service to trapping the enemy in the valley, APACHE SNOW was the first
allied attack plan to make a systematic attempt to actually shut down
enemy escape routes from day one. On the west side of Route 548, four

battalions would combat assault on D–Day, while to the east, a series of firebases (BRADLEY, AIRBORNE, BERCHTESGADEN) would serve as patrol bases while also providing artillery support. At the north end, the 9th Marines at FB ERSKINE and the ARVN at FB TURNAGE/ TIGER* were to put a cork in the Route 548/922 escape route, while to the south at Ta Bat, LZ CURRAHEE would perform that role for Route 548 and the trails leading west along the Rao Lao Valley. It was hoped that the enemy had been herded into and corralled in the center of the valley by the closure of the northern link to Base Area 611 during Operation DEWEY CANYON and the shutdown of the southern end of Route 548 into Base Area 607 by Operation MASSACHUSETTS STRIKER.

At 0845 on the morning of 8 May 1969, D/1/506 Infantry air assaulted into the A Shau Valley, west of Ta Bat and about 2 km from Route 548, along the west bank of the Rao Lao where it exits the valley on its journey to join the mighty Mekong River. The company secured the area for the construction of FSB CURRAHEE (YC399949).[2] In the meantime, the ARVN were moving to build FSB BRADLEY (YD278119), and A/2/501 Infantry relieved the 3/187th at FSB AIRBORNE (YD355071).

The formal beginning of Operation APACHE SNOW came on 10 May. The 1st Battalion, 9th Marines (LTC Thomas J. Culkin) air assaulted to reestablish FSB ERSKINE (YD164107) in the Da Krong Valley. The Marines were to patrol the upper (southern) Da Krong Valley as far south as Route 922, covering the same terrain they had been in in February and March. The 2nd Battalion (LTC George C. Fox) was similarly patrolling the Da Krong Valley out of FB RAZOR to the north. Neither unit encountered any significant enemy contact, as it appeared that the NVA had not yet attempted to reassert their dominance over the area. On 25 May, Marine engineers placed two major cuts in Route 922, each 12 feet wide and 20 feet deep, and a third 12 feet wide and 10 feet deep, and on the following day caused two 50-meter landslides. The interdiction plan completed, the Marine engineers, infantry, and artillery were withdrawn from the APACHE SNOW AO during the period 27–31 May and the Marine participation was terminated.[3]

---

* The XXIV Corps AAR refers to this as FSB TURNAGE, while the 3rd Brigade, 101 Abn Div AAR calls it FB TIGER.

The last fire base to be recommissioned on 10 May was at Tiger Mountain (Hill 1228), where a reinforced company of the 2/1 ARVN helicoptered in to FSB TURNAGE (aka TIGER) to close the Route 548 connections to Route 922. This location was about 5 km north of the Tam Boi area, and thus constituted the final link in the series of LZs about 5,000 meters apart forming (in theory) a net across the western side of the A Shau Valley.

There were four combat assaults on D-Day (10 May) into landing zones spaced 4–5 km apart and all within 1,000 meters of the border with Laos. First in was the 1/506 Infantry (LTC James Bowers) at 0730 to the southernmost LZ, 8 km due west of FB CURRAHEE and 1 km north of Hill 996. From there, the insertions proceeded in sequence to the north: 3/187 Infantry (LTC Weldon Honeycutt) at 0800, 2/501 Infantry at 1000, and 4/1 ARVN at 1020. A fifth combat assault, at 1300, placed the 2/1 ARVN on Tiger Mountain, 5 km north of its sister unit. The U.S. LZs were cold; the ARVN LZs were hot. The 4/1 continued in contact for several hours and a CH-47 Chinook bringing in an ARVN 105mm howitzer to Tiger Mountain was shot down, but there were no casualties.[4]

For the next several days, Companies A, B, and C of the 1/506th Infantry (the Currahees) patrolled an area bounded on the west by the Laotian border and on the south by the Rao Lao. They operated separately, but always within a few hundred meters of each other, and discovered the usual A Shau scattering of enemy caches and camps, while finding the enemy elusive. NVA mortars inside Laos bombarded C Company on the western side of Hill 996 on 11 May, wounding 22 paratroops. At 1220 hours on 13 May, the first alert was received that the 3/187th to the north was in contact, and the following day the battalion turned back to the north, where it was designated to join the battle for the Dong Ap Bia hill mass on 15 May.[5]

The 3/187th Infantry (the Rakkasans) landed at 0800 less than 2,000 meters to the northwest of Hill 937, the highest peak of the Dong Ap Bia hill mass. Initially, two companies moved to the north and west, searching the assigned AO, while the rest of the battalion (B Company did not arrive until 1600) moved to the south. The battalion's AAR paints a rosy picture:

The insertion had caught the enemy by surprise. D Company . . . moved its 1st Platoon up the ridge to the southeast. It was from this position, the key terrain to the immediate northwest of Dong Ap Bia (Hill 937), that the Battalion Headquarters controlled and supported operations throughout the period 10 to 21 May. Occupation of this terrain also resulted in an effective blocking position, as it stood on the major ridgeline and trail complex between Dong Ap Bia and the Laotian border.[6]

The air assaults on 10 May were preceded by artillery and air "prep" of about 30 possible landing zones throughout the area, in order to deceive the NVA about the insertion locations. The buildup at FSB AIRBORNE, combined with the establishment of FSB CURRAHEE and FSB BRADLEY on 8 May, had certainly alerted the enemy in the valley to the impending allied incursion, and the LZ prep signaled the imminence of the attack, so only the actual locations of the combat assaults came as a surprise. While it is true that the position occupied by the 3/187th was astride a trail complex between Dong Ap Bia and Laos, that position was not "effective" in isolating the 29th Regiment. Enemy reinforcements and evacuations took place during the ensuing days between Hill 937 and Laos via ravines to the south and southwest of the 3/187th and between that battalion and the 1/506th Infantry approaching from the south.

Just before dark, the NVA ambushed the B/3/187 point as it moved southeast just over 1 km from the peak of Dong Ap Bia. Three men were wounded by AK-47, RPG, and machine gun fire. F-105 fighter bombers were on station, and a FAC directed them onto the enemy bunkers where they placed accurate 20mm cannon fire and then napalm. By the time the air strike and medevac were completed, it was too late to advance, and the troops settled in place for their first night on the mountain. According to the III MAF intelligence report, 9 NVA were KIA in this encounter.[7] The next day (11 May), LTC Honeycutt's troops continued to RIF to the south. Their planned mission was to cover the Dong Ap Bia area and then move to the northeast toward A Luoi. Alas, that plan was soon aborted: "At 1625H, B Co 3-187 Infantry began to receive heavy small arms fire and machine gun fire from YC325982. This action marked the beginning of the ten-day battle for Dong Ap Bia Mountain."[8] Before taking up that story, consideration must be given to actions elsewhere in the APACHE SNOW area of operations.

The Other Major APACHE SNOW Battle: Firebase AIRBORNE Under Attack

Third to go in along the border were the troops of the 2/501st Inf (-), assaulting into LZ Green (YD288015). From there, they conducted RIF operations to the northwest during the period 10–13 May. On 11 May, scouts of the 2/17 Cav led the ground troops to five rafts hidden near the LZ. A and E Companies remained at FB AIRBORNE, conducting local patrolling and guarding two batteries of artillery.

Prior to its relief at FSB AIRBORNE on 8 May, the 3/187th Infantry had conducted search and destroy patrols in the area for two weeks. Many enemy caches were located, including one less than 200 meters from the new American base, containing 1,000 rocket propelled grenades. A preliminary report (6 May) of the caches found by the 3/187 Infantry included, among other items, 28 60mm mortars, 478 122mm rockets, over 2,500 mortar rounds, over 2,200 RPG rockets, and almost 24 tons of rice.[9]

Even as the reconnaissance around the base continued, AIRBORNE was rapidly developed into a 12-gun artillery complex supporting the elements of the 3rd Brigade operating on the South Vietnam/Laos border. By 12 May, four 155mm howitzers of C/2/11 Artillery, six 105mm howitzers of C/2/319 Artillery, two 105mm howitzers of B/2/319 Artillery, and Company A/2/501 Infantry were present. The base, originally blasted out of the hilltop jungle by a "daisy cutter" bomb, occupied a position on top of the ridge approximately 25 meters wide and 200 meters long. About 50 meters higher and separated from the fire base by the helicopter pad was a separate perimeter about 50 meters in diameter containing two 81mm mortar pits and defended by E/2/501 Infantry. Three strands of concertina wire laced with trip flares protected the base, and the jungle came right up to the edge of that wire. The enemy was still very much present in the area, and incoming mortar and small arms fire were regular occurrences.[10]

The NVA defending Warehouse 54 (at some point shortened to "the Warehouse") had been driven away, or so it appeared, and a number of equipment, weapons, ammunition, and food caches proximate to FSB AIRBORNE were captured and either destroyed or evacuated by the 3/187th Infantry. But the enemy had not gone far and received orders on about 9 May to attack the American artillery, which must have appeared exceptionally vulnerable. The 806th Battalion of the 6th NVA Regiment and the K12 Sapper Battalion, supported by elements of two other battalions, conducted the usual meticulous reconnaissance of AIRBORNE before launching their assault at 0330 on the morning of 13 May. There were only 113 men (46 sappers and 67 infantry) in the assault force. They moved to their assault positions at last light on 12 May:

> The attack was conceived, planned and executed in the typical thorough fashion of a well-trained sapper unit: a stealthy approach, a violent execution lasting approximately 90 minutes, and then a hasty withdrawal leaving a covering force in contact. Most of the trip flares in the wire had been neutralized by tying down the striker with a bamboo strip. The enemy dead were well equipped with satchel charges, dressed in loin cloths or shorts, though several were completely naked. They also wore the head band to prevent sweat from getting in their eyes.[11]

The sappers were inside the American position before the defenders were even aware that an attack was underway, having silently cut the concertina and neutralized the trip flares in a half-dozen approach lanes. Keeping the defenders' heads down with mortar and heavy RPG fire, the sappers quickly overran several of the artillery pieces. One 155mm howitzer was destroyed in a direct hit by an 82mm mortar round, and four 105mm howitzers were put out of action by sappers tossing satchel charges. A Conex container occupied by the Fire Direction Center of the 2/319th Artillery was hit with an RPG and riddled with AK-47 fire. The battery commander, CPT Moulton L. Freeman, was killed in action during the battle at the FDC. Soldiers in a number of other bunkers were killed or wounded by satchel charges thrown in before the defense could be organized, as the sappers ran throughout the base leaving devastation and death in their paths. Eventually, the American defense became more organized, as small clusters of GIs fought back. By 0500, an AC-47 Spooky flare ship and other gunships arrived and placed heavy minigun fire on the enemy positions just outside the base.[12] Although unable to strike directly at the enemy already inside the fire base, artillery from other fire bases assisted in silencing the enemy mortars and hitting likely assembly areas and withdrawal routes.

When the enemy withdrew around 0500, they left behind 39 bodies. Twenty-seven Americans were killed in action, and 61 were wounded.[13] It was one of the most devastating attacks on an American firebase during the war. "On the basis of body count alone, this constituted an American victory, and that is exactly how the spokesman at division level portrayed it to the press."[14] The battle narrative prepared by the 22nd Military History Detachment was less sanguine:

> Lessons are being learned and relearned in the Vietnam theater, and the one lesson that seems to crop up time and time again is the increased need for security. Security must be active in the form of patrols and ambushes, and passive in the form of wire, claymores, listening posts and increase alertness during the hours of darkness. . . . All evidence pointed toward the fact that the NVA forces would attempt to defend the area. Therefore, the need for security was even greater than in a normal exposed fire base. The enemy suffered heavily in this attack. If he had been detected prior to entering the perimeter, his losses may not have increased, but our losses would have been less severe.[15]

The brigade commander, COL Joseph Conmy, commented after the war, "It was the worst result from a sapper attack that I have ever seen."[16] One may ponder whether the enemy attack on FSB AIRBORNE during the early morning hours of 13 May had anything to do with the obduracy with which the 101st Airborne Division would prosecute the assault on Hamburger Hill during the coming days.

The rest of the 2/501st Infantry, then patrolling the border several kilometers north of Dong Ap Bia, would be prematurely withdrawn from that mission and flown back to AIRBORNE on 13 May to hunt for the enemy attackers and renew the search-and-destroy patrols in the Warehouse. Even as the 101st Airborne claimed to write the script in the valley, it was the NVA piper who called the tune.

The ARVN 1st Infantry Division during APACHE SNOW

As noted previously, the last allied element to air assault into the Laotian border area on D-Day was the 4/1 ARVN Battalion, inserted at YD241041 on the Tam Boi ridge, where major caches had been discovered during DEWEY CANYON. On D+1, a third battalion of the ARVN 1st Regiment, 3/1, air assaulted to an LZ at YD197044, 4,000 meters farther west and just north of Lang Ha, another position well known to the U.S. 9th Marines who had been here earlier. Also on D+1, the rest of the 2/1 ARVN arrived at LZ TIGER. The three ARVN battalions of the 1st Regiment proceeded to patrol their respective areas. In the Ta Bat vicinity, the 3/2 ARVN Infantry Battalion also conducted a reconnaissance in force.[17]

The 4/1 ARVN Infantry Battalion was in the field almost four weeks, working its way gradually to Hill 1022 on the west wall of the A Shau Valley about 1,500 meters south of LZ TIGER, where it was extracted on 5 June. On 14 May, there was a brief firefight with an enemy unit of unknown size.[18] Numerous caches were found, including one with 320 mortar rounds and 500 cases of rifle ammunition on 22 May (YD266030).

On 12 May along the Laotian border, the 3/1 ARVN discovered a cache containing 8 individual and 12 crew-served weapons, and on 13 May a larger haul of 8 vehicles, 25 crew-served, and 78 individual weapons. It would continue to discover enemy caches and graves in this area, which was right where the Marines had crossed back into Vietnam from Laos in early March, for the next several days. On 22 May, the 3rd Battalion captured 5 trucks and 3,000 pounds of rice (YD227082). On 4 and 5 June, two more enemy caches were taken (YD233113 and YD248117) north of Tiger Mountain, including 4 damaged trucks, four 12.7mm AA guns, and ten 122mm rockets. On 14 May, less than 1 km southeast of FSB TURNAGE (YD257081), the 2/1 ARVN bagged a cache including a bulldozer, 3 Molotava trucks, and 66 SKS rifles. On 17 May west of FSB TURNAGE (YD232082), the same battalion found and destroyed 4 Russian trucks, 5 bulldozers, and 300 lbs of TNT. The 2/1 Battalion had 8 men WIA when the enemy fired ten 122mm rockets on 19 May. On 29 May, the ARVN FSB BRADLEY, defended by the 2/1 ARVN Infantry, was attacked by NVA sappers in company strength. The results were 13 ARVN KIA, 24 ARVN WIA, 4 US WIA, and 22 NVA KIA.[19]

The 2/3 ARVN Infantry Battalion was committed to the battle at Hill 937 pursuant to an agreement on the evening of 18 May between CG, XXIV Corps (LTG Stilwell), MG Zais, and General Truong, commander of the 1st ARVN Division. The arrival of that unit on the eastern slopes of Dong Ap

Bia on the afternoon of 19 May and its participation in the final assault on Hamburger Hill on 20 May are covered in chapter 19. After participating in the final assault on Hamburger Hill, on 21 May, the 2/3 was assigned to conduct a search and destroy operation in an area of eight square kilometers to the east and southeast of Hill 937 and east of the area that had been traversed by the 1/506th Infantry on 13–20 May. One large ammunition and weapons cache was found. On 26 May, the 2/3 was sent southwest to Hill 996, which had been occupied by the 1/506th Infantry on 11–12 May.[20] On 1 June, the battalion in this area (YC302944) was hit with 70–80 mortar rounds.[21]

APACHE SNOW – The Air Cavalry[22]

The Divisional Cavalry Reconnaissance Squadron of the 101st Airborne Division was the 2/17th Cavalry (LTC William W. Deloach). By 1 May, its conversion from ground cavalry to an air cavalry squadron was complete, with three air cavalry troops and one ground troop (D). Attached was Company L, 75th Infantry (Ranger). The unit deployed helicopter "Pink" teams consisting of one OH-6A light observation helicopter (LOH, pronounced "Loach") and one AH-1G Huey Cobra gunship, aero-rifle platoons known as "Blues," "Sniffer" human-smell detection helicopters, and six-man Ranger teams, in constant and wide-ranging reconnaissance operations through the AO. The Blues also served as the Ready Reaction Force which could be inserted to rescue downed helicopter crews, to reinforce ranger teams in contact, or to exploit enemy camps and caches that were not readily reachable by the airmobile infantry companies. It is incredible that in spite of the vulnerability of the Loaches and the small size of the blue platoons, only two men from the 2/17th were KIA during APACHE SNOW. Some of the air cavalry engagements are listed here (there were also innumerable sightings and minor engagements involving enemy huts, bunkers, and small groups of NVA soldiers):

10 May – Pink teams scouted all of the landing zones used on the morning of D-Day, remaining on station until the infantry (U.S. and ARVN) was inserted and providing gunship support at the red (ARVN) LZs.

10 May (YC319142) – Less than 1 km southeast of the 1/506 LZ, destroyed a machine gun or light AAA weapon at a cave entrance

10 May (YC3171000) – In the Warehouse 54 area, engaged and damaged an enemy bulldozer (no indication of whether the bulldozer returned fire)

11 May (vicinity of LZ PEPPER [YD337026]) – The Blue platoons of A and B Troops were inserted at abandoned LZ PEPPER, atop Dong So Ridge, where they conducted RIF until the following day.

11 May (YD200030) – Inside Laos and south of the 3/1 ARVN LZ, a C Troop gunship destroyed one truck and one bulldozer.

11 May (YD275080) – About 2 km southeast of Tiger Mountain on the west wall of the valley, "spotted hill [Hill 1062] honeycombed with spider holes, tunnels & bunker complexes"

12 May (YD354024) – The Blues destroyed an NVA 12.7mm heavy machine gun.

12 May (YD308084) – Just off Route 548 near Warehouse 54. Found and destroyed 2 bulldozers and 1 truck.

All three air cavalry troops were committed to supporting the 3rd Brigade during the early days of the operation, but that commitment tapered off as the month of May wore on, and no ranger team patrols in the A Shau were identified.

Hamburger Hill (Dong Ap Bia/Hill 937) – 11–20 May 1969[23]

The battle for Dong Ap Bia, "The Mountain of the Crouching Beast" in Vietnamese mythology, is reported in detail in Samuel Zaffiri's book and also in articles by Charles Willbanks, et al.[24] Frank Boccia, a platoon leader in the 3/187th Rakkasans, has also written an outstanding first-person narrative of his experience at Hamburger Hill.[25] The battalions committed included the 3/187 Infantry (11–20 May), 1/506 Infantry (12–20 May), 2/501 Infantry (19–20 May), and 2/3 ARVN Infantry (19–20 May). While the NVA bastion on the Dong Ap Bia massif, including hills 937, 916, 900, and 800, occupied an area over 2 km from north to south and 3 km east to west, the heaviest fighting took place within just a few hundred meters of the peak of Hill 937 (see Map 14).

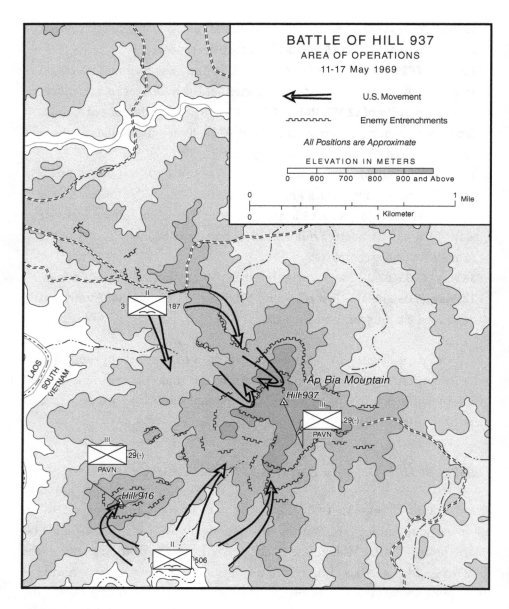

MAP 14

Source: Traas, Adrian G. *Transition, November 1968 – December 1969* (Washington, DC: Center of Military History, United States Army 2018) p. 32.

B/3/187th Infantry – 11 May

On 11 May, B/3/187 Infantry (CPT Charles Littnan) renewed its move up the ridgeline. It was assumed that the action the evening before involved only enemy trail watchers, and LTC Honeycutt expected to be atop Hill 937 by the afternoon. As the point element moved with extreme caution up the trail and through the area of the previous afternoon's battle, they found four enemy bodies (killed by the air strike just before dark on 10 May), weapons, equipment, and many blood-stained bandages. In the pocket of one of the dead soldiers were some military documents. These were sent with a courier back down the trail to the battalion headquarters, where before the morning was over, they had been translated, revealing the presence on the mountain of the 29th NVA Regiment, which was spoiling for a fight. The translator, a Hoi Chanh (rallier from the enemy side), was terrified—a sensible response, as it turned out.

After working through the battleground, B Company made relatively rapid progress following enemy blood trails in the general direction of the mountaintop, and by 1230 hours was only 800 meters from the objective. At a little after 1600 hours, the point platoon encountered a bunker complex and was stopped short. The fire was much heavier this time around, including a .51-cal machine gun and at least four command-detonated claymore mines. Before contact was broken, there were 3 American KIA and 19 WIA. A first attempt to rescue the wounded failed, and more men were hit by enemy fire. Finally, after the battalion commander instructed that the company use maximum fire with maneuver to establish a perimeter in front of the wounded, this was accomplished. SP4 John E. McCarrell was posthumously awarded the Silver Star for the action on this day, when after braving heavy enemy fire to save one wounded comrade, he was killed when attempting to retrieve a second injured soldier from an exposed position.[26]

Helicopter gunships and air strikes were called in to hammer the entrenched enemy as Bravo Company pulled back with its casualties and called for medevac. The Bravo Company LZ a few hundred meters back down the trail was too small for a Huey, even after an engineer team worked to enlarge it, so a LOH shuttled the wounded paratroops down to the battalion base, where they were transferred to UH-1 medevac helicopters. That same LOH, piloted by "Crazy Rairdon," would continue to perform this lifesaving trip many times in the next 10 days.

In the meantime, while the medical helicopters were conducting their mission, Huey Cobra gunships were called to attack the NVA position. Six rockets were fired, hitting instead the 3/187 battalion headquarters location, resulting in 2 killed and 35 wounded (15 requiring medevac), including the battalion commander, operations, intelligence, and artillery liaison officers. This was just the first of at least four friendly fire incidents that would plague the Rakkasans during the battle for Hill 937. Medevac for both B Company and the friendly air strike casualties was completed without enemy fire by 1842 hours.[27] LTC Honeycutt, now suspecting that there was at least an NVA company on top of Hill 937, instructed the commanders of A and C companies, then operating to the west near the Laotian border, to begin moving in the direction of the mountain. He also planned to move D Company east and then south, so that by 13 May he could attack in a broad arc from the north, northwest, and west.[28]

3/187th Infantry – 12 May

On 12 May, all companies of the 3/187th had reoriented to move to the east and southeast. After a rainy night (it rained most afternoons and/or evenings while the battle was taking place, turning the hillside into a slippery morass late each day), A Company (CPT Gerald Harkins) moved out early, reaching the battalion headquarters by 0800, to take over security from D Company. An engineer team was requested to blow a landing zone near the B company location, but when a helicopter with that team approached around 1000, it was shot down. Three engineers and three helicopter crewmen were wounded in the crash.

B Company maneuvered against the enemy bunkers encountered on 11 May. A very cautious approach to the site of the ambush paid off, as only one man was wounded when the enemy hit the advancing troops with the same heavy fire as before, now accompanied by mortars. Bravo Company pulled back and allowed air and artillery to try again to silence the enemy guns. At around 1230 hours, enemy mortar fire fell once more on Bravo Company's position, but no one was hit. Gunships and counter–mortar fire were directed against the suspected enemy location. During the day, a total of eight air strikes would augment gunship and artillery fire against the enemy on Dong Ap Bia. Nine wounded, including the six from the downed helicopter, were medevaced during the afternoon. In the early afternoon, B Company approached the bunkers one more time, expecting to find that the heavy ordnance that had been pounding the area for several hours would have killed or driven off the NVA, now estimated at two companies.[29] They were still there. The careful RIF caused the NVA to reveal their positions without resulting in any Screaming Eagle casualties.

C Company (CPT Dean Johnson) moved south and east, approaching Hill 937 from the southwest. During the day, the company received frequent sniper fire. Charlie Company was ambushed at 1820, with eight WIA. After pounding the enemy with mortar, artillery, and gunship fire, the company pushed through as the enemy fled. Just 75 meters farther on, another barrage of RPG fire stopped the column and CPT Johnson pulled back a short way to establish his NDP (night defensive position).† D company (CPT Luther Sanders), moving east, reported no enemy contact on this day but encountered a deep ravine that almost brought movement to a standstill.[30]

---

† Although the command historian's narrative on the battle shows C company approaching to the north (left) of B Company, the grid locations and Zaffiri both indicate that C Company was to the south (right) of B Company by the late afternoon of May 12. Also, while I have endeavored to reconcile casualty data from various sources, that is not always possible. On 12 May, for example, the 3/187 AAR casualty summary (Para 12.b.) reports a total of 1 KIA and 2 WIA for the day, but Appendix 3 of that source states that 9 WIA were medevaced from the B Company LZ (this would include 6 helicopter crash victims), and Zaffiri reports 8 WIA in Charlie Company at 1820 hours. If there was a KIA, he is unidentified, and not listed at the Virtual Wall.

3/187th Infantry – 13 May

On 13 May, B and C Companies moved out before dawn and were soon in contact once again. Air support was used throughout the day, including a total of 10 airstrikes and multiple teams of helicopter gunships. Medevac support was also critical. D Company, now moving northeast toward the hilltop, also ran into resistance and had a number of men wounded. Around 1500 hours, a medevac helicopter from the 326th Medical Battalion arrived and was in the process of extracting the wounded by basket when it was hit with an RPG and crashed straight down. Three crewmen and two previously wounded passengers from D company were killed.

> Throughout the afternoon of the 13th, B and C Companies had continued maneuvering toward Hill 937, attempting to gain a favorable position for a major attack on the 14th. It was not an easy task. Both units maintained contact throughout the afternoon. The company's contact with the enemy ceased at 1730 hours. C Company received mortars and small arms fire as late as 1935 hours . . . the enemy broke contact at 1941 hours.[31]

For the day, the 3/187th had 4 KIA and 33 WIA. During the afternoon, shrapnel from a 500-lb. bomb dropped by an F-105 killed PFC Myles D. Westman of Bravo Company and wounded one other trooper.[32] Westman's official cause of death was listed as "enemy fire"; this incident was not reported in the AAR, and there were no recorded "friendly fire" losses for the 3/187th on this date. But it was what it was.

At 1220 hours, the brigade commander alerted the 1/506 Infantry, operating several kilometers to the south, to "plan for the immediate movement of 1-506 units N and NW to reinforce the 3-187 Abn Inf. . . ."[33] The battalion immediately began to move toward Hill 937.

3/187th Infantry – 14 May

On the night of 13–14 May, the three companies of 1/506 Infantry were still over 2.5 km south of Hill 937, and although they would be on the move early, there was no way that they could be expected to join the attack on the 14th. In fact, movement was not rapid, and when they went into their NDPs that night, they were still 2,000 to 2,500 meters from the top of Dong Ap Bia. B Company's point man was killed by a sniper.[34]

LTC Honeycutt decided to launch the first full-blown assault against the enemy on 14 May.

> Although the colonel was not positive Captain Sanders could get Delta in position fast enough to have an impact on the fight, he simply could not risk postponing the attack for fear the NVA might use the respite to further strengthen their position on the mountain. Honeycutt knew for a fact that the NVA were using the large draw on the southwest side Hill 900 to shuttle fresh troops and supplies up to the mountain at night. He had requested that 1/506th attempt to cut this draw, but so far the Currahees had not advanced close enough to the mountain to do so. . . . For now the 3/187th was going to have a go at the mountain alone. It was not a situation to relish, but he had no other choice."[35]

Well, of course he did have a choice. He could have continued to pound the enemy with air strikes and indirect fire while awaiting reinforcements to cut the enemy line of communications with Laos and bring pressure from 360 degrees instead of just a two-column front of 150 meters. Or he could launch a frontal assault against a heavily entrenched force that likely outnumbered his own, while the battalion had no reserve company that could be called upon to reinforce the assault (A Company was the "palace guard" for battalion headquarters, and would later provide one platoon to assist with casualty evacuation). As early as 12 May, the enemy force was estimated at two companies.[36] It was two battalions, with regimental heavy weapons and support troops in Laos, two kilometers to the west.

Bravo and Charlie Companies, only 150 meters apart on separate ridge fingers coming from the summit of Hill 937, supported by heavy (13) air strikes and the usual mortar, artillery, and gunship support, kicked off the assault around 0800. Resistance was staunch and included the mass employment of command-detonated claymore mines placed in trees, but both companies overran the areas where they had been stymied before, knocking out multiple NVA bunkers and fighting positions. By 0900, C Company reported that one platoon was almost at the top of the hill, and at 0930, B Company's lead platoon was also atop the ridge. Both units were under heavy fire. CPT Johnson reported that the enemy was surrounding his unit, and LTC Honeycutt gave permission for him to pull back. "Charlie Company was, in fact, slowly coming apart at the seams. . . ."[37] As this would expose CPT Littnan's right flank, he was also authorized to withdraw from the peak. The NVA had one thing the paratroops did not have: reserves, and they counterattacked Charlie Company from the front and flanks. The rear echelon of Bravo Company, at the small LOH LZ still being used for medevac, also came under ground attack from a small enemy force.

At noon, LTC Honeycutt reported the opposition at two companies to a full battalion. Appendix 5 (14 May) to the 3/187th AAR: "At 1216 hours, B Company commander reported that the top of the ridge was covered with blood, pieces of bodies, and enemy dead, but that he would have to wait to give an accurate figure on enemy KIAs." No further information from any company on enemy losses is recorded for 14 May. The historian's narrative reports that a "rapid count on the battlefield resulted in [sic] forty-seven NVA killed by small arms fire and twenty-nine killed by air strikes." The "Enemy Losses" section of the 3/187th AAR records 60 "KIA by 3/187th small arms only."[38]

Delta Company never came close to joining the attack on the peak. The official narrative for 14 May on APACHE SNOW by the command historian reports: "B, C, and D Companies . . . began a deliberate fire and maneuver on the Dong Ap Bia complex," and that "B, C, and D Companies were in the attack echelon."[39] As far as Delta Company is concerned, this is poppycock. Most of the day was spent in the vicinity of the deep ravine where the medevac Huey had gone down, trying to recover the bodies.[40] They were constantly harassed by enemy snipers and small ambushes; the NVA were *all over* Dong Ap Bia. Finally, at around 1400, when the company had reached the crash site, an NVA platoon was already there and waiting, and opened up with RPGs and machine guns.

Ten more men were wounded, and the enemy had to be suppressed with a heavy dose of gunships and artillery. It was late afternoon before the company was reassembled and out of the ravine. At no point during the day was Delta anywhere close to maneuvering up, let alone attacking, the north side of the mountain.

By evening, the 3/187th Infantry this day had 12 KIA and over 80 wounded;[41] all but one of the KIA were in Charlie Company. B Company reported foxhole strength of 86, and C Company had just 70 men.[42] This was barely over 50% of the 134 and 135 they had at the start of the operation. COL Conmy flew in for a chat with LTC Honeycutt just before dark. Honeycutt expressed his concern about the enemy ability to reinforce: "They're movin' fresh troops up those draws from Laos every night." He also complained about the slow approach of his sister battalion: "The 506th has got to get their asses in gear and get involved in this fight."[43] The enemy could also reinforce from other parts of the mountain, as there were no allied troops anywhere on the eastern half of Dong Ap Bia.

The division commander, MG Melvin Zais, was also getting worried about what was going on in this remote corner of the A Shau Valley. He had never before seen the enemy stand up to such a concentration of firepower for such an extended period of time; Dak To came to mind as the most recent similar fight. He spoke with his assistant division commander, BG Jim Smith, whose long experience in Vietnam also failed to provide an explanation for the NVA's stubborn refusal to give up the fight. Zais agreed, in any case, with Smith's recommendation that they continue until the battle could be brought "to a successful conclusion."[44] That night, CPT Sanders of Delta Company observed at least a hundred enemy campfires on the slopes of Dong Ap Bia. The "successful conclusion" was not going to come easy.

In spite of the concern of LTC Honeycutt, the 1/506th Infantry continued to move slowly, with its closest companies still 1,500 meters away by the evening of the 15th. The battalion had only two WIA for the day.[45] The 2/501st Infantry, searching the area immediately to the north, had of course been withdrawn to the FB AIRBORNE area after the attack there on the morning of the 13th and was unavailable.

3/187th Infantry – 15 May

The plan for the morning of 15 May was for A Company to switch places with the broken C Company. D Company would attack from the north, B Company would attempt to move back up to the top of the ridge along the same line as the previous morning, and A Company would be on the right flank a bit farther to the south than C Company on the 14th.[46] It would be another 10-air-strike day. Once again, however, D Company was still occupied with the downed helicopter site and would not join in the attacks on this day,[47] and once again the 1/506th could manage only slow progress against light opposition.

During the night of 14–15 May, enemy troops were detected moving down a draw between B and C Companies. It was anticipated that these troops would attempt to attack the American companies from the rear when they launched their assault positions the following morning. When morning came, Alpha and Bravo Companies slowly began to advance toward the top of the mountain. Before reaching their assault positions, on order, both companies reversed direction and faced the draw where the enemy was hiding. They waited silently. The NVA infantry launched their attack into a hail of rifle and machine gun fire. Fighter-bombers, artillery, and gunships finished the job, and an entire company of North Vietnamese troops was wiped out.[48]

After the defeat of the enemy enveloping force, A and B Companies moved up the mountain against relatively light opposition. By 1400, each company was within 150 meters of the top and had no more than a dozen WIA. Then, disaster struck again in the form of another "friendly fire" attack by aerial rocket artillery gunships. The strike hit the command group of Bravo Company, with 2 KIA and 14 WIA. Company Commander Littnan, his first sergeant, artillery forward observer, and RTOs were all wounded. It would be about two hours before a replacement commander, CPT Butch Chappel, could reach the company. The enemy continued to demonstrate its omnipresence and depth, as sensing the chaos following the elimination of its command group, NVA infiltrators began to harass the company from the draw on its right flank and from the rear. Two squads attacked the LZ where the wounded were awaiting medevac;

they were turned back by an alert machine gun crew. Enemy RPGs also hit the battalion headquarters area at 1600, wounding 13.

One abiding myth about the battle for Hamburger Hill is that the battleship *New Jersey* fired its 16-inch guns in support of the Rakkasans. This myth may have originated with a quote in Samuel Zaffiri's book from an interview with Frank Boccia![49] In Boccia's own book, published years later, he clarified, correctly, that "the *New Jersey* left Vietnamese waters for Korea in late April, weeks before Dong Ap Bia."[50] In any case, the *New Jersey's* big guns had a range of 39 km, not nearly enough to reach any part of the A Shau Valley.

At 1500, the A Company commander reported that he was just 75 meters from the crest, but B Company was out of contact due to the loss of the company radios. After consulting with COL Conmy, LTC Honeycutt ordered A Company to return to the previous night's laager, not wishing to leave a lone, weakened unit to hold the hill. When CPT Chappel joined B Company soon thereafter, he expressed the need to "sort out what he had" and retreated to form a joint NDP with A Company, but not until one final joint abortive attempt to assault up the hill was made under orders from the battalion commander.

LTC Honeycutt avowed that the helicopter gunship strike on Bravo Company was responsible for the failure to take the hill on 15 May. A Company lost 1 KIA and 15 WIA, and still had 100 men at day's end. B Company lost 2 KIA and 19 WIA, all but 5 WIA due to the "friendly fire" incident; it had only 65 men when the sun went down.[51] Placing all of the blame on "friendly fire" conveniently avoids asking the question, why was such a risky assault against such a determined foe made with absolutely no provision for a reserve? Even a single platoon, sent forward at the right moment, could have replaced all of the casualties in both assault companies and maintained the attack momentum, if led forward by an infantry officer of stature to assume command of B Company, perhaps from the battalion staff, or even the C Company commander.

It may also be an open question whether the Screaming Eagles could have held the top of Dong Ap Bia. There were still large numbers of unengaged enemy soldiers on the mountain; NVA reinforcements could easily be brought in from Laos, and they knew the bunker and tunnel system intimately, so occupying the enemy's former bastion in late after-noon might have placed the Rakkasans in an untenable position. Around midnight, an NVA sapper attack on the battalion headquarters area was thwarted by alert troops of C Company; at dawn, 14 enemy bodies were found, with many more likely dragged away.

3/187th Infantry and 1/506th Infantry – 16 May

The 3rd Brigade plan for 16 May called for a coordinated attack by the 3/187 and the 1/506. The Rakkasans were to be the anvil, advancing only slightly and holding the enemy in place, while the Currahees would be the hammer, attacking from the south.‡ It didn't happen that way, as the two lead companies of 1/506 Infantry were constantly harassed and ambushed by squad- to platoon-size enemy forces.[52] After three days of slow advances against minimal opposition, but across incredibly tough terrain, the 1/506 at last reached the southern end of the Dong Ap Bia massif and promptly ran into a much more numerous and tenacious enemy.

At 0830, A Company (CPT Linwood Burney), 1/506th Infantry was ambushed approaching Hill 800, about 1,500 meters due south of Hill 937 on the same ridge. Contact with four or five enemy bunkers quickly resulted in 1 KIA and 7 WIA, and the company pulled back to bring in air and artillery. After the softening up, A Company swept forward, only to see the enemy bug out just as the second attack was launched. Stranger still, when the paratroops were consolidating their position, a platoon of NVA came out of the jungle and launched a frontal assault. They were stopped by the fire of the Currahees, and then driven off with help from the not-always-friendly Huey Cobras. Only one NVA KIA and one AK-47 rifle were left behind. The survivors fled over the top of Hill 800. There were no American casualties, but a severe thunderstorm hit, and as sometimes happened to infantry units on hilltops, lightning struck, severely injuring two men. A Company would advance no farther this day.

---

‡ The 1/506 had only three companies in the field, as D/1/506 was on the perimeter at FSB CURRAHEE.

B Company left its NDP atop Hill 916, went only 150 meters toward Hill 900, and was hit from front and rear at 1045. It, too, lost 1 KIA and 8 WIA before the firefight ended two hours later. Heavy sniper fire prevented normal medevac, and a LOH was able to complete that mission. When the company tried to resume forward movement late in the day, it was stopped again by well-bunkered NVA soldiers. By dusk, all elements of the 1/506 Infantry were still at least 1,000 meters from the top of Hamburger Hill.[53]

In the 3/187 sector, the companies were reshuffled, with B Company joining C Company at the battalion HQs, A Company remaining in place, and D Company moving to B Company's old ridge finger. LTC Honeycutt had finally abandoned the idea of having D Company move to the north side of Hill 937, as the huge ravine had made untenable any link between the rest of the battalion and a unit sent to the east. After Alpha and Delta Companies occupied assault positions during the morning, the brigade commander called off the attack when 1/506 stalled to the south.[54] The 3/187 Infantry made no contact with the enemy, and all elements pulled back several hundred meters and established strong defense perimeters. Air strikes and indirect fire continued to be directed at Hamburger Hill, including delayed-fuse 1,000-lb. bombs to destroy the deep enemy bunkers.

Hamburger Hill – The Controversy

Before concluding the narrative on the battle of Dong Ap Bia, a pause is necessary to consider the controversy that arose over the high U.S. casualties and the tactics that begat them, when the press became aware of the battle and "Hamburger Hill" became a media and political firestorm.

On 16 May, an Associated Press reporter named Jay Sharbutt[§] flew into the LZ on Hill 937 and spoke with a number of Rakkasans who had taken part in the battles over the previous several days. He then flew on to FSB BERCHTESGADEN, where he interviewed the division commander, MG Melvin Zais. Sharbutt's first question was: "Why don't you just pull back and hit [Hill 937] with B52 strikes?"

---

§ Sharbutt died of a heart attack in 1992, aged 51.

The general's response was, in his own eyes, diplomatic, explaining that in order to use B-52s, the troops would have to be pulled back at least two miles, that tactical air strikes were more accurate and at least as effective, and that pulling back was a bad tactic because the same ground would have to be retaken later on. However, behind the scenes, Zais was "stunned" by Sharbutt's question, because the answer was so obvious and there was, in his eyes, no flaw in the tactical management of the battle. The general prided himself on his tact, having suppressed his first reaction "to tell Sharbutt 'to confine himself to writing about the battle rather than second-guessing the commander.'"[55] American generals have been subject to "second-guessing" by the press and the public since at least the times of Ulysses Grant and Douglas MacArthur, and ought never to believe that their actions are beyond question, let alone reproach. General Zais's remarks about not giving up "ground you've gained" also ring hollow, as the assault companies of the 3/187 Infantry pulled back at dusk every single day during the first week of the attacks, abandoning at least some of the ground they had just sacrificed so much to conquer. On 16–17 May, once the brigade commander decided to use the 3/187th as a blocking force, that battalion withdrew several hundred meters to await the arrival of the 1/506th Infantry to join the battle for the top of Hill 937.

Sharbutt's first article appeared in the 18 May issue of the *New York Times* ("Mountain Battle Tough, Bloody for GIs"), was followed by two more and was widely distributed under the Associated Press byline. Zais's evasiveness was based, according to a later confidential memorandum, on his concern that Sharbutt was trying to create controversy. This became a self-fulfilling prophecy, as the media and the U.S. Senate nurtured a debate that would not die down for weeks. What began as a simple, legitimate query arising from Jay Sharbutt's trip to the battle scene and effort to understand why the battle was being fought as it was soon became a national furor over the American role in Vietnam in 1969.

The 101st Airborne Division officers in the chain of command for APACHE SNOW, Zais, Conmy, and Honeycutt, believed that Sharbutt's stories were sensationalized and inaccurate. In fact, other than a liberal use of journalistic prose, that was not the case.

The reports that followed described the devastation on Dong Ap Bia. The mountain was "almost bare," according to AP reporter Jay Sharbutt, "its heavy jungle cover blasted apart by artillery, rockets, bombs, and napalm," and the anger and frustration of the men fighting to gain the top were unmistakable. "After all these air and artillery strikes," one soldier told Sharbutt, "those gooks are still in there fighting. All of us are wondering why [U.S. forces] . . . can't just pull back and B-52 that hill." Another soldier, badly wounded, told CBS News correspondent Richard Threlkeld that the hill was "absolute suicide." As fighting continued, the term Hamburger Hill seemed to appear out of nowhere, the product of some soldier's cynicism or of a reporter's morbid wit.[56]

Indeed, the origin of the "Hamburger Hill" moniker is uncertain. While William Head says Sharbutt "coined the term," the source of that conclusion is not identified.[57] The Joint Chiefs of Staff history implies that the name came from the men who fought there: "During the operation, troops had complained of the difficulty of the battle, nicknaming the mountain "Hamburger Hill" because it "chewed men up like meat."[58] When Senator Edward Kennedy took up the cause on 20 May, even as the infantry battle on Hill 937 was in its final hours, the political battle was just being joined: "It was, he said, 'senseless and irresponsible to continue to send our young men to their deaths to capture hills and positions that have no relation to ending the conflict.' American lives were being wasted, he added, merely to preserve military pride.'"[59] And yet, even as the question about whether the infantry attacks on Dong Ap Bia were "senseless and irresponsible" moved the public debate to a higher plane, it also had the effect of sidetracking any serious consideration of whether or not LTC Honeycutt's repeated frontal assaults were, if unsuccessful, nonetheless appropriate applications of American tactical doctrine.

The question, then and later, about the actions of LTC Honeycutt and his battalion at Hamburger Hill boils down to: "Should American infantry have engaged in a ground attack on the mountain?" But here the discussion becomes muddied, because the debate really has both *military* (tactical and operational) and *political* connotations. Noted Vietnam War historian Harry G. Summers, Jr., writing on the 30th anniversary of the battle, addressed the political aspect by quoting the famous military theoretician, Carl von Clausewitz:

> "Once the expenditure of effort exceeds the value of the political object, the object must be renounced." And that's exactly what happened. The expenditure of effort at Hamburger Hill exceeded the value the American people attached to the war in Vietnam. The public had turned against the war a year and a half earlier, and it was their intense reaction to the cost of that battle in American lives, inflamed by sensationalist media reporting, that forced the Nixon administration to order the end of major ground operations.[60]

LTC Honeycutt and his immediate superiors could not be held responsible for the *political* implications of the battle, as that was well above their pay grades, though General Abrams could.

The 101st Airborne Division commanders in 1969 and most subsequent military analysts justify the assault(s) with the rationale, "That is where the enemy was and we were in the A Shau Valley to kill the enemy." But that is a very simplistic and evasive answer to a military decision-making process involving both operational and tactical questions.

Operationally, the command level at which the designation of the destruction of the 29th NVA Regiment as the objective was first adopted or assumed is unclear, although brigade (Conmy), division (Zais), corps (Stillwell), and MACV (Abrams) all concurred. There were numerous instances in the past where U.S. infantry, facing a strongly defended enemy position in the A Shau Valley, simply turned and walked away, leaving it to air or artillery to achieve maximum destruction at minimum human cost.** This would occur again on 24 May in the immediate aftermath of Hamburger Hill, when the 1/506th Infantry turned its back on the enemy fortifications on Hill 916. The "flight, not fight" decisions were probably

---

** E.g., the 5/7 Cavalry west of Tiger Mountain and the 1/7 Cavalry in the box canyon during Operation DELAWARE, or the 1/327 Infantry on Dong So Ridge during SOMERSET PLAIN.

driven primarily by the necessity of adhering to tight operational timetables, where weather played a role in the unwillingness to become deeply committed to possibly prolonged engagements, but the question remains as to why Dong Ap Bia was selected for a battle royal, when there was no evidence that the mountaintop was occupied by anything other than combat troops, while other enemy defensive positions that may have shielded large caches or heavy weapons were left alone.

What the field commanders, Zais, Conmy, and Honeycutt, *were* responsible for was the direction of the battle. Hence, the *tactical* question should be "Were the tactics appropriate to the objective, which was the destruction of the NVA 29th Regiment?" The essence of the entire "Hamburger Hill" controversy *at the field command level* basically boils down to these points:

1) **Were the attacks justified by the possibility that the 3/187th Infantry could, without reinforcement or supporting attacks, have destroyed the enemy on Dong Ap Bia?**

2) **If a single-handed victory was not probable, were those attacks necessary and effective to "hold" the enemy in place until a brigade-strength assault could be planned?**

The daily plans of LTC Honeycutt, the man on the spot, were routinely approved by his superiors. Military plans, as the old adage goes, rarely survive first contact with the enemy. The 3/187, during the first few days of the battle (11–13 May) as the enemy's strength was being repeatedly reassessed, persisted in frontal assaults against an enemy bastion that was gradually revealed to consist of heavily fortified bunkers, spider holes, tunnels, and command-detonated claymores. The position was massive in size and well manned, with sufficient reserves to send out company and smaller elements to attack the 3/187 from the flanks and rear. The attacks by A, B, and C Companies, essentially reconnaissance in force, were executed with the expectation that the enemy would likely soon hightail it for the border, as was its usual practice when pummeled by air and artillery. The probes on 11–13 May were thus necessary and appropriate to fix the enemy's strength and deployment, and the increase in casualties on the latter date (4 KIA and 33 WIA) was a logical culmination of the recon process.

The 3/187th Infantry blocking force role on 16–17 May (there were no casualties on either day) was ordered by brigade, and the final multibattalion push on 20 May was well planned. But the two-company frontal assaults of 14–15 May have more in common with the tactics of Burnside at Fredericksburg or the Huertgen Forest campaign in World War II than with modern fire and maneuver tactics. Those assaults were planned to be part of a pincer attack, with D/3/187th Infantry on the left and the 1/506th Infantry on the right, but were carried on in spite of the enemy's delaying actions having forestalled those supporting maneuvers, a situation that was known to the 3/187th commander in advance of commitment of the assault forces. In actuality, on May 14–15, 12 men were KIA, including "friendly-fire" losses, 139 were wounded, and 114 required evacuation. On 18 May, when there was still no effective support available from the 1/506th, LTC Honeycutt's attack cost the lives of 13 of his men, while 63 others were wounded and required evacuation. There was a reserve company on the 18th, but that element had to fight its way to the front, and the overall condition by the battalion by this date was such that all three companies could do little more than extricate themselves from their advanced positions when the assault stalled. It is a valid question as to whether the battalion commander should have persisted with those attacks. If the battalion could serve as a blocking force on 16 May, awaiting the arrival of the rest of the brigade, why could that not have been done on 14 May, and why was it necessary to attack on the 18th, prior to the commitment of reinforcements? "Military Pride?" It should be remembered that throughout the struggle, the NVA demonstrated the ability to detach counterattack and spoiling attack elements to harass the 3/187th from all sides, and the absence of any allied pressure on the eastern side of the enemy fortress and of any ground interdiction of its lines of communication with Laos ensured that the NVA could control the ultimate outcome of each day's combat by employing uncommitted troops, a luxury LTC Honeycutt did not have.

The unasked question most pertinent to the direction of the battle by COL Conmy and MG Zais is why there was no reinforcement of LTC Honeycutt's battalion, until finally on 18 May a single company (A/2/506) was flown in to augment his command (at that time, he specifically asked for one, and only one company). The 3/187 suffered well over 50% losses against an enemy that received fresh troops every night, and yet the Rakassans were on their own. Tactically, in the attack, reserves are committed to reinforce success, not mitigate failure. If Honeycutt's attacks were succeeding, why weren't his attacks supported by Brigade? If he was failing, why weren't the attacks stopped?

In summary, the frontal assaults of the 3/187 from 14 May on went forward each day as planned the night before and were pursued in spite of the failure of the other parts of each day's plan to fall into place. Why? General Zais's rationale, as explained by Harry Summer,[61] was that offense is the very best defense,†† but the enemy could have been held in place (if it *wanted* to remain) as well by using 3/187 as a blocking force as by crawling up the mountain, until the forces were at hand to isolate the enemy and launch a truly coordinated multibattalion attack. An offense by an outnumbered unit against a defense that could reinforce at will was not a "best defense."‡‡

There is also a question of why the 1/506 Infantry was so slow to reach the battle area. That battalion was supposed to cross Hill 916 to the southwest and simultaneously come up along the Hill 800–Hill 900 ridge due south of Hill 937, cut the NVA line of communications running southwest of Dong Ap Bia, and then join the attack on the mountain. While it may have been a bit slow getting off the mark on 13–14 May, its movement thereafter was seriously impeded by the terrain and the enemy. In a 1990 interview with *Vietnam* magazine, COL Conmy (who received a

---

†† The North Vietnamese commanders who led the attack on the rear of the 3/187 assault force on 15 May and the sapper attack on the battalion headquarters the night before might question this.

‡‡ See the appendix for further discussion on the application of the Principles of War at Hamburger Hill.

Silver Star for action in the A Shau Valley), spoke about the 1/506th (LTC James Bowers):

> The terrain that Bowers's men had to face on the south side of Hill 937, in my opinion, was even more difficult than what Honeycutt had to contend with. . . . there is some condemnation of the 1/506. I don't agree with it. As the brigade commander, I felt they did the very best they could. And they were in the final assault to take the hill. They had very rough terrain to traverse in reaching the top of the mountain. [The terrain was] triple-canopy jungle, very deep ravines, knifelike ridges, and so heavily vegetated that you could blunder into an enemy bunker and never see it until you were almost on top of it.[62]

Its fourth company (D) might have been relieved by Brigade from securing FB CURRAHEE to provide additional punch (or to reinforce 3/187). While we may accept the brigade commander's explanations for the tardy arrival of the 1/506th Infantry at the slopes of Hill 937, LTC Honeycutt's belief that the movement of that battalion lacked the urgency demanded by the intensity of the battle likely carries some merit.

In the aftermath of the publication of Jay Sharbutt's articles, the story acquired a life of its own. "The paratroopers came down the mountain, their green shirts darkened with sweat, their weapons gone, their bandages stained brown and red – with mud and blood. . . . 'That damned Blackjack [LTC Honeycutt's radio call sign] won't stop until he kills every one of us,' said one of the 40 to 50 101st Airborne troopers who was wounded.'"[63] Other senators took up the cry raised by Ted Kennedy. What inadvertently, but certainly, compounded the public outrage, was a *Life Magazine* article on 27 June, displaying photographs of the 241 American KIAs the previous week. Only 5 of those pictured had died in the A Shau, but the photos were accompanied by a quote from a final letter written by one of the Hamburger Hill casualties: "You may not be able to read this. I am writing in a hurry. I see death coming up the hill." Readers quickly concluded that all of the dead were from the battle in the A Shau Valley. In fact, total KIA in the 3/187th Infantry were 46, and 113 Americans died during the entire operation.

Samuel Zaffiri has written an impassioned account of the course of the battle, especially the courage and fortitude shown by the young Screaming Eagle infantrymen at Hamburger Hill. He supports the contentions of those in the chain of command, at the time and later (LTC Honeycutt – COL Conmy – MG Zais – LTG Stillwell – GEN Abrams), that the continued frontal assaults by the 3/187th Infantry were the appropriate tactic for engaging the 29th NVA Regiment at Dong Ap Bia. The subsequent writings of military historians and analysts generally concur with this position. The United States Army certainly endorsed the decisions of Honeycutt and Zais, as they went on to attain 2-star and 4-star rank, respectively. LTC Honeycutt also received the Distinguished Service Cross, America's second-highest military decoration, for his actions from 10–20 May 1969.

The most notable exception to the almost universal endorsement of the U.S. command decisions is a little-known 1992 work by James W. McCoy. McCoy, a "technical writer and specialist in the field of maneuver warfare," according to his book jacket, wrote what may well be the most scathing overall critique of the American military during the Vietnam War. While some of his facts are wrong, particularly concerning U.S. forces employment, the information on NVA/VC tactics is unsurpassed.

One of the enemy stratagems revealed by McCoy was known as the "hill trap." He describes at length the application of that deployment at Hamburger Hill.[64] The key aspects of the enemy "trap" into which LTC Honeycutt and his superiors fell with breathless willingness included the following:

1) **"Trails and other avenues of approach up the mountains were defended in depth with multiple bunkered trench lines and sniper positions."**

2) **The use of stream beds and draws "as avenues of approach to seek rear or flanking attack positions" (the hill trap maneuver, a repeated part of the NVA tactics on Hamburger Hill), and also for resupply, reinforcement, and casualty evacuation nightly from Laos.**

3) While enemy defense lines were incredibly well prepared, they typically did not extend entirely around the mountain. "Large [allied] units, deployed with widely probing columns, could have got[ten] behind the enemy lines."

4) By attacking up the west slope of the mountain, LTC Honeycutt maximized his men's exposure to enemy artillery in Laos. An attack from the eastern side would have provided "reverse slope" protection, minimizing the effectiveness of NVA indirect fire.

McCoy commented:

American leadership during the battle of Hamburger Hill was hardly competent. U.S. commanders knew what they were up against on the second day after the first American paratroop battalion arrived on the scene. The Americans discovered the enemy resupply roads and trails leading west out of Laos straight up the mountain, but they didn't interdict them. They discovered, from captured documents, that they were facing an entire NVA regiment, but they didn't reinforce.[65]

There is evidence that the Brigade commander of the "Apache Snow Operation" against Hamburger Hill realized that with two or more battalions moving simultaneously up the mountain, the reds could not easily concentrate against one or the other. Carrying out maneuvers based on an understanding of that simple military fact could have thwarted the enemy's consistent locally superior troop concentrations against strung out American forces.[66]

3/187th Infantry and 1/506th Infantry – 17 May

By the night of 16–17 May, the companies of the 3/187th Infantry all occupied positions several hundred meters west of the battle sites of the previous days.

On 17 May, A and B Companies, 1/506th Infantry, were in contact with NVA troops in bunkers for the entire day. Bravo Company, on the left, advanced from its position on the northeast slope of Hill 916 through a saddle and toward a small knoll west of Hill 900. They soon ran into another strongly fortified NVA bunker system, losing one man killed and seven wounded. For the rest of the day, that is where they remained, still over 1 km southwest of the top of Dong Ap Bia.

Alpha Company began the morning atop Hill 800 and had moved less than 50 meters toward Hill 900 before running into the next fortified line. An attempt to flank the enemy was unsuccessful. The 3rd Brigade had planned to use CS riot control gas during the final push up the mountain, and gas masks had been delivered to the 3/187th and 1/506th. A Company now called for gunships to place the CS on the enemy bunkers. The wind was slack, so that when the aerial rocket artillery gunships placed about 50 rounds of riot control agent on top of the NVA, it worked with great effectiveness. Fifteen to twenty enemy soldiers came swarming up out of the ground like fire ants from a disturbed mound. Ten of the enemy were killed and the others fled, finally clearing the northern end of Hill 800. Taking advantage of the enemy retreat, A Company quickly moved forward, only to encounter yet another strong enemy position, including multiple 12.7mm heavy machine guns, in the saddle between Hills 800 and 900. Hue Cobra gunships were directed at the enemy bunkers but had little effect against the strongly constructed fortifications, and one GI was seriously wounded by shrapnel from their rockets. As the afternoon wore on, the NVA sent squad-strength flankers against A Company, and it was only after a prolonged disengagement that the Currahees were able to make their way back up Hill 800, where they would once again spend the night. They had lost five men killed in the battle on the ridge.

While the 1/506th was fighting hard but unable to advance, LTC Honeycutt and his troops were anxiously awaiting the order to head back up the hill and rejoin the battle. On 17 May, that order never came. He did receive word that positive identification of the enemy had been confirmed. He was facing one to two battalions of the 29th NVA Regiment, estimated at 600 men each. When the battle resumed on 18 May, the overmatched 3/187th Infantry at long last would begin to be joined by other allied battalions in the battle for Dong Ap Bia.

19

OPERATION APACHE
SNOW – PART II

Hamburger Hill – 3/187th Infantry and 1/506th Infantry – 18 May

The terrain over which the two American battalions had been attempting to advance was incredibly close. Heavy jungle, impenetrable bamboo thickets, razor-sharp elephant grass, and steep inclines made maneuver all but impossible in most areas. While the battle descriptions give the impression of platoon or company assaults, these were most often conducted, at least at initial contact, on a front of one man. The units moved along very narrow trails atop almost equally narrow ridgelines, in some cases true "razorbacks," until approaching a clearing of varying size that was almost always protected by heavily camouflaged fighting positions, bunkers, tree-mounted claymore mines, and snipers. There the enemy would stay, holding out in its reinforced A-frame structures against everything but a direct hit from a bomb or the heaviest artillery, renewing its screen of claymore mines each night. After several days of combat, these battle sites were covered in craters and devoid of overhead vegetation, but the ground was an indescribable jumble of uprooted and splintered trees, bodies, and used and mangled military detritus. While the nice blue arrows on military battle maps may seem to indicate that the American assault columns (for columns they were) worked in coordination, in fact there was little or nothing each attacking element could do to support its neighbor. Support by fire was impossible. About the best that could be expected was that the enemy might be dissuaded from engaging from the flank (which it often did, with small counterattacks a constant threat) if a friendly unit was nearby, and that NVA soldiers held in place on one ridge finger would be unable to move to and reinforce an adjacent battle.

On 18 May 1969, both of the U.S. battalions on the Dong Ap Bia massif would finally be engaged within about 1,000 meters of each other. Yet each fought its own war. Eight air strikes were delivered by 0830. A lengthy artillery prep followed for almost an hour, including howitzer-delivered CS gas. This time, the CS was ineffective, as it either missed the target or was dissipated in a brisk breeze. Then five companies attacked.

In the Rakkasan sector, with Delta Company on the left and Alpha Company on the right, progress was initially rapid, and by 1000 they were within 200 meters of the crest. According to Samuel Zaffiri, these two companies were under brigade orders to advance to contact. "They were not, however, to become decisively engaged until the 1/506th was ready to go for the mountain."[1] The Delta Company attack soon became snared in a vicious dogfight. The NVA threw everything in their arsenal at the Americans: claymores, small arms, light and heavy machine guns, rocket-propelled grenades, and 61mm and 82mm mortar fire. The D Company commander was severely wounded, and all of the officers in the company were hit during the day. Before long, A Company was also engaged.

LTC Honeycutt ordered Charlie Company, back at the battalion base camp with Bravo, to load up with ammunition and move up to support the attack and help evacuate casualties. On their way there, the C Company troops were themselves pinned down by enemy RPGs and mortars. As the attacking troops were running low on ammunition, a platoon of B Company, loaded with ammo and water, was sent forward to follow in the trace of C Company. Then the "friendly-fire" demon struck once more. Again, it was an ARA Huey Cobra, this time hitting the B Company resupply platoon with just a few seconds of 7.62mm mini-gun fire. One man was killed and four wounded.

The battle near the top of the mountain raged for several hours. Multiple enemy bunker lines were overrun, and Alpha Company was relatively unscathed (10 WIA). Delta Company was terribly punished, with reported 10 KIA and 23 WIA. Charlie Company made its way forward, and Captain Johnson was ordered to take command of the combined C/D company force. C company had taken even heavier losses, 11 KIA and 27 WIA.*

By 1430, with the companies near the top of Hamburger Hill engaging North Vietnamese troops on all sides and continuing to take casualties, and with no help in sight from the 1/506th Infantry, LTC Honeycutt asked for and was granted permission to withdraw. With the usual help of air strikes, gunships, and artillery, the battered remnants of the 3/187th Infantry slowly pulled back down the mountain. Sixty-three men were medically evacuated. All of this was accomplished in the midst of a rainstorm of biblical proportions that struck in midafternoon, sometimes reducing visibility to 20 meters.

At 1645 hours, Company A, 2/506th Infantry arrived at the battalion LZ. The rest of that battalion would move to FB AIRBORNE, relieving the 2/501 Infantry to join the fight at Dong Ap Bia the next day.

At 2332 hours that night, the 3/187th Infantry companies reported these headcounts: A – 83, B – 78, C – 76, and D – 57.[2] These numbers include an undetermined number of green troops; the 3rd Brigade AAR reports that the battalion received 9 officers and 118 enlisted replacements during the course of APACHE SNOW. Casualties now approximated 50% in A and B companies and 80% in C and D companies.

A medic serving with Bravo Company, a conscientious objector named Nikko Schoch, received our nation's second highest award for valor, the Distinguished Service Cross, for his heroism with Bravo Company during the battle for Hamburger Hill on this and the preceding days:

Specialist Four Schoch distinguished himself by exceptionally valorous actions during the period 10 to 18 May 1969 while serving as a medical aidman during a mission to capture the enemy citadel of Dong Ap Bia Mountain. On 10 May his company engaged an entrenched North Vietnamese force, and Specialist Schoch rushed to the area of fiercest conflict and

---

* Casualty figures cited are from the 3/187 AAR daily appendix; the casualty summary elsewhere records 13 KIA and 60 WIA total, exclusive of the B Company friendly-fire losses.

began to administer medical aid to the wounded. Once, he moved to aid three seriously wounded men lying in an area completely devoid of cover. While treating one of the men, he became the target of a sniper in a nearby tree. Taking the weapon of the man he was treating, he killed the sniper. On 13 May his unit assaulted the enemy stronghold and again came under heavy concentration of hostile fire. As Specialist Schoch was applying first aid to the wounded of the lead element, the medical aidman of another platoon sustained serious wounds and could not breathe. Braving hostile fire, he skillfully performed a tracheotomy on his wounded comrade who resumed breathing and was evacuated. Later as he was treating a casualty, an enemy fragmentation grenade fell near him and the wounded man. He instantly grabbed the grenade and threw it into a nearby bomb crater and then eliminated the enemy soldier who had thrown the device. After completing treatment, he carried the American to a landing zone for evacuation. On the following day, Specialist Schoch treated and evacuated four wounded soldiers who had been well forward in a maneuver toward the summit of the mountain. On 15 May, as the battle for the hill still raged, a helicopter carrying ammunition was downed by hostile fire. Despite the fact that the burning aircraft might explode at any moment, Specialist Schoch ran to the wreckage and retrieved an unconscious survivor and carried him through a barrage of sniper fire to safety where he administered first aid, saving the man's life. For the remainder of the day and until 18 May, he took charge of medical treatment and evacuation on the emergency landing zone.[3]

For the hapless 1/506th Infantry, the 18th was much like the 17th. B Company started off well but then was mousetrapped in a large ambush. Retreating 200 meters to prepare for medevac, the unit was hit from all sides by an NVA counterattack. In firefights that continued until after dark, Bravo Company had 4 KIA and over 20 wounded. Its night position was at most 200 meters closer to Hill 937 than in the morning. A Company was able to advance a bit farther, reaching the vicinity of Hill 900 and destroying a number of enemy bunkers and a mortar position in the process. At one point, the company commander reported that he was "just south of Hill 937," but that was a map-reading error, as his unit dug in for the night on the ridge south of Hill 900, still 700–800 meters short of the goal. A Company lost two men during the day's combat.

Hamburger Hill – Four Battalions Close In – 19 May

Sometime on the afternoon of 18 May, MG Zais "was once again faced with the decision of whether to call off the attack on Dong Ap Bia or order yet another assault."[4] This question was one that had been pondered daily for the previous week, and it is not surprising that the intention to continue attacking was once more the answer. What changed on this day was the decision to strongly reinforce the attack force. In addition to bringing forward the 2/506th Infantry (LTC Gene Sherron) to replace LTC Honeycutt's depleted command, 2/501st Infantry (from the 2nd Brigade of the 101st Airborne Division) and the ARVN 2/3rd Infantry would be air assaulted in to help corral the NVA infantry on the slopes of Dong Ap Bia.

When LTC Honeycutt learned of the plan to replace his unit, he was incensed. General Zais flew to the 3/187th perimeter on the afternoon of the 18th and met with Honeycutt, who had just learned of his planned relief from LTC Sherron. After a brief discussion, Honeycutt laid it on the line: "General, if there is anybody that deserves to take that sonofabitch [Hill 937], it's the Rakkasans—and you know that as well as I do. And there just is no goddamn way in hell I wanna see Sherron and the 2/506th come in here and take that mountain after all we've been through. And if it ain't gonna be that way, then you just better fire my ass right now. Right this minute."[5] Honeycutt asked that he be reinforced with one company for the final assault. He had his way, and A/2/506th came in to help with the final assault, but the rest of that battalion would go to FB AIRBORNE.

For most of the battalions that would take part in the next big assault, now scheduled for 20 May, the 19th was a day to move to their attack positions. The 3/187th Infantry had minimal contact with the NVA, and there was only one WIA during the day. The commanders of Alpha and

Charlie Companies, plus the commander of A/2/506th Infantry, met with LTC Honeycutt in the morning to go over the attack order, and then moved cautiously to the next day's line of departure. Between 1300 and 1430, the 2/3 ARVN air assaulted into a two-ship LZ about 1,000 meters southeast of Hill 937 and moved to a position about 500 meters from the objective. The ARVN infantry would wear red and yellow armbands to help prevent any friendly fire accidents as they joined with the American grunts in a tightening circle. Three companies of the 2/501st Infantry inserted into a landing zone about 1.5 km northeast of the mountain top, completing their lift by 1515. Both of the landing zones were green.

The 1/506th Infantry also was moving toward its planned assault positions for the 20th but was doing so against stiff resistance. B Company continued to move northeast with only light contact, but A Company (CPT Bill Womble) encountered a defense in depth on Hill 900. At one point, there were over 20 enemy bunkers active, manned by at least two men each. At 1715, the Currahees at last reached the top of Hill 900. There, they found a large command bunker equipped with field telephones and maps—probably a battalion headquarters, if not that of a regiment. They reported 18 NVA KIA by body count. C Company (CPT Will Stymiest), which had been following A Company for several days, was ordered forward to join in the attack, and killed 10 NVA. The battalion lost 3 men KIA and 19 WIA during these attacks. While still several hundred meters from the top of Hill 937, the 2/506th was finally approaching the defenses of Hill 937 proper.

Hamburger Hill – The Final Assault – 20 May

On the morning of 20 May 1969, multiple air strikes were called on top of the mountain, and the tube and aerial rocket artillery (gunships) worked it over for the 10th day. All four battalions began to advance up the slopes of Hamburger Hill at 1000 hours (see Map 15).

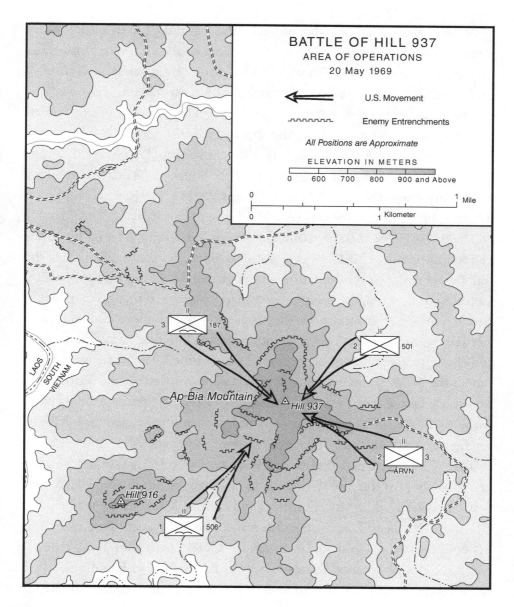

MAP 15

Source: Traas, Adrian G. *Transition, November 1968 – December 1969* (Washington, DC: Center of Military History, United States Army 2018) p. 38.

What happened, and in what order, on the top of the hill that day is likely forever shrouded in the fog of war. According to the command historian's narrative, "At 1100H 3-187th with A Company on the right, C Company in the center and attached A Company, 2-506th on the left, moved southeast for a coordinated attack on the Hill. . . . [The assault continued] with A Company, 3-187th reaching the top of Hill 937 first at 1145H."[6] The 3/187th Infantry AAR has a slightly different version: "All the 3/187 units were on the military crest of the hill by 1127 hours. . . . The first company to actually reach the top of Hill 937 was C/3/187. He [sic] accomplished this at 1144. . . ."[7] But there is yet a third side to this story.

The 2/3 ARVN Infantry Battalion (Pham Van Dinh), advised by Major Harvey Zimmerle, USMC, was responsible for the southeast quadrant of the hill. Before dawn on the 20th, from its designated line of departure about 500 meters from the crest, the battalion commander observed that his men would have to cross open ground when they began to advance after the air and artillery prep were completed. "Fearing heavy losses Dinh made a fateful decision and chose not to wait until the proposed assault time." The move in darkness caught the NVA napping, and the lead company "moved quickly through an undefended route to the summit of Hamburger Hill. . . . Around 1000 hours, while 3/187 began to batter through heavy enemy resistance, 2/3 ARVN reached the crest of Hamburger Hill, compromising enemy positions."[8]

The ARVN marked their front line with purple smoke and consolidated their position, while trying to arrange medevac for 3 KIA and 5 WIA. Overhead, in the command and control (CC) helicopter of COL Conmy, LTC Cecil Fair, Senior Advisor to the 3rd ARVN Regiment and his counterpart, Regiment Commander COL Pham Van Hoa, coordinated the ARVN troops' maneuvers with the American battalions. "Fair reported to Conmy that the summit of Dong Ap Bia was in friendly hands and that 2/3 awaited further orders."

Fair was soon shocked to receive a new directive from Conmy, who told him "to get your people off the hill, because we are going to fire an artillery preparation on top of the hill." Fair could not understand the order, and Colonel Hoa was so frustrated that he "almost jumped out of the helicopter."

Fair reminded Conmy that there was a full ARVN battalion atop Dong Ap Bia that could launch an attack on the rear of the enemy forces locked in battle with 3/187 and requested that he adjust the fire plan accordingly. Conmy refused Fair's advice and said, "if your guys are not out of there they are gonna get blown to hell."[9]

The story told by Andrew Wiest, although refuted by Conmy and Honeycutt (the latter "vehemently" in a phone interview with Wiest), is well documented in his book. The ARVN went back down the mountain, returning several hours later when the top had been "taken" by the 3/187th. It must be said that the decision of Dinh to advance ahead of the scheduled time for the coordinated attack, and without first notifying his superior and obtaining approval, would be regarded as a serious breach of discipline for an American commander. In moving early, he avoided possible heavy casualties in a ground assault, but exposed his troops to the high risk of friendly fire losses due to the constant air, gunship, and artillery fire at the mountain top, and also greatly increased the chance of a firefight between his unit and the 3/187th Infantry. It is to be regretted that COL Conmy did not display flexibility in taking advantage of the ARVN's surprise attack, and it would have been interesting to know if he had some reason for ordering the 2/3 off the mountain other than preserving the pride of the Rakkasan commander.

The 1/506th, still facing the heaviest enemy fire, advanced little during the day but kept up the pressure. As early as 13 May, when the 1/506th was first ordered to march to the sound of the guns on Dong Ap Bia, an important part of its mission was to close the enemy LOC running down the ravine southwest of Hill 937. Regrettably, that was *never* effectively accomplished. Excerpts from the 3/187th AAR, Appendix 11 to ANNEX C (20 May 1969) read:

At around 1120, LTC Honeycutt "told A/3/187 to watch his flanks, because the 'gooks' are up and running around up here and streaming off the western side of the hill into the draws. At this time, he enjoined the Bn CO 1/506 to move quickly to catch the enemy spilling into the draw *several hundred meters to the front* of the 1/506 [emphasis added]."

"At 1252 hours the Bn CO reiterated to Brigade that he could see the enemy retreating toward the 1/506. The location of the retreating elements was marked and struck innumerable times by ARA, assault teams, airstrikes, and mortars. By 1255, with the noose tightening . . . the Bn CO reported enemy running in every direction from the hill in their confused attempt to escape. The majority however were spilling into the draw on the west in a desperate rush for the border less than two kilometers away.[10]

The final American battalion, 2/501st, approaching from the northeast, had the farthest to go but encountered no enemy resistance, although it came across many abandoned bunkers and huts. By midafternoon, all elements were primarily involved in mopping up enemy stragglers and conducting body count. For the day, the 3/187th and attached A/2/506th had 2 KIA and 60 WIA. In the 2/506th, 4 men were killed and 15 wounded.

Hamburger Hill – The Final Count

In the final analysis, the number of Americans killed in action on Dong Ap Bia was remarkably low. This may appear to be an incredible conclusion. Here are the numbers according to the 101st Airborne Division Fact Sheet issued on 24 May:

	KIA/MIA	WIA/ Evacuated	WIA/Returned to Duty	TOTAL
3/187th INF	38	255	50	343
1/506th INF	23	91	3	117
2/506th INF	1	11		12
2/501st INF	1	10		11
2/3 ARVN	UNK			UNK

While the KIA figures are indeed low compared to other major attrition battles earlier in the war, the ratio of WIA to KIA is extremely high. This was probably because most of the casualties were caused by enemy RPGs, claymores, and mortar fire, resulting in a large number of shrapnel compared to gunshot wounds.

Reported enemy body count was pegged at 630, of whom 355 were claimed by the 3/187th Infantry, 51 attributed to the 2/3 ARVN, and 47 credited to tactical airstrikes. It is interesting that no enemy losses were identified from artillery or gunship fire, and the Tac Air figure is remarkably low for the 271 sorties that were flown. One million pounds (not a typo) of artillery ammunition, 513 tons, was used during the battle, including almost 17,000 rounds of tube artillery and 2,622 2.75" aerial rockets. There were undoubtedly many enemy KIA whose bodies were either buried under the pulverized hilltop or carried off by their comrades. Mass graves were found for days afterward as the enemy escape routes were searched (the uncovered bodies are included in the 630 figure). Only 25 crew-served and 152 individual weapons were captured, most of those in two caches found by the 1/506th Infantry on 21 and 24 May.

Based on an alleged report from a Special Forces Team operating in Laos after the battle, it has been claimed that as many as 1,000–1,100 additional enemy casualties were incurred.[†] At the end of the battle, two badly wounded enemy soldiers were captured. They reported that 80% of their company was killed, and that the 7th and 8th battalions of the 29th Regiment were decimated.

The 3/187th Infantry was also rendered combat ineffective and would require several months of replacement infusion and reorganization before it would be ready for combat again. It left the hill on 21 May. On 5 June, the new commander of the 101st Airborne Division, MG John M. Wright, withdrew all American troops from Hill 937.

APACHE SNOW – 2/501st Infantry – 21 May – 7 June

After taking part in the final assault on Hill 937 on 20 May, the 2/501st Infantry participated in the systematic search of the enemy bunker complex on the mountain the following day. On 22–23 May, sweeping west toward the border, more bunkers and enemy bodies were found.

---

† I have found no documentation to support this claim, although much of the MACVSOG documentation is still missing or heavily redacted.

On 23 May, the D company point was ambushed (YC320988) in the same area where the 3/187th had been just before the start of the big battle. The enemy platoon hit hard with small arms fire, rocket-propelled grenades, satchel charges, and grenades. Gunships and artillery were called; one man was killed and eight wounded. Three NVA bodies were found when the battle ended.[11]

On 24 May, the 2/501st reached the border and turned north and then east. B Company ran into an enemy squad on 25 May and was in contact through the afternoon. One NVA was reported KIA and there were no American losses.[12] The following day, the battalion was redirected to move in the direction of Ale Ninh (1) and (2) (YD344003 and YD323013). These were the same objectives as in the original Operational Narrative! Late on the night of 30–31 May, Bravo Company ambushed an NVA column just 200 meters from where the 3/187th landed on 10 May, but the enemy reaction was strong, and when the battlefield was finally secured the next morning, 4 Americans were KIA and 6 WIA. Enemy casualties were unknown.[13]

APACHE SNOW – 1/506th Infantry 21 May – 7 June[14]

On the morning of 21 May, the mopping up began on the Dong Ap Bia massif. The 1/506th Infantry would have to retrace much of the ground it had moved across previously, as it hastened to close the ring around Hill 937. Hill 900, about 1,000 meters south of 937, was replete with bunkers, many filled with enemy ammunition, supplies, weapons, and the bodies of former defenders. There were also mass graves, booby traps, and prowling NVA soldiers. Four 1/506th troops were KIA on 21 May,[15] three in C Company by a claymore mine and the fourth in B Company. These were the last fatalities the battalion would suffer during the operation.

On 21–22 May, the three companies of the Currahees combed through one abandoned bunker complex after another, compiling long lists of captured booty and adding many notches to the enemy body count for Hamburger Hill.

Meanwhile, a short distance to the west, Hill 916 loomed. On 22 May a FAC reported heavy anti-aircraft machine gun fire from that peak, and air strikes and artillery were liberally used. By 1000H on the morning of 23 May, A and B Companies were moving toward Hill 916. By noon, both companies were in contact with NVA defenders, and artillery and ARA were called in. C Company joined the move, and it, too, was stopped by a bunker. The first enemy position was overcome, but further advance was stalled by a large bunker complex. The ARA was again called, and misdirected rockets struck the company, wounding five men. Three men in B Company also received shrapnel wounds from the aerial artillery. The companies in contact settled in for the night in three perimeters about 800 meters east of Hill 916. D Company, guarding the perimeter at FSB CURRAHEE, was not spared from the enemy's attention. At 1345 hours, six 122mm rockets were fired at the base. All landed outside the perimeter and only one man was wounded.

On 24 May, the battalion reversed direction, declining to assault Hill 916. It swept back to the east, ending the day about 400 meters farther from Hill 916. Over 60 bunkers were destroyed, and another long list of captured equipment and ammunition was drawn up. The 25th and 26th of May brought the battalion back to the vicinity of Hill 800, the southernmost peak of Dong Ap Bia. There, the search litany continued. On the east side of Hill 800 on the 27th, high-tension wires with three types of insulators were found, illustrating the sophistication of the enemy complex on the western side of the A Shau Valley. The bunkers were big and comfortable, and had 2–3 feet of overhead cover.

And so the story went. On 28 May, A Company relieved D Company at FSB CURRAHEE, and Delta joined Bravo and Charlie to the west. On 29 May, the NVA launched six more 122mm rockets at CURRAHEE, but again they landed outside the base; there were no casualties. The battalion continued to RIF due east. After a day of rest on 30 May in honor of Buddha's birthday, D Company ran into a small enemy force on the 31st. The enemy broke contact, then reinitiated later in the afternoon. Five D Company troops were wounded. Enemy casualties were not known.

On the first day of June, the companies in the field were ordered to move to FSB CURRAHEE, 2 km to the east, to prepare for a combat assault on 3 June. The entire battalion was back at the firebase by 1045H on 2 June, gaining a brief respite, with time for showers, ice cream, a change of clothes, and equipment cleaning and repairs. At 0830 on 3 June, Companies B and D were inserted about 2 km northeast of Hill 937 (YC347990). Bravo Company conducted a RIF to the southwest, and Delta Company did the same in the opposite direction. On 4 June, B Company came under RPG fire. Six troops were wounded, and artillery was called on the enemy position with unknown results. Both companies found abandoned enemy positions, but there was no further contact on 5 or 6 June. From 6–8 June 1969, the companies of the 1/506th Infantry were rotated to Eagle Beach on the coast for a quick rest, and then returned to firebases in or near the A Shau Valley. When the battalion's participation in Operation APACHE SNOW ended at midnight on 8 June,[16] A Company was patrolling north of the EAGLES NEST, B Company was providing security for FSB EAGLES NEST and FSB BERCHTESGADEN, and C Company and D Company were headed to FSB GEORGIA (YD420033) for RIFs in that area. The battalion claimed 159 enemy KIA, while losing 22 KIA, 1 MIA, and 136 WIA.

APACHE SNOW - 2/506th Infantry in The Warehouse[17]

With the extraction of the badly battered 3/187th Infantry from Hamburger Hill on 21 May, the 2/506th Infantry (LTC Gene T. Sherron) was brought in from the coast to fill the gap in the roster. Alpha and Echo Companies arrived at FSB AIRBORNE on 21 May, and the rest of the battalion came in on the 22nd. Patrolling began immediately, with no enemy contact on the first day. On 23 May, first B Company patrolling to the southeast and then D Company to the northwest ran into enemy troops as the skirmishes in the Warehouse area were renewed. B Company had two KIA and five WIA. D Company also had 2 KIA. The NVA fired a few 60mm mortar rounds at the firebase just after midnight, to no effect.

Before dawn on 24 May, the NVA again hit the firebase with a barrage of mortar fire. Thirty-eight rounds were counted, but the light mortar bombs again did not cause any casualties. Delta Company made contact with the enemy three times during the day, but the fire was light and neither side took any casualties. Charlie Company, operating just east of Delta, had two brushes with the enemy, also resulting in no known injuries. Bravo Company's turn was next, when it was ambushed by RPG-firing enemy soldiers on 25 May, just 200 meters north of its battle on the 23rd. On 26 May, skirmishes between Bravo and Delta Companies and the NVA resulted in one U.S. WIA and one NVA KIA.

The enemy mortar attack on FSB AIRBORNE at 0425 hours on 27 May was small but exceptionally accurate. Only seven or eight rounds were fired, but direct hits were scored on two bunkers and a listening post just outside the perimeter wire. The result was four KIA and six WIA. In the field that day, C Company found three caches, the largest supply stash yet discovered by the 2/506th, including 300 82mm mortar rounds, 600 60mm mortar rounds, and 200 RPG rounds. The 28th was uneventful until evening. As D Company was establishing its NDP, a booby trap consisting of two or three claymore mines was tripped. Fourteen D Company troops were wounded.

On 29 May, none of the elements of the 2/506th engaged the enemy, but another mishap befell D Company. While in its night laager, a single white phosphorous shell impacted the perimeter, wounding five soldiers. The round was believed to have been a 105mm howitzer shell fired by C/2/319th Artillery at AIRBORNE. The last two days in May, all three patrolling companies found enemy caches but made no contact with NVA troops. They captured 315 mortar rounds, 900 rounds of 12.7mm ammunition, and other munitions. On the first day of June, the wreckage of a UH-1 Slick and the body of the pilot were found. The first of several large rice caches was also discovered.

After five days with no casualties due to enemy fire, 2 June brought the first A Company losses. While on a short-range RIF to the southeast of the firebase, an A Company platoon ran into 10 NVA soldiers who were setting up a claymore booby trap (what GIs would call an "electronic ambush"). In the ensuing battle, the enemy detonated the mines, resulting in one U.S. KIA and four U.S. WIA. Four NVA soldiers were also killed, including two whose bodies would be found the next day. B, C, and D Companies continued to patrol their areas on 2–4 June, finding more rice, mortar ammunition, bunkers, and huts.

The 5th of June brought more of the same. C and D Companies each had brief contacts with squad-size enemy forces and each sustained three WIA. B Company, having crossed over Route 548 and patrolling almost 2 km to the west of the road and north of the Dong So ridge, made contact twice on 6 June. In the first firefight, the NVA hit the company with mortar and RPG fire at 0630H, killing 1 and wounding 12. In the afternoon, small arms fire wounded another U.S. trooper. There were no reported enemy casualties in either battle.

Operation APACHE SNOW ended for the 2/506th Infantry on 7 June. D Company found several small caches that day and killed 1 NVA soldier. A total of 33 contacts and 30 enemy KIA were reported by the 2/506th Infantry for the operation. Twenty-seven individual weapons were captured, along with large amounts of ammunition and over 10 tons of rice; almost all of the captured supplies were found within what is loosely described as the Warehouse 54 area. Friendly casualties were summarized as 11 KIA and 60 WIA.

APACHE SNOW – Results and Resurgence of the NVA

There are multiple Combat After Action Reports on Operation APACHE SNOW, issued by U.S. Headquarters from battalion to corps level. The casualty and enemy equipment figures do not necessarily reconcile with each other, in part because they were prepared at different times. Using the latest and highest-level report, that of the XXIV Corps dated 27 August 1969, the statistics are as follows:

KIA: U.S. – 90; ARVN – 31; TOTAL – 121
WIA: U.S. – 582; ARVN – 137; TOTAL – 719
Enemy KIA by U.S. – 792; ARVN – 229; TOTAL – 1,021
Enemy significant equipment losses
    Crew-served weapons 141
    Individual weapons   613
    Vehicles   53

The U.S. casualty figure is certainly too low. Sixty-six men died at Hamburger Hill, including the 3 medevac helicopter crewmen lost on

13 May, and another 27 during the battle at FS Airborne on 12 May. Here are the most accurate KIA/MIA figures I have been able to develop:

3/187 Inf	36
2/501 Inf	19 (includes 13 at FSB Airborne)
1/506 Inf	23
2/506 Inf	11
2/17 Air Cav	2
2/319 Art	14 (FSB Airborne)
326 Med Bn	3 (medevac helicopter loss 5/13)
TOTAL	108 (this includes friendly-fire but not noncombat losses)

Air support was heavy: 905 tac air sorties were flown, 777 directed by FAC and 128 Combat Sky Spot (radar-vectored). A total of almost 1,400 tons of bombs and over 200 tons of napalm were delivered, resulting in 29 enemy KIA, 72 secondary fires, and 87 secondary explosions. There were 7 B–52 ARC LIGHT strikes in preparation for the operation and an additional 12 during APACHE SNOW. These resulted in approximately 25 secondary explosions.[18]

As per the script, the allies again claimed to be "King of the Valley," if not the (Hamburger) Hill, and to have swept the North Vietnamese from the A Shau.

RESULTS: The objectives of Operation Apache Snow were successfully carried out. A series of mutually supporting fire support bases . . . was established. . . . The result of this operation was the destruction of the combat effectiveness of the maneuver battalions of the 29th NVA Regiment . . . which forced the broken elements to retreat across the border into Laos. It also denied them use of the northern A Shau Valley as either a staging area for attack or as a storage area for supplies.[19]

Operation APACHE SNOW was directed toward defeating the North Vietnamese Army in the Northern A Shau Valley. Through intense close-in fighting, the 101st Airborne troopers contributed another gallant victory to the Vietnamese War effort by almost completely eliminating one North

Vietnamese Regiment and putting the Valley under Free World control for the first time since 1965. Though APACHE SNOW is over, the 101st Airborne Division continues their aggressive hard-hitting attack to return control of Thua Thien Province to the Republic of Vietnam.[20]

The author of the narrative does, in the end, at least attempt to indicate that the victory may be transitory. After first concluding that the valley is "under Free World control for the first time since 1965," he then hedges that claim by noting that the 101st Airborne Division "continues [its] . . . attack to return control of Thua Thien Province to the Republic of Vietnam."

Even before the termination of APACHE SNOW, the NVA swarmed back into the valley. The 2/17th Air Cavalry, which had been first on the scene at FSB AIRBORNE, was back in combat, this time at Dong Ap Bia, as the operation came to a close. On 6 June at 1144H, the ARP of Alpha Troop was inserted at YC328978, 300–400 meters south of the top of Hill 937 and right in the heart of where the "final" battle of 20 May had taken place. The king-of-the-mountain U.S. and ARVN infantry had departed. The NVA were still there. The LZ was green, but within an hour, the platoon was in contact. The battle lasted throughout the afternoon, with air strikes, artillery, and gunships employed against at least 20 NVA bunkers. At 1533H, the 1st Platoon of D/2/17th Cavalry (the ground troop) was brought in to reinforce the Blues as they fought toward enemy bunkers 150 meters away. One NVA .51-cal heavy machine gun and one light machine gun were destroyed and 5 NVA KIA. The 2/17th lost one man KIA and three WIA before the battle ended, and the U.S. troops were extracted at 1800 hours.

A message from XXIV Corps to the CG, III MAF on 18 June included the following intelligence:

During the period 1-18 June there were numerous VRS [visual reconnaissances] indicating construction of new huts and bunkers on the northern and western slopes of Hill 937. . . .

Enemy 122mm rockets were fired on FSB BERCHTESGADEN from the vicinity of Hill 937 on 14 June. On 10 June FSB CURRAHEE received several 122mm rockets from the same area. . . . PWS captured following a sapper attack on FSB BERCHTESGADEN on 14 June indicated that they were to return to the vicinity of Hill 937 following the attack. PWS captured after a ground attack on FSB CURRAHEE on 15 June indicated that they were from the C20 sapper company of the 29th Regt. According to the PW reports, there were an estimated 1000 personnel in the immediate vicinity of Hill 937. 101st ABN DIV OB [order of battle] holds the 29th Regt at an estimated strength of 1200 personnel following losses incurred during the battle against the regt in May 69. VRS of the trail complex to the NW of Hill 937 leading from BA 611 in Laos indicate continuous enemy use of the trail network to infiltrate into the northern and western slopes of the hill.[21]

COL Joseph B. Conmy, Jr., commanded the 3rd Brigade of the 101st Airborne Division during APACHE SNOW. In an article in 1989, commemorating the 20th anniversary of Hamburger Hill, he avowed the success of APACHE SNOW and the victory on Hamburger Hill, in the process epitomizing the "We were winning when I left" syndrome: "The 3rd Brigade stayed in the A Shau for months, keeping it clear of the enemy. . . . "We and our successors controlled the area until ordered out of Vietnam three years later."[22] As the coming battles during Operation MONTGOMERY RENDEZVOUS would substantiate, the allies did not control the A Shau and the enemy had not abandoned the valley for its sanctuaries in Laos. The enemy was still there, still attacking, and still rapidly rebuilding.

Aftershock: General Abrams and Hamburger Hill

At his weekly intelligence meeting with his staff on May 3 (i.e., 10 days *before* the battle at Dong Ap Bia began), GEN Abrams commented (re the inevitability of being criticized): "It's like the 29th Regiment. You fight them in the A Shau and they piss on you for that. You fight them in Hue and they raise hell about that. You can't find a good place to fight them. The places aren't good" (emphasis in original).[23] A premonition? Abe's preference for going after the enemy's "logistical nose"[24] is well established. Ditto for his understanding of the need to get away from the "big battalion" operations of the Westmoreland years and to strengthen the ARVN and pacification. Both of those operational approaches moved away from the attrition strategy of his predecessor. Less has been written about his views on engaging in knock-down, drag-out fights with large, heavily fortified enemy units, perhaps because his ready acceptance of those engagements might tinge our belief in his enlightened leadership or perhaps because Hamburger Hill was the first, and last, such battle under his command.

During the battle for Dak To, in November 1967, one of the last major protracted slugging matches with NVA regulars in the western badlands, DEPCOMUSMACV was sent up to the battle area to review the situation. He looked at the enemy defenses:

> I viewed one position today consisting of 3 concentric trenches making 3 complete circles around the top of one mountain. I am informed that at the bottom of the trench are some horizontal holes dug back into the mountain where the enemy stays during bombing and artillery.[25]

There were at least four NVA regiments in the Dak To area, and fighting had been going on for over two weeks at the time of Abe's visit. In the final tally of the fight for Hill 875, the 173rd Airborne Brigade lost 122 KIA/MIA and 253 wounded. For the campaign as a whole, the brigade had 208 KIA and 645 WIA, over 50% of the infantrymen in its line companies. Enemy losses were variously estimated as 1,000–1,644 killed.[26] In summarizing his report on Dak To to Chairman of the Joint Chiefs Wheeler (cc to GEN Westmoreland and ADM Sharp), Abrams endorsed this full-blooded application of the doctrine of attrition:

> The battlefield in Kontum resembles a chess board at play. We have units and fire bases on peaks and ridges and the enemy has too. We shift ours by helicopter to check him or get behind him or assault him. Extensive use is made of LRRP, mike forces, CIDG, the people sniffer, air observers and all other means to locate the enemy. When found maximum fire power is put on him. *I believe when the enemy comes forth from Cambodia or Laos with his principal formations looking for a fight we must go out and fight him* [emphasis added].[27]

It does not come as a surprise, therefore, that Abe stood 100% behind MG Zais and LTG Stillwell in their decisions to pursue the battle of Hill 937 to the bitter end. Discussing Hamburger Hill at the 31 MAY 1969 WEIU, while the public outcry was still going on "back in the world," Abe reiterated his opinion about where to fight the 29th Regiment:

> After all, you know, the <u>29th</u> that they tangled with up there, the 29th was in <u>Hue</u> last year. And this year the 29th fought out there on Hamburger Hill. They weren't within – Christ, they weren't within <u>V-2</u> range of Hue. So I don't know whether it's better— Well, I <u>do</u> know [laughing]. I think it's better to fight the 29th in the A Shau. We wouldn't get any less casualties fighting them in Hue. We took a lot of casualties in Hue [Tet 1968] [<u>emphasis in original</u>].[28]

There is a very interesting quotation from General Abrams at the 21 MAR 1970 WIEU. He is speaking about a conversation he had with the 4th Infantry Division leaders concerning tactics:

> Then I got talking with them about this matter of working against the <u>system</u>. I've sort of become wedded to preaching about the system. And even going so far as to say, You can go out there and cut one of these battalions down to 60 men, but in three or four months, five months, they'll build it back again—so long as their <u>system</u> is functioning—supply, medical, infiltration, recruits—you know, all of those things. So the way to—one of the things—if you can get at that <u>system</u> and start putting inefficiency, failures, into the <u>system</u>, they're just going to have to move back [i.e., across the border into Cambodia or Laos] [<u>emphasis in original</u>].[29]

How badly punished was the 29th Regiment at Dong Ap Bia? For months, MACV Intelligence had carried the unit on the monthly ORDER OF BATTLE SUMMARY at a total strength of 1,690 men, including 450 each in the 7th and 8th Battalions. In June 1969, citing the presence of those two battalions and the regimental headquarters in the vicinity of Hill 937, the estimate dropped to 970 men, including 200 in each battalion. In July, the regiment was estimated at 1,400 men, with 300 in the 7th and 360 in the 8th battalions.[30] That estimate would hold until the end of the year.‡ So the regiment lost an estimated 720 men, but in less than two months had recovered about 60% of those losses, whether through replacements or recovery of wounded. Within a year, the regiment would again be heavily engaged in the preliminaries for the RIPCORD battle. Conmy's later assertion that "[w]e destroyed the 29th NVA Regiment as a fighting force"[31] does not hold up. Similar assertions about "destroyed" enemy units by American officers throughout the war invariably underestimated the enemy's recuperative powers. On the other hand, according to John Roberts, a year later, "the 3-187 'Rakkasans' were completely committed to the pacification program in the villages along the coast since the battalion was almost destroyed on Dong Ap Bia" (ROBEJG01, 36). So the PAVN lost two battalions for a few months and the U.S. one battalion for a year.

While we'll never know, it is interesting to speculate on whether, looking back after another year in command, Abe might have begun to comprehend the ultimate futility of big-battle attrition.

Arguably, William C. Westmoreland's greatest failure as a theatre military leader was his inability to accept that the outcome of a battle could have political reverberations far exceeding in scope the military result. Not only could the effect on the political situation be greater than what a victory or defeat might imply for future military plans, but also, that effect might even operate in the opposite direction. Thus, Hue. Yet, while some allowance may be made for Westy's failure to understand what Tet 1968 meant to the future course of the conflict, the surprise at the political aftermath of that event may be the only thing that exceeded the surprise of the offensive itself. Hamburger Hill was not a surprise, and General Abrams should have grasped that another bloody, prolonged battle for a worthless piece of real estate simply wasn't going to wash with the American public, regardless of the body count, the kill ratio, and whether we were king of

---

‡ The III MAF PerIntRep figures on the 29th Regiment show an even stronger recovery, listing it at 1,800 before the battle, 1,150 in June, and back up to 1,820 by mid-July.

the mountain when it was all over. Abe had understood that Khe Sanh was not worth fighting over; he somehow missed the signs at Dong Ap Bia.

Just as Tet 1968 had marked the turning point in the War in Vietnam, so too would Hamburger Hill be the last big attrition battle and turn U.S. public opinion even more sharply against a continued conflict. As noted Vietnam War scholar James Willbanks wrote: "For much of the American public, Hamburger Hill crystallized the frustration of winning costly battles without ever consummating a strategic victory. The hill had been won but at a very high price—then only to be abandoned for the Communists to reoccupy."[32] General Abrams failed at the time to perceive that U.S. casualties, regardless of the enemy losses or the alleged importance of the "objective," were the true metric by which the American public would gauge military strategy in 1969. As the commanding general, he was responsible for taking that into consideration—perhaps even the most important consideration—as he led his army.

In the wake of Hamburger Hill, on 19 June, President Nixon issued a specific order to General Abrams: "He is to conduct the war with a minimum of American casualties."[33] It is unfortunate that Abe did not recognize this before Dong Ap Bia.

For the second time in two years, the American people signaled sharp opposition to a continuing commitment to the conflict in Vietnam. In the first instance, the A Shau Valley played a critical role in the siege of Hue in early 1968. The valley was again the locus in May 1969 at Hamburger Hill: "The truth is that it was one of the most significant battles of the war, for it spelled the end of major American ground operations in Vietnam,"[34] wrote Harry Summers, Jr. James Willbanks is even more succinct: "Before Hamburger Hill, the U.S. forces were still seeking victory on the battlefield; after Hamburger Hill, they were only seeking a way out."[35]

20

OPERATION MONTGOMERY RENDEZVOUS

KENTUCKY JUMPER Phase III: MONTGOMERY RENDEZVOUS

The operations of the 101st Airborne Division in Thua Thien Province from March through August 1969 were conducted under the umbrella code name KENTUCKY JUMPER. The third and final operation in the series commenced simultaneously with the termination of APACHE SNOW.* MONTGOMERY RENDEZVOUS was controlled by the 3rd Brigade, 101st Airborne Division, with brigade headquarters at FSB BERCHTESGADEN and COL Conmy in command. The primary mission now was the interdiction of the enemy lines of communication; "search and destroy" was running out of known enemy base camps to exploit. The Area of Operations was centered on the Rao Lao and northern A Shau Valleys (see Map 16).[1] The 3rd ARVN Regiment also participated, with the 1/1 ARVN, 2/3 ARVN, and 4/3 ARVN battalions under command.

---

* KENTUCKY JUMPER was a division-wide operation and included other named operations outside the Valley.

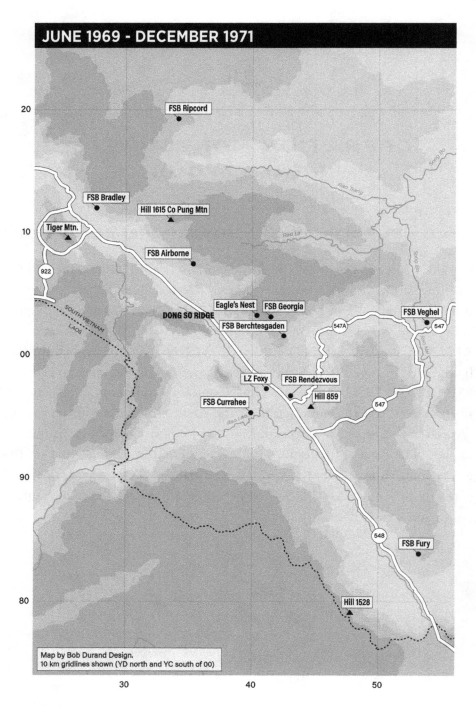

MAP 16

Source: Bob Durand Design

As APACHE SNOW ended, the 1/506th and 2/506th Infantry occupied positions from which they would segue directly into MONTGOMERY RENDEZVOUS. Alpha Troop, 2/17th Cavalry, was the reconnaissance unit. The infantry battalions involved changed several times during the operation; these are the combat units initially assigned:

FSB EAGLES NEST	B/1/506 Inf (part)
	A/1/506 Inf operating to the northeast
FSB BERCHTESGADEN	B/2/319 Art
	B/1/506 Inf (-)
FSB AIRBORNE	A/2/319 Art
	B/2/11 Art
	A & B/2/506 Inf
	C & D/2/506 Inf operating to the southwest
FSB CURRAHEE	C/2/319 Art
	E/1/506 Inf
FB GEORGIA	C & D/1/506 Inf (to RIF to northeast)
Route 547A	3/5 Cav

9 June 1969 – MONTGOMERY RENDEZVOUS Begins

An important part of the master plan for KENTUCKY JUMPER was the construction of an all-weather road into the A Shau Valley. From the division logistical hub at FSB BIRMINGHAM near Hue, the "A Shau Expressway" would follow Route 547 as far as FSB BLAZE, 20 km to the southwest. From there, the alternate route first constructed by North Vietnamese Army engineers in early 1968 that branched west down the Rao Nho Valley and then south and southwest to a junction with Route 548 about 1 km north of the abandoned Ta Bat airstrip would be upgraded. This road was known to the allies as Route 547A. In late March, when MASSACHUSETTS STRIKER was still in progress, Tiger Task Force from the 45th Engineer Group set out from FSB BASTOGNE on the road-building mission. This force was assembled around the 27th Combat Engineer Battalion, reinforced with seven specialized engineer companies, including the 59th Land Clearing Company with giant Rome Plow bulldozers.

By 9 April, the task force had reached FSB BLAZE, and thereafter the road work began in earnest. First, a 160-foot Bailey bridge was constructed to span the Song Bo River; from there to the A Shau Valley, the existing road required extensive work, as it was heavily cratered by ARC LIGHT strikes and mined by the NVA. Two months later, when MONTGOMERY RENDEZVOUS began, the engineers were within 5 km of Route 548; the juncture would be completed to great fanfare on 20 June. To provide security for the engineers, with the departure from Vietnam of the U.S. Ninth Infantry Division (south of Saigon) imminent, the division's reconnaissance squadron, the 3rd Squadron, 5th Cavalry, was sent north to I Corps in the spring of 1969, minus its air cavalry troop. As MONTGOMERY RENDEZVOUS commenced, the armored cavalry occupied several firebases along Routes 547 and 547A and escorted the engineer teams daily from the secure base camps to their work sites.[2]

The first day of Operation MONTGOMERY RENDEZVOUS served as a stark reminder that the North Vietnamese Army was still present throughout the Area of Operations. At 0740H on 9 June, three 5-ton engineer trucks from FSB CANNON were ambushed along Route 547, as they neared FSB BASTOGNE. Killed in the action were 3 men from D Company, 27th Engineer Battalion, and 13 others were wounded. Ten enemy soldiers were KIA in the counterattack by C Troop, 3/5th Cavalry and A Company, 2/327th Infantry, and 3 more GIs were wounded.[3]

Earlier that same morning, Team Alaska from L/75th Infantry (Ranger) was inserted on the top of Dong So Ridge (YD343025), north of FSB CURRAHEE. The LZ was green. This insertion was typical of the use of ranger patrols in economy of force missions to keep an eye out for enemy activity in areas not currently being swept by infantry companies.[4] On the other side of Route 548 from Dong So, in the Warehouse 54 area, D/2/506th Infantry ran into an enemy force in late afternoon. Two troopers were killed in this action. SP4 Guadalupe Prado, Jr. was posthumously awarded the Silver Star.

At approximately 1700 hours on the cited date, Company D, while moving in a reconnaissance in force operation, was engaged by an unknown size enemy force. During the initial contact Specialist Prado was situated on a rockshelf above the rest of his element. Disregarding his unprotected position, he leveled intense M-16 rifle fire against the enemy ranks, thus allowing the remainder of his platoon to take cover. His courageous act allowed the element to organize an effective volume of fire against the insurgent force and hold casualties to a minimum. Continuing to place effective fire upon the enemy from his open position, Specialist Prado was struck by small arms fire and fatally wounded.[5]

In the evening, the enemy delivered one more reminder of its intention to contest possession of the valley. Forty 82mm mortar rounds, accompanied by RPG and small arms fire, struck FSB CURRAHEE starting at 1945 hours. 1LT Eddie L. Smith of A Battery, 2/319th Field Artillery, was killed in this action, and 23 men were wounded.[6]

The die was cast – the North Vietnamese Army would continue to ambush the allied patrols and to attack firebases with indirect fire and sappers for the duration of MONTGOMERY RENDEZVOUS.

10–30 June: The Americans Build and the North Vietnamese Bombard

At first light on 10 June, Ranger Team Wyoming from Company L, 75th Infantry inserted on Tiger Mountain, where a firebase (TURNAGE) occupied by the ARVN 1st Regiment during APACHE SNOW had been recently abandoned. Teams Alaska and Wyoming would continue for several days to scout the mountains and trails for signs of the enemy, but neither made contact. Alaska was extracted on 12 June and Wyoming soon thereafter.[7]

An AH-1G Cobra gunship was shot down over Hamburger Hill (YC318982) by small arms and RPG fire in the late afternoon of 10 June while making a firing run at an enemy position. The C Troop Blues were on call as the ready reaction force and were inserted near the crash site at 1713H. The LZ was green, but as the troopers made their way to the downed helicopter, they were ambushed. AK-47s and RPGs opened fire at the rescue force, killing aerorifleman SP4 Robert L. Howard. The rest of the C Troop Blues arrived, and after suppressing the enemy fire and killing three NVA soldiers, reached the crash site. A platoon of D Troop infantry arrived to reinforce the air cavalry at 1800H. Both pilots were found injured and a medevac helicopter with a jungle penetrator hoist was requested; the Cobra was a total loss and was destroyed in place. The extraction of the injured helicopter crew and then of the reaction forces was completed without further enemy contact, although this was done under the light of artillery flares and was not completed until 2150H.[8]

As Tiger Task Force was completing the last section of Route 547A into the valley, another important construction project was started and completed in just 10 days by the 326th Airborne Engineer Battalion. On 3 June, with APACHE SNOW winding down, the first engineer teams arrived at LZ FOXY (YC405975), about 3 km northeast of FSB CURRAHEE, the location selected for a new Ta Bat airstrip. In a test of recently developed techniques to break down heavy construction equipment for transport by CH–47 Chinook and CH–54 Flying Crane helicopters, an astonishing inventory of engineer equipment was disassembled, flown into the valley, and reassembled to build the new airfield: ". . . four D5A bulldozers [a 15,000-lb item!], two M450 bulldozers, two TD–6 bulldozers (Seabees), two MRS-100's, four graders, two backhoes, five ¾ ton dump trucks, three sheepsfoot rollers, two Vibrapace, and one 13-wheel roller." A 36-inch culvert 90 feet long was also flown in to help with standing water, the "chief concern" in the construction of the field. On 5 June, C/326th Engineer Battalion began initial land clearing. Construction of the 1,500-foot airstrip was completed in just 54 hours. "On 13 June, runway marking panels were placed and the strip was checked by Air Force representatives. A C7A Caribou, 'Flying Virginia,' landed at LZ FOXY at 1520 hours."[9] That same day, Tiger Task Force and Route 547A reached the edge of the A Shau Valley about 3 km to the east, and a LOH from A/2/17th Cavalry went down along the Laotian border 10 km to the south. The crew was rescued, but the aircraft was a loss.[10] The ARVN 3/3 Infantry would arrive to secure the airfield on 22 June.[11]

The enemy did not passively accept the apparent allied endeavor to establish a permanent presence in the valley. In the early morning hours of

14 June, in an attack eerily reminiscent of the assault on FSB AIRBORNE one month earlier, the C2 and C3 Sapper Companies of the K-3 Battalion, 6th NVA Regiment, attacked FSB BERCHTESGADEN with devastating results. The NVA sappers first completed their usual careful and comprehensive reconnaissance of the American installation. They then cautiously and without detection cleared three paths into the base, cutting concertina wire, deactivating claymore mines, and tying down trip flares. When the assault started at 0345 hours, the firebase was subjected to heavy 82mm mortar and RPG fire, pinning the defenders in their bunkers, as the assault troops rushed into the base with satchel charges and AK-47s. BERCHTESGADEN was defended by elements of the 2/327th Infantry, B/2/506th Infantry, B/2/319th Artillery (105mm howitzers), and C/2/11th Artillery (155mm howitzers), and included the command bunker of the 2/327th Infantry and the Tactical Command Post of the 3rd Brigade, 101st Airborne Division.[12]

The battle lasted until 0530 and included hand-to-hand combat. COL Conmy was wounded in the satchel charge attack on his command bunker. The defenders called on artillery illumination from nearby firebases, as well as artillery and helicopter gunship fire, Air Force AC-47 gunships, and U.S. Navy A-7s from the USS Enterprise.[13] When the enemy withdrew, they left behind 33 bodies and 3 prisoners, out of an attack force of just 96 men. American losses were 12 KIA and 47 WIA.[14] One of those killed was CPT Robert M. Snell, USMA Class of 1966, of the 2/320th Artillery, who received the Distinguished Service Cross posthumously for his bravery during the battle:

Captain Snell distinguished himself on 14 June 1969 while serving as an artillery liaison officer to the 327th Infantry. In the early morning hours, an intense mortar attack followed by a ground assault was launched against Fire Support Base Berchtesgaden. When the first enemy rounds hit, several personnel were injured and in need of immediate medical attention. Captain Snell, with complete disregard for his own safety, moved fifty meters from his protected bunker through the heavy volume of enemy mortar fire to assist the wounded. He carried one wounded soldier back to the safety of the bunker and was going out the bunker door to retrieve another man when a round impacted at his feet and mortally wounded him. Captain Snell's unselfish dedication to his fellow soldiers was directly responsible for saving the life of the man he had carried to safety.[15]

The total surprise achieved by the attackers was the more remarkable, as on three occasions within the week preceding the attack, recon teams of two or three NVA were seen near the base, and sappers had recently hit FSBs AIRBORNE and BRADLEY with early-morning assaults. Several trip flares were set off between midnight and 0300H as the enemy prepared the assault paths, but the defenders assumed that gusty winds caused these incidents. Neither a "mad minute" of firing by the entire perimeter at 0300H nor the use of grenades in the areas where trip flares were ignited caused the enemy to reveal his presence prematurely.[16] In the late 1960s, the North Vietnamese Army was regarded by many as having the finest light infantry in the world. The battle at FSB BERCHTESGADEN demonstrated the reasons for this designation.

Two mornings later,[†] the enemy had FSB CURRAHEE in its sights. This time, however, a vigilant sentry spotted a sapper in the wire at 0125H, alerting the defenders and causing the NVA commander to prematurely attack through the wire, without having breached the concertina and infiltrated the base in advance. The result was a major defeat for the enemy. At the time of the attack, the firebase was occupied by Companies B and E and the command post of the 2/502nd Infantry (LTC George D. Moore, Jr.), Battery A/2/319th Artillery (105mm howitzers), Battery B/2/11th Artillery (155mm howitzers), and an ARVN 105mm howitzer battery. Using illumination from its own 81mm mortars and an air force flare ship to expose the enemy to rocket and minigun attacks by Cobra gunships, several waves of attackers were routed. At 0200, 75 NVA attacked the northwest side of the perimeter; at 0440 it was the western side, and at 0545 they came from the north. With the help of an aerial resupply of ammunition, none of the enemy reached the bunker line.

---

† Sources conflict on whether this attack occurred on 15 or 16 June.

When dawn came and the battle subsided, 51 NVA bodies were found and 3 prisoners taken. American losses were just 7 WIA, of whom only 3 required evacuation.[17] The attackers were from the C-20 Sapper Company of the 29th NVA Regiment,[18] the regiment allegedly "destroyed" at Hamburger Hill on 20 May. This battle demonstrated the other side of the coin – when poor execution by the attackers was coupled with an alert defense, it resulted in a lopsided defeat of the NVA.

On 17 June at 1700H, enemy mortars again hit FSB CURRAHEE. Forty-five rounds of 82mm fire struck the base, detonating a 105mm ammunition bunker and wounding 10.[19] C Company, 1/506th Infantry had been on patrol 5–6 km northwest of FSB BERCHTESGADEN since the beginning of MONTGOMERY RENDEZVOUS, finding some signs of enemy presence but no action until 18 June, when one trooper cleaning by a stream was shot by a sniper; he died shortly after evacuation. In the afternoon, the company was picked up and delivered to FSB CURRAHEE, in the vicinity of which it would patrol for the next four days, looking for the elusive NVA mortar teams who continued to harass the firebase. The NVA welcomed C Company to CURRAHEE with a small mortar attack at 1640H. There were no casualties. From 19–22 June, C/1/506th Infantry continued to patrol and conduct short-range combat assaults within 2 km to the south of FSB CURRAHEE, finding some signs of past enemy occupation but making no contact.[20]

20 June was a red-letter day in the A Shau Valley. The A Shau Expressway from the lowlands (Routes 547/547A) was completed as it linked to Route 548. The event was celebrated with the presence of LTG Richard Stillwell, XXIV Corps CG (to be replaced by General Zais just six days later), and marked by the arrival of allied armor in the valley for the first time ever.[21] An 80-vehicle convoy of the 3/5 Cavalry and the ARVN 3rd Squadron, 7th Cavalry, rolled into the A Shau Valley. As subsequent events would show, the NVA would generally avoid contact with the armored cavalry during its tenure there. A bit later, M-113 armored personnel carriers from the 3/5 Cavalry with attached tanks from the 2/34 Armor would scale Hamburger Hill, to what end besides publicity photographs is not recorded.[22]

On 23 June, Charlie Company, 1/506th Infantry, was lifted to a landing zone about 8 km southwest of the FSB CURRAHEE and just south of the Rao Lao (at YC322917). The area centered on this LZ, in a circle 3 km in diameter on both sides of the Rao Lao and bordering Laos from the northwest to the south, would be a hot spot for enemy engagement over the next several weeks. Throughout the day on the 23rd, C Company patrolled to the southeast of the LZ, finding many freshly used trails. At 1800H, the 1st Platoon was ambushed, the enemy immediately killing the platoon leader and shortly thereafter his RTO. A squad leader took over, leading a charge that broke the ambush, but the damage was done.

An LOH from A/2/17th Cavalry was shot down a little over 1 km south of C/1/506th Infantry on the morning of the 24th. C/2/506th Infantry secured the site, and one crewman with a gunshot wound and one with minor injuries were evacuated. There was sporadic enemy fire but no further casualties. The Loach was a loss and was destroyed. On 27 and 28 June, several large enemy bridges were found by scouts of the air cavalry near the Laotian border and 5–6 km south of the Rao Lao. The bridges (one was 70 meters long) were destroyed by air strikes. C/1/506th Infantry was withdrawn on 28 June and sent to FSB Berchtesgaden, which it would secure until 10 July. On 3 July, the 1/506th received a new commanding officer, LTC Arnold Hayward.[23]

Hill 996 (YC313939): 11 July 1969

The first week in July was unusually quiet in the A Shau. When the lull came to an end, it would be with another battle for a hilltop and the U.S. would incur its heaviest one-day casualties of Operation MONTGOMERY RENDEZVOUS. As the month began, the 1/506th Infantry secured FSBs BERCHTESGADEN and CURRAHEE and, together with the 3/3 ARVN Infantry Battalion, conducted RIFs to the southwest of CURRAHEE. The 2/506th Infantry guarded FSBs AIRBORNE and EAGLES NEST, patrolling southwest and northeast of those bases. The armored cavalry, 3/5 Cavalry, and the ARVN 3/7 Cavalry were doing RIF southwest of A Luoi. The 2/327th Infantry was along Route

547A, just outside the valley. The first week in July, Team Wyoming from L/75th Infantry (Ranger) was patrolling near the valley floor, between Tiger Mountain and Dong So Ridge; it was extracted on 8 July having made no contact.[24]

On the afternoon of 8 July, LOH scouts from A/2/17 Cavalry received fire on several occasions from an area between Hill 996 and the Rao Lao, within 2 km of Laos. The next morning, an FAC reported a similar experience. At 1215H on 9 July, the A Troop ARP was inserted into a green LZ 1,500 meters south of Hill 996 and was reinforced at 1800H by a platoon of D/2/17 Cavalry. Patrols late in the day failed to find the enemy, but just before midnight the enemy found the cavalry, as small arms fire was received. As the night wore on, grenades were intermittently thrown into the NDP (YC312923) and mortar fire was also incoming. ARA, artillery, and Spooky were in support of the troopers, and contact was not broken until after dawn on 10 July. B/1/506th Infantry arrived at 1115H to help evacuate the casualties and guard the LZ for extraction of the A Troop Blues and D Troop infantry. Five men were KIA in these exchanges, four from D troop and one from A Troop, and four were WIA. Three men from D Troop were briefly MIA but were later found alive in a cave. By midafternoon, all cavalrymen had been extracted.[25]

10 July 1969 was a dreary, cold, rainy day in the Rao Lao Valley west of FSB CURRAHEE. Delta Company (CPT Owen Ditchfield), 1/506th Infantry, was inserted into what turned out to be a hot landing zone at 1115H, some distance east of its objective, Hill 996 (YC313939). Hill 996, just over 1 km from the Laotian border and less than 5 km southwest of Hamburger Hill, had been visited by the 1/506th Infantry at the beginning of Operation APACHE SNOW. On 10 May, B Company had killed an NVA officer there who was found to be in possession of plans for attacks on U.S. bases.[26] Delta Company called in artillery fire on the mortar position that was bombarding the LZ, and after silencing the enemy, began its trek toward Hill 996. That night was spent short of the objective, and all was quiet.[27]

The next morning, D Company moved out again. Around 1230H, while crossing a saddle about 400 meters from the top of the hill, the unit came under heavy fire. The fire struck the battalion command group, traveling with the company, severely wounding the new battalion commander, LTC Hayward, the artillery forward observer, LT Len Griffin, and one of the radio operators. The other RTO was killed and both of the battalion HQs radios were destroyed. What happened next in D Company is hazy, but it appears that the company commander lost control of the situation, at least temporarily. One of CPT Ditchfield's RTOs, SP4 George H. Fry, was instrumental on his own initiative in reestablishing radio links with the company's platoons and also with the battalion operations officer. This made it possible to coordinate the movement of Company B toward the opposite side of Hill 996 and eventually to overrun the enemy bunkers. SP4 Fry was posthumously awarded the Silver Star.

At approximately 1230 hours on the cited date, Company D was advancing along a trail when it encountered a heavy volume of enemy small arms, automatic weapons, and rocket propelled grenade fire from an unknown size enemy force in well-fortified positions. During the ensuing battle, two battalion radio-telephone operators were killed and their radios destroyed. Then Company D lost contact with one of its platoons because of the uneven terrain. Realizing that control was impossible without effective communication, Specialist Fry climbed a nearby ridge in an attempt to make communication with all the company's platoons. He remained fully unprotected from the enemy fire and effectively established radio contact within the company and to the Battalion Command Post, allowing the Company Commander to maneuver his elements and to coordinate with the Battalion S-3 element. Disregarding the hostile fire, Specialist Fry held his ground in order to relay messages. It was while relaying one of Company D's messages that he was mortally wounded by the intense enemy fire.[28]

After the battalion command group was put out of action, D Company fought for a while and then moved toward the hilltop, failing to leave a security element with the wounded troops. The NVA came through the area from the rear several hours later, executing LTC Hayward and PFC Curtiss Fernhof. LT Griffin survived by playing dead.[29]

As Delta Company struggled uphill against heavy opposition, Bravo Company (CPT Harold J. Erickson) was sent in to attack Hill 996 from the opposite direction. The top of the hill was about 100 meters long by less than 50 meters wide, with the long axis running southwest to northeast. Steep slopes on the eastern and western sides channeled attackers to approach along the long axis, and the crest was defended by 13 well-camouflaged bunkers. While D Company kept up pressure from the north, by about 1730 hours B Company was approaching from the south. Leaving most of the company to establish a night defensive position, the 2nd Platoon moved uphill in driving rain, until about 50 meters from the top when enemy fire from hidden bunkers wounded five men. At this point, then–SP4 Gordon Roberts launched a one–man attack against the enemy force, for which he would receive the Medal of Honor.

Sergeant [then SP4] Roberts's platoon was maneuvering along a ridge to attack heavily fortified enemy bunker positions which had pinned down an adjoining friendly company. As the platoon approached the enemy positions, it was suddenly pinned down by heavy automatic weapons and grenade fire from camouflaged enemy fortifications atop the overlooking hill. Seeing his platoon immobilized and in danger of failing in its mission, Sergeant Roberts crawled rapidly toward the closest enemy bunker. With complete disregard for his safety, he leaped to his feet and charged the bunker, firing as he ran. Despite the intense enemy fire directed at him, Sergeant Roberts silenced the two-man bunker. Without hesitation, Sergeant Roberts continued his one-man assault on a second bunker. As he neared the second bunker, a burst of enemy fire knocked his rifle from his hands. Sergeant Roberts picked up a rifle dropped by a comrade and continued his assault, silencing the bunker. He continued his charge against a third bunker and destroyed it with well-thrown hand grenades. Although Sergeant Roberts was now cut off from his platoon, he continued his assault against a fourth enemy emplacement. He fought through a heavy hail of fire to join elements of the adjoining company which had been pinned down by the enemy fire. Although continually exposed to hostile fire, he assisted in moving wounded personnel from exposed positions on the hilltop to an evacuation area before returning to his unit. By his gallant and self-less actions, Sergeant Roberts contributed directly to saving the lives of his comrades and served as an inspiration to his fellow soldiers in the defeat of the enemy force.[30]

SP4 Roberts's squad leader provides the rest of the story:

After his one-man assault, we received very little fire from the hill, and within twenty minutes the hill was ours. We found four dead enemy soldiers on top of the hill, and a varied assortment of weapons used to defend the hill. The most important factor [in] taking the hill was the fact that the company [D] we were helping was caught in a saddle with enemy soldiers coming in from behind them. If it wasn't for Specialist Four Roberts's outstanding individual effort, I believe that the company would have received a greater number of casualties.[31]

SP4 Roberts passed entirely through the enemy perimeter and assisted with the evacuation of Delta Company's wounded, before returning to B Company at 2200H that night. Gordon Roberts would go on to received two Silver Star Medals, including one (see below) for action in the A Shau Valley just 10 days later. Barely 19 years old in July 1969, he would leave the Army in 1971, return to the service in 1989, and be the last surviving Vietnam War Medal of Honor recipient on active duty, before retiring with the rank of Colonel. He was one of the most decorated American soldiers of the war.

As implied above, while the actions of a few individuals in the battle at Hill 996 are well documented, the overall picture of the maneuvers of the companies involved throughout the afternoon and evening of 11 July is murky at best. According to the Virtual Wall, 20 men were killed in action: 6 from HHC/1/506 (the battalion commander, 2 RTOs and 3 medics), 8 from D/1/506, 1 from B/1/506, and 5 from C/2/506. Twenty-six troopers were wounded. NVA losses were reported at 10 KIA.[32] There is no mention in the medal files of C/2/506 being committed to the fight, nor do the maps in the Roberts file show that unit. On the other hand, I have found no record of any other action involving C/2/506th Infantry on 11 July 1969.

According to Len Griffin, CPT Ditchfield was relieved of command of D Company on 12 July. For five days after the battle, elements of 1/506th Infantry remained on the hill, constructing a three-ship LZ and conducting local patrols.

"Heavy and Continuing Contacts:" 13–31 July 1969

"In heavy and continuing contacts west of FSB CURRAHEE after 9 July, 101st Abn Div (AM) elements identified at least two battalions of the apparently recently reinfiltrated 803d Regt. The regiment may have been called upon to fill the void left by the badly defeated 29th Regt. As with the 29th Regt, the 803d has been aggressive in its offensive activities."[33] On 13 July, in the Warehouse 54 area (YD367076), A/2/506 Infantry captured a cache on high ground containing 1 machine gun, 65 SKS rifles, over 400 mortar rounds, and other ammunition. The next day in the same area, D/2/506 Infantry bagged another 1,000 mortar bombs.[34] On 16 July, D/1/327th Infantry fought with an enemy company 10 km south southwest of A Luoi, killing 10 NVA while suffering 1 KIA and 7 WIA.[35] On 18 July elsewhere in the MONTGOMERY RENDEZVOUS AO, a battle at a hot LZ resulted in 14 U.S. WIA and 1 NVA KIA.[36]

The new kid in town in July for the 3rd Brigade was the 1/327th Infantry, committed to MONTGOMERY RENDEZVOUS on 13 July.[37] "Tiger Force," the battalion recon platoon, flew into Hill 996 on 17 July. Two days later they were ambushed. In a series of actions, initially as the platoon sergeant and then, after the platoon's officer was wounded, as acting platoon leader, SSG John G. Gertsch became the second Screaming Eagle in less than 10 days to earn the Medal of Honor in the vicinity of Hill 996.

During the initial phase of an operation to seize a strongly defended enemy position, Staff Sergeant Gertsch's platoon leader was seriously wounded and lay exposed to intense enemy fire. Forsaking his own safety, without hesitation, Staff Sergeant Gertsch rushed to aid his fallen leader and dragged him to a sheltered position. He then assumed command of the heavily engaged platoon and led his men in a fierce counterattack that forced the enemy to withdraw. Later, a small element of Staff Sergeant Gertsch's unit was reconnoitering when attacked again by the enemy. Staff Sergeant Gertsch moved forward to his besieged element and immediately charged, firing as he advanced. His determined assault forced the enemy troops to withdraw in confusion and made possible the recovery of two wounded men who had been exposed to heavy enemy fire. Sometime later his platoon came under attack by an enemy force employing automatic weapons, grenade, and rocket fire. Staff Sergeant Gertsch was severely wounded during the onslaught but continued to command his platoon despite his painful wound. While moving under fire and encouraging his men, he sighted an aidman treating a wounded officer from an adjacent unit. Realizing that both men were in imminent danger of being killed, he rushed forward and positioned himself between them and the enemy nearby. While the wounded officer was being moved to safety, Staff Sergeant Gertsch was mortally wounded by enemy fire. Without Staff Sergeant Gertsch's courage, ability to inspire others, and profound concern for the welfare of his men, the loss of life among his fellow soldiers would have been significantly greater.[38]

Five troopers were lost and nine wounded in the action on 19 July.[39] Like Gordon Roberts, SSG Gertsch was one of the most highly decorated soldiers of the Vietnam War, receiving, depending on the source, somewhere between two and five Silver Stars. He did not survive to be presented with the Medal of Honor.

On 20 July, about 5.5 km southwest of A Luoi, C and D Companies of the 2nd Battalion, 506th Infantry were ambushed by heavy small arms, grenade, and rocket-propelled grenade fire. Three troopers were KIA and 10 were WIA in this firefight; 3 NVA bodies were found.[40]

The next day, B/1/506th was in another firefight, and SP4 Gordon Roberts, who would receive the Medal of Honor for his actions just 10 days earlier, was once again a paragon of courage.

As Company B began to ascend the slope of a ridge north of Dong Ap Bia Mountain, the lead element came under intense enemy small arms, rocket-propelled grenades, and automatic weapons fire from an undetermined size enemy force situated in well-fortified bunkers. During the initial burst of fire, the lead man was seriously wounded and remained in the line of enemy fire. The company commander called for smoke grenades in an attempt to mark the location so that aerial rocket artillery could be employed on the entrenched enemy. Specialist Roberts then moved up from the rear of the company with a load of smoke grenades. He threw the smoke grenades between the wounded man and the enemy positions. As enemy small arms rounds impacted all around him, he dragged his wounded comrade to safety. He then moved to a good location and began firing an accurate, heavy volume of fire at the enemy, killing one of the insurgents. It was his courageous efforts that kept friendly casualties to a minimum.[41]

At 0500H on 26 July, A/1/506th Infantry, still operating near Hill 996 about 2 km from the Laotian border, was attacked by two platoons of NVA sappers in its NDP. The sappers penetrated the perimeter and used small arms and RPGs; the Currahees fought back with M16s, M79s, M60s, and grenades, supported by artillery and a USAF C-47 gunship. When the battle was over, 8 GIs were dead and 11 wounded. The enemy left 16 bodies behind.[42] The final tally for the month of July 1969 was 47 KIA and 157 WIA friendly casualties, and 180 NVA KIA.[43]

Late in the month, a prisoner of war captured one kilometer south of Dong Ap Bia confirmed that the 9th Battalion of the 29th PAVN Regiment arrived too late to participate in the battle for Dong Ap Bia, but was now operating in the area.[44]

The Hill 996 Area: 1–15 August 1969

In early August, the NVA continued to launch hit and run attacks against allied NDPs throughout the MONTGOMERY RENDEZVOUS AO. Allied intelligence placed three enemy regiments in or near the A Shau Valley. "The 29th and 803rd Regiments, with about 2,200 men each, were believed to be located in Base Area 611 west of the A Shau Valley. The 806th Battalion, 6th NVA Regiment, was located in the vicinity of YD3310 [Warehouse 54]. Other 6th NVA regiment forces were unlocated."[45] Troop A, 2/17th Cavalry, continued to search for the enemy throughout the AO, with a primary focus on the area west of Hill 996 and along the Rao Lao Valley west of FSB CURRAHEE. On 1 August, the A Troop Blues inserted just north of Dong So Ridge, finding only an abandoned classroom and cave. On 4 August, after scout helicopters took ground fire in the area, the Blues were inserted along the border 1 km northwest of Hill 996. They accounted for 2 NVA KIA before extraction later that day.[46] On the evening of 3 August, an element of the 3/5th Cavalry, including tanks, was hit just east of Dong Ap Bia. The armored unit suffered no casualties, but killed 9 NVA and captured 2 crew served and 4 individual weapons.[47] On 4 August, the ARVN 3rd Regiment FSB BRADLEY received twenty 122mm rockets; casualties were light.[48] That same day, the 1/506th Infantry and 2/3 ARVN at FSB CURRAHEE were hit with twenty-two 122mm rockets. Casualties were light, but 3,000 gallons of fuel were destroyed.[49]

The 1/506th Infantry continued to operate west of FSB CURRAHEE. On 5 August, Charlie Company (CPT Reg Moore) was working the area between Hills 916 and 996 in platoon patrols and staying mainly along ridgelines to avoid being trapped by an ambush. From the time one of the point men first spotted and killed two NVA soldiers at 0725 until the last enemy mortar shells fell over 13 hours later, the company was in on-and-off firefights throughout the day. By the time it ended, 10 enemy soldiers had been killed and C Company had 1 KIA and at least 6 WIA. The company commander decided to make a rare night move on the night of 5–6 May in order to consolidate his scattered platoons and to avoid the enemy attacking him in fixed positions. After the arrival of the 4th Platoon from FSB BERCHTESGADEN and a bit of rest time for the sleepless troopers, the company moved out at 1100H on 6 August, just a few hundred meters northeast of Hill 996.

Just a few weeks after the battle for Hill 996, the area between Dong Ap Bia and the Rao Lao was once again teeming with NVA troops spoiling for battle. As on the previous day, C Company fought a prolonged series of firefights, several initiated by enemy ambushers in bunkers, with "business hours" starting just after noon and lasting until the final medevac was complete at 0345 hours the next morning. Four Currahees were killed and at least eight wounded in a day that saw many acts of individual courage and superb leadership under fire.[50]

The 1/506th Infantry's sister battalion, 2/506th Infantry, was also battling the enemy near the Laotian border. On 5 August, B and C Companies both were in contact (YC312955), losing 5 men KIA and 18 WIA, while claiming 6 NVA KIA. Two days later, B/2/506th Infantry again made contact with an enemy force of unknown size, killing 2 but losing 1 KIA and 16 WIA in a short, vicious firefight from 0730 till 0815.[51]

On 10 August, the last significant battle of the operation occurred when B/1/327th Infantry destroyed an enemy mortar position near the crest of Hill 996 (YC311938). Eleven NVA soldiers were KIA in this fight and the Currahees had only five wounded. Later that afternoon, as they searched toward the border with Laos, B Company was ambushed. This time, the result was much worse. Five Americans were killed, including platoon leader 2LT Carl A. Peterson (Citadel, '68),[52] and six were wounded, with only one NVA body found.[53]

One of the final skirmishes of MONTGOMERY RENDEZVOUS also involved the 1/327th Infantry, when Charlie Company fought a 20-minute dawn battle on 12 August with an enemy force of unknown size just 1,500 meters from Laos near Hill 996. When he withdrew, the NVA commander left five bodies behind. The Screaming Eagles had five WIA and none KIA.[54] The other battle on 12 August was fought one km northwest of Hill 996, right on the Laotian/South Vietnamese border. While crossing a stream, C/1/506th Infantry was ambushed with small arms and machine gun fire by a well-camouflaged and fortified enemy unit and took one KIA and two WIA early in the fight. Enemy casualties were not reported.[55]

FSB CURRAHEE continued to be frequently attacked by indirect fire, receiving thirty-four 82mm mortar rounds on 11 August, twenty 122mm rockets on 15 August, and thirteen 122mm rockets on 16 August.[56]

MONTGOMERY RENDEZVOUS and KENTUCKY JUMPER: Recap

The basic mission of the KENTUCKY JUMPER operations in the A Shau Valley was to "locate and destroy enemy forces, caches, and LOC,"[57] and to "establish a firm Allied presence in the A Shau and cork up that traditional infiltration route."[58] When KENTUCKY JUMPER was conceived in early 1969, there was no plan for the withdrawal of American forces from South Vietnam. When that plan was first formalized in National Security Decision Memorandum 9 on 1 April 1969, it was with the expectation that "there will be no de-escalation except as an outgrowth of *mutual* troop withdrawals" (emphasis added).[59] Thus, the concept, culminating with the completion of the upgrade of Route 547/547A as an all-weather ground supply route from the coast and the construction of a new airfield north of Ta Bat, was for a *permanent* allied presence in the center of the A Shau Valley.

But while allied control was postulated, a contested AO was all that could be achieved by the time MONTGOMERY RENDEZVOUS was terminated. As the last two weeks of operations had evinced, all allied battalions in the AO were still opposed by enemy forces. And as the operations throughout 1969 had revealed, wherever in the A Shau watershed the allies went, from Route 922 and Tam Boi in the north, to Tiger Mountain, to Dong So Ridge, to Warehouse 54, to A Luoi, Dong Ap Bia and Ta Bat, to the Rao Lao Valley draining into Laos and guarded by Hill 996, to south of the old A Shau airstrip, the North Vietnamese were there. There when the allies arrived; there when the allies departed.

Yet the valley was indeed a very different place than it had been just 18 months earlier. Gone were the batteries of North Vietnamese anti-aircraft artillery that had greeted the First Cavalry Division. Gone were NVA truck convoys that paraded down Route 548 with their lights on. Gone were the enemy tanks and bulldozers. In place of the engineers, anti–aircraft batteries, rear area service troops, and counter–reconnaissance teams that denoted its status as a secure NVA base prior to Tet 1968, the valley was now defended by an entire division of enemy infantry. The A Shau was in the front line. To be sure, the combat support and service units assigned to Binh Tram 42 (formerly BT 7) had not gone far. In early August, four engineering battalions, two anti–aircraft battalions, one artillery battalion, and two transportation battalions were still assigned but were mainly just across the border in Base Area 611, awaiting the time when they could resume work in the valley. "Route 922 was . . . the only road the NVA tried to keep open in the I Corps area during the 1968 and 1969 summer wet seasons."[60] Route 922 led directly to Route 548 at Tiger Mountain.

Whereas prior to Tet 1968, the enemy had sought to draw the allied maneuver battalions away from the lowlands and cities and into the hinterlands, where they could not support pacification and protect the populace, now the NVA leaders saw their LOC threatened and tried to draw the 101st Airborne Division back to the coast!

> Since early July there [had] been [a] significant increase in activity by the 5th Regt in central THUA THIEN (P). In the first three weeks of July, all subordinate units of the regiment, with the exception of the 810th Bn, were identified in contact. Of particular importance was the activity by the CHI THUA I and II Sapper Bns. . . . The increased offensive action by the 5th Regt may [have been] an attempt to create enough activity in the lowlands to force the allies to divert their attention and forces from the A SHAU Valley, as continued interdiction of the A SHAU suppresses the bulk of logistical activity for the 4th and 5th Regiments.[61]

It wasn't only the NVA regiments in central Thua Thien that were discomforted by the allies in A Shau.

> An even more drastic alteration of LOCs occurred by mid-1969 when permanent Allied control of the valley forced the NVA to supply Base Area

112/127 [the staging point for attacks on Da Nang] by detouring south down the Ho Chi Minh Trail to Chavan, east by route 165/966 to the I Corps/II Corps Border, and then northeast along Route 14. This lengthy detour to the south more than doubled the distance to Base Area 112.[62]

Patience. This had always been (with Tet 1968 the highly notable exception) a hallmark of the North Vietnamese strategy for the unification of Vietnam. With the withdrawal of U.S. combat units from the country already underway, just one more operation and six more weeks would reveal the *impermanence* of the allied positions in the A Shau Valley.

21

OPERATION LOUISIANA LEE

Operation LOUISIANA LEE

Hard on the heels of KENTUCKY JUMPER came LOUISIANA LEE. It might have been named "SHADOW BOXING" for all of the blows that were struck against the North Vietnamese Army. When the books were closed on 28 September, the final official tally was 67 enemy KIA and 3 captured, at an official cost of 7 Americans KIA and 34 WIA.[*1]

As LOUISIANA LEE commenced on 15 August, the 3rd Brigade of the 101st Airborne Division had three battalions committed in the vicinity of the A Shau Valley. The 1/506th Infantry (LTC David R. Pinney), with three companies, conducted reconnaissance in force (RIF) patrols out of FSB CURRAHEE (YC399949), while B/1/506 manned the perimeter at FSB BERCHTESGADEN (YD425011). The 3/187th Infantry protected the allied line of communication along Route 547, from FSB BLAZE (YD530020) to FSB RENDEZVOUS (YC432961). Armored cavalry of the 3/5th Cavalry (LTC Joseph L. Hadaway) carried out RIF operations in the southern half of the A Shau Valley. The helicopters and aero-riflemen of A Troop, 2/17th Air Cavalry, continued to scout the length and breadth of the valley and the surrounding mountains for signs of the increasingly invisible enemy. The ARVN 2/3 Infantry also operated out

---

* The official U.S. KIA number is too low by at least 100%. At least 5 men from 1/506th Infantry, 2 from the 3/187th Infantry, 4 from the 59th Engineer Company, 2 from the 326th Medical Battalion, and 1 from A/158th Assault Helicopter Battalion were killed during LOUISIANA LEE. Details are included in the text.

of FSB CURRAHEE, while Troop 2 of the ARVN 7th Armored Cavalry worked with its U.S. counterpart.

The NVA forces identified in the area included the 29th and 803rd Regiments, which were believed to be rebuilding in Base Area 611 a few kilometers west of Hamburger Hill (YC2998), and the 806th Battalion of the 6th NVA Regiment, still based in the Warehouse 54 area about 8 km east of Tiger Mountain (YD3310).[2]

LOUISIANA LEE – 1/506th Infantry Area of Operations

The NVA presence around Hill 996 persisted. C/1/506th Infantry patrolled to the north and west of CURRAHEE and inserted a six-man ranger-type patrol near Hill 996 after dark on 15 August to monitor enemy activity in the wake of the departure of larger U.S. forces from that area. The team remained in place until 20 August, when it was withdrawn under fire. Although one or two men were wounded, the extraction was completed. On 19 August, the closure of FSB CURRAHEE was begun, with the backhauling or demolition of anything and everything that might be of use to the North Vietnamese.

On 21 August, 1/506 Infantry sent an RIF element into the hills on the east side of the valley (YC5091). There were no contacts reported, but two days later, a UH–1H (67-17644) was downed in the same area. The door gunner, SP4 William W. Ellis, III of A/158 Assault Helicopter Battalion, was KIA.[3]

While the Americans began to withdraw their first units from Vietnam (the 9th Infantry Division in August 1969) and to shut down their last bases on the valley floor, the NVA provided reminders that as the heirs–apparent of the possession of the A Shau Valley, they would not delay in reoccupying their bases there. On 23 August, twenty 60mm mortar rounds fell on the soon–to–close FSB CURRAHEE, killing SSG Juan L. G. Duenas and medic SP4 Philip J. Vevera of the 1/506th Infantry.[4]

At 0320 hours the following day, a combined mortar and sapper attack by the 806th NVA Battalion and K–12 Sapper Battalion hit the 3rd Brigade headquarters at FSB BERCHTESGADEN. Defended by Bravo and

Echo Companies of the 1/506th Infantry, with air support from ARA and "Spooky" USAF gunships, the CURRAHEES got the best of the enemy once again. Thirty-one NVA were killed, with 8 RPG launchers and 6 AK-47s captured. U.S. casualties were 3 KIA (CPL William F. Golliday, PFC Lenard F. Moeggenborg, and PFC Richard J. White) and 8 WIA.[5]

While A and D Companies continued to RIF farther afield with no contact, C/1/506th Infantry was at CURRAHEE from 19–26 August, closing the base and sending out local patrols up to 4 km to the west and southwest. On 20 August, members of the company took a final swim in the nearby Rao Lao, one of the few pleasant memories that would endure from the battalion's long tenure in the valley. On 24 and 25 August, daily indirect fire attacks on CURRAHEE continued, with nine 122mm rockets counted on the 24th and six on the 25th.[6] On 26 August, leaving behind two ambushes, the company abandoned the remains of FSB CURRAHEE and moved to FSB RENDEZVOUS. Two days later, the company was returned to CURRAHEE for a final "police call."[7] Also, on 28 August, a B-52 ARC LIGHT strike hit the area northeast of Hill 996, through which the 1/506th Infantry had fought its way to Hamburger Hill three months earlier.[8] The 1/506th Infantry continued to work in the FSB RENDEZVOUS / FSB BERCHTESGADEN / Route 547 area until on or about 21 September, when the battalion was withdrawn to Camp Evans.

LOUISIANA LEE – 3/5th Cavalry Area of Operations

The allied armored cavalry, running free on the floor of the A Shau Valley at last, soon found that the NVA had no stomach for a fight with tanks and armored personnel carriers. On D-Day (15 August), "Task Force 3-5 Cav conducted a road march from FSB Rendezvous to the A-Shau Airstrip vicinity grid YC495836. From this point, detailed reconnaissance-in-force and land clearing operations were conducted from 15 August to 7 September 1969." Attached to the 3/5th Cavalry for this operation were B Company, 3/187th Infantry; 2nd Platoon, C Company, 2/34th Armor (M-48 tanks); engineers from the 45th Engineer Group with Rome Plows; and the 2nd Company, 7th ARVN Armored Cavalry (replaced by the 3rd Company on 24 August). The Cavalry Squadron also took along its own artillery as it marched down the A Shau: the self-propelled 105mm howitzers of A Battery, 1/40th Artillery.[9]

As was often the case when allied troops entered an area previously dominated by the enemy, the first few days brought frequent contact with small enemy elements. Subsequently, it became harder and harder to "find, fix, and finish" enemy units as they left the area to await the withdrawal of the allied troops. On 15 August, A/3/5th Cavalry reported one enemy KIA after that individual took on the armor with his RPG-7. Aerial scouts from A/2/17th Air Cavalry directed the ground force to several enemy sites, but no further contact was made. B/3/187th Infantry, providing a light infantry component for the task force, ran into another NVA soldier the next day, whose single-handed attack also resulted in his death and the capture of his AK-47 assault rifle.

On 17 August, about 1.5 km north of the A Shau airstrip, B/3/187 was ambushed by an unknown-sized enemy force. This firefight resulted in the elimination of two more NVA soldiers and the capture of two 12.7mm heavy machine guns. Two GIs were KIA (PFC Curtis Bowman and SP4 Jayson F. Ulrich) and two wounded.[10] Probably while en route to pick up the wounded from this engagement, an OH-6A medevac helicopter from the 326th Medical Battalion, Eagle Dustoff 905, was shot down on the east side of the valley (YC546847).[11]

This LOH is thought to be the only OH-6A to carry DUSTOFF red crosses during the Vietnam War. The air ambulance platoon borrowed this LOH from the 3rd Bde, 101st ABN as an experiment to see if they could eliminate some UH-1H host missions by taking advantage of the LOH's small size. It served as an air ambulance for just three weeks prior to being destroyed.[12]

The pilot, CPT George L. Miner, and the aircraft commander, WO1 Gerald L. Caton, were killed and the LOH was destroyed. The NVA soldiers killed were the last reported enemy casualties recorded by the 3/5th Cavalry and attached units during the operations. While no soldiers in the

3/5th were lost, four more troops attached to the squadron would die on 31 August, for a total of six Americans vs. four NVA KIA for Task Force 3-5 Cav.

In the ensuing days, the armored cavalrymen, assisted by the air cavalry, operated throughout the southern part of the valley, capturing a number of abandoned NVA vehicles and anti-aircraft weapons. On 20 August, A/3/5th Cavalry took possession of a Russian-made bulldozer and two 2½ ton trucks (YC507806). On 22 August, at the very southern end of the valley less than 1 km north of the Laotian Border, the same troop found a 37mm anti-aircraft gun carriage and three 2½ ton trucks. The next day, nearby, the 2/17th Air Cavalry found two additional 37mm anti-aircraft gun carriages and three more 2½ ton trucks. No major caches were uncovered, but caves, small ammunition caches, and a few lighter enemy weapons were taken. Enemy soldiers in small numbers were sometimes sighted and engaged, with unknown results.[13]

B/3/187th Infantry conducted an RIF to the border with Laos, some five kilometers west of the A Shau airstrip, on 25 August, finding little except "five boxes of .51-cal ammo at least 3 months old." It patrolled in the vicinity of the border the next day but found nothing more.

On 31 August, while trying to recover a damaged bulldozer one kilometer north of the A Shau airstrip, an engineer from the 59th Engineer Company (Land Clearing) stepped on an anti-personnel mine, resulting in the deaths of four combat engineers (SP5 Douglas A. Bennett, SP4 Jerome E. Bowers, SGT Michael Preslipski, Jr., and SFC Edwin B. Ryder).[14] Two Americans and one ARVN soldier were also wounded in this incident.

From 1 September on, "negative enemy activity" was the usual report. On 6 September, the cavalry squadron packed up and move north back to FSB RENDEZVOUS. On 7 September it was ordered to return to Camp Evans for a maintenance stand down.[15]

Unlike some After Action Reports, the 3/5th Cavalry summary was realistic in its self-evaluation: "Throughout the operation, enemy contact was light and sporadic. . . . The land clearing operations opened large areas of the valley floor which facilitated the surveillance of enemy infiltration routes. Continuous and widespread movement of TF 3-5 Cav denied the enemy full use of his infiltration routes in the southern portion of the A Shau Valley."[16]

LOUISIANA LEE – Blues and Rangers

On 19 August, the Aerorifle Platoon of A/2/17th Air Cavalry with a scout dog team attached was inserted several kilometers northeast of Dong Ap Bia. It was extracted 2½ hours later, having made no contact with the enemy. A similar insertion took place on 21 August two kilometers to the west, followed by a patrol one kilometer to the east to a pick-up zone. This likewise resulted in a "dry hole." Following up on aerial recon reports, the A Troop Blues were dropped just north of Dong So Ridge on 24 August. Spending four hours on the valley floor, the air cavalrymen destroyed an enemy bridge over the A Shau and a few huts but did not encounter any live enemy.[17]

In early September, the 2/17th Air Cavalry moved its reconnaissance focus briefly to the Warehouse 54 area once again. On 11 September, the A Troop Blues were inserted (YD365010) at 1357 hours and extracted at 1605, with no enemy contact. Two days later, their landing zone was about 1 km north of the abandoned CURRAHEE firebase (YC386961), and again they found no sign of the enemy. The recon of old hot spots of North Vietnamese activity continued on the 14th, when the platoon was inserted just off the eastern end of Dong So Ridge (YC373016), where the 1st Cavalry Division's arrival in the valley had made so much trouble for the enemy in April 1968. This patrol, too, came up empty handed. On 16 September, the Blues were put down 2½ km southeast of FSB RENDEZVOUS (YC443938). They trooped west five kilometers to the old CURRAHEE location (YC397942), where they were picked up six hours later, after yet another walk in the A Shau sunshine. With little rest, the aerial rifle platoon was reinserted at 1220 hours on 17 September about 5 km northwest of their pickup zone from the previous day (YC371974). They were extracted at 1750 hours, "negative sitrep," completing their fifth patrol in seven days. After a few days' well-earned rest, the Blue Platoon of A Troop, 2/17th Cavalry, wrapped up its participation in LOUISANA LEE with a double-header on 21 September, revisiting the LZs used five days earlier, but this time by inserting on the valley floor (YC442936) at 1217 and extracting at 1340, and then repeating the "quickie" with an insertion at CURRAHEE (abandoned) (YC396943) at 1344 and extraction at 1521 hours.[18]

The Rangers of L/75th Infantry returned to the A Shau Valley on 14 September, when Recon Teams Burma (YC477880) and Dahomy (YC435910) were inserted 3 km southeast and west of Kon Tom, respectively. These teams would observe the area recently vacated by TF 3-5 Cav as it moved north to depart the valley. At 1035 hours on 18 September, both teams were extracted from the mountains west of the valley, having spent four days in the field and seen no enemy.[19]

None of the ARP or Ranger Team patrols got into fights with the shadowy North Vietnamese.

LOUISIANA LEE – *Other Activities*

In the latter part of the operation, the 3/187th Infantry was relieved of its security role on Route 547 by 2/506th Infantry and tasked with conducting RIF in the northern part of the A Shau Valley and the southern Da Krong Valley (vicinity of former FSBs ERSKINE AND CUNNINGHAM). "Elements were inserted and one company encountered CS type agent and small arms fire on one LZ. The remainder of the operation yielded no significant contact, but units found numerous signs of recent activity."[20] In order to facilitate operations in the northern valley, FSBs TIGER and ERSKINE were briefly reopened by B Company, 326th Engineer Battalion on 14 September.[21]

"In late September, a plan was developed to close all fire bases in the A Shau Valley."[22] This process had actually begun in August, with the closure of FSB CURRAHEE on 26 August and the closure also by 15 September of FSB BERCHTESGADEN. The enemy shifted the focus of indirect fire attacks to FSB RENDEZVOUS, which received fourteen 122mm rockets on 15 September.[23] On 26 September, RENDEZVOUS was closed by the 3/187th Infantry.[24]

The Success of DEWEY CANYON, KENTUCKY JUMPER, and LOUISIANA LEE

Allied operations in the A Shau Valley in 1969 had resulted in the destruction of large amounts of enemy equipment and munitions. The Rome Plows of the 59th Engineer Company cleared 1,100 acres along Route 548 on the valley floor, facilitating improved aerial observation and targeting of the enemy should it try to reopen that infiltration and supply route.[25] More importantly, the constant presence of allied forces, extended into September 1969, denied the NVA the critical bases and communications arteries in and through the A Shau Valley. The allies briefly reopened the oft-used firebase at Tiger Mountain at the northern extreme and occupied the abandoned A Shau airstrip at the southern end of the valley with armored cavalry, while maintaining firebases in the central valley (FSBs CURRAHEE and RENDEZVOUS). By late 1969, this resulted in the dropping of Base Areas 101 (south of Quang Tri) and 114 (the staging area for enemy attacks on Hue) from the list of active NVA bases in I Corps.[26] Not only did the occupation of the valley place a chokehold on enemy supply of his operations in northern I Corps (Quang Tri and Thua Thien Provinces, except for the DMZ area), but it also had a major impact on the NVA activity directed at Da Nang (Quang Nam Province). In 1968, the enemy had constructed a truck route along Route 614 in western Quang Nam Province, connected to the southern terminus of Route 548 through A Shau, facilitating a much-improved logistical situation for Base Area 112, its forward base for offensive action directed at Da Nang. But "there was no evidence of truck traffic after November 1969."[27] The ground interdiction of the southern A Shau Valley thus disrupted, for as long as it could be maintained, North Vietnamese operations over the entire northern part of South Vietnam.

The American Withdrawal Begins

"In late September, all Allied troops [in the A Shau] were withdrawn when the 101st Airborne redeployed to Quang Tri Province to replace the

3d Marine Division, being pulled out of Vietnam. For a time, the enemy was cautious about reentering the A Shau Valley in strength, but it was only a matter of time before NVA logistics would again move through terrain the Allies had once securely held."[28]

On 8 June 1969, during talks at Midway Island with President Thieu, President Richard Nixon made the initial announcement that he had

... decided to order immediate redeployment from Vietnam of a division equivalent of 25,000 men. This troop replacement [sic] will begin within the next 30 days and still be completed by the end of August. During the month of August and at regular intervals thereafter, we shall review the situation, [bearing] in mind the three criteria that I have previously mentioned with regard to troop replacement [sic]:

First, the progress insofar as the training and equipping of South Vietnamese forces;
Second, progress in the Paris peace talks;
Third, the level of enemy activity.

I will announce plans for further replacements (sic) as decisions are made. As replacements (sic) of United States forces begin, I want to emphasize two fundamental principles: no actions will be taken which threaten the safety of our troops, and the troops of our allies, and second, no action will be taken which endangers the attainment of our objective, the right of self-determination for the people of South Vietnam.

The initial increment of U.S. troop withdrawals, completed in August, included two brigades of the 9th Infantry Division in the southern part of the country and the 9th Marines from I Corps. "'We drew out the 9th Marines,' recalled Major General William K. Jones [CG, 3 Mar Div], 'because they were the Swing/Ready regiment; the regiment that was sort of a Division reserve, or not occupying a fire support base.'"[29] The battalions of the 9th Marines departed Vietnam on 14 July, 1 August, and 13 August.[30]

The initial U.S. troop withdrawals had little impact on the overall allied situation, drawing on combat battalions that were either in a reserve role or in areas where enemy activity was low. That would not be the case as the "redeployments" continued, and one of the first areas to be affected, in a very profound and prolonged way, was the A Shau Valley. This occurred not because the 101st Airborne Division was tagged to leave Vietnam (it would in fact be the last complete U.S. Division to remain in country) but because the 3rd Brigade was withdrawn from the A Shau to screen the AO of the 4th Marines when that regiment headed for Okinawa in October and November 1969. The sudden and complete departure of American ground troops from the valley was a direct contradiction of the KEN-TUCKY JUMPER plan, which foresaw the requirement for a *permanent* ground troop commitment in the A Shau.

Why did COMUSMACV select the A Shau, of all areas in the Republic of Vietnam, to be one of the first places in the country to be for all intents and purposes ceded to the enemy, after struggling for 17 months to wrest it away from the NVA? There appear to have been several considerations. As he pondered his recommendations for the next round of "redeployments" (due to be submitted by 3 August 1969), on 24 July 1969, General Abrams briefed RVN Defense Minister LTG Nguyen Van Vy (later removed from office after the alleged embezzlement of several million dollars from the ARVN pension fund) on "assessment at this time relative to a further reduction of U.S. forces," and asked his input concerning, in particular, the importance of border areas to the GVN (Government of Vietnam). When asked about Kontum, Pleiku, Phuoc Long, Tay Ninh and other "border areas" in II and III Corps, General Vy responded that "the GVN regards these border provinces, these border areas, as *extremely* important and would regard the loss of one or any of them as an extremely serious political setback" (emphasis in original).[31] As these areas were generally in the AOs of United States forces, the American divisions committed there (1st, 4th, and 25th Infantry) were, at least for the time being, taken off the table in selection for the next increment of troop withdrawals. Why wasn't A Shau included in this discussion? Presumably because, unlike the "border areas" farther south, there was no Vietnamese population to protect. The *military* value of the

A Shau was defined by its importance to the enemy, whereas the military significance of Kontum, Pleiku, et al. was due to their *political* importance to the Republic of Vietnam.

Within I Corps, there were of course other units besides the 3rd Brigade, 101st Airborne Division, that might have been used to screen in the north for the 3rd Marine Division's departure. But the 1st Cavalry Division had already been relocated to III Corps, and the 3rd Brigade, 82nd Airborne Division would itself depart in December 1969. By January 1970, the 1st Brigade, 5th Infantry Division (Mechanized) would be in Quang Tri, the 101st Airborne Division in Thua Thien, the 1st Marine Division in Quang Nam, and the Americal Division in Quang Tin and Quang Ngai. The position of MACV in the fall of 1969 was that none of the U.S. units operating in the populated areas along the coast could be spared for operations in the hinterlands. The Commanding General, 101st Airborne Division from May 1969 to May 1970, MG John M. Wright, Jr., discussed this decision in his Senior Officer Debriefing Report:

In August [1969], the division staff began detailed studies on the expected effect of the pending monsoon season on the division's ability to sustain operations in the mountainous regions of the AO, particularly in the A Shau Valley. Although weather would be a major factor influencing the division's ability to keep open the air and ground lines of communication into the A Shau Valley, the decision was made in early September to maintain forces in the A Shau Valley. This decision was altered shortly thereafter, when the division was assigned the mission to screen the redeployment of the 3rd Marine Division from Quang Tri Province. A study of the mission, requiring a brigade of two infantry battalions, dictated a revision of the earlier estimate and decision. *To maintain forces in Quang Tri Province and in the A Shau Valley would have major impact on the division's ability to support pacification in the populated lowlands* [emphasis added]. . . . The operations of the 1st and 2nd Brigades in the lowlands and piedmont areas were characterized by small unit patrols and ambushes. These operations, designed to prevent NVA/VC units from using the population as a source of supplies and recruits, effectively stopped the flow of rice from the lowlands into the mountains.[32]

Pacification, which was supposed to have been a burden the burgeoning South Vietnamese military forces could take on (including the various popular, regional, and self-defense forces), still required almost three full divisions of American troops in I Corps, with no battalions available to disrupt the enemy LOC. The similarity between MG Wright's Abrams-era explanation for the failure of the XXIV Corps to contest the A Shau Valley in late 1969, and the "pacification first" policy of General Westmoreland in I Corps in 1967 (pending the redeployment of major U.S. Army units to I Corps) is striking, and would appear to support the position that it was primarily, if not solely, the radical tilting of the balance of forces from the enemy to the allies resulting from the very heavy casualties of the Tet Offensive that enabled the apparent change in operational focus under the new COMUSMACV. Once the enemy recovered strength in I Corps sufficient to threaten the rice-producing coastal areas while simultaneously protecting and expanding its bases in the jungles, and abetted by the American "redeployments," GEN Abrams, like his predecessor, opted to concentrate his forces in the lowlands, ceding the A Shau to the NVA. Almost as soon as the drawdown of U.S. Forces commenced, the number of U.S. maneuver battalions in the northern part of the country was deemed inadequate for both the pacification support and LOC RIF missions, and the latter were sacrificed. There was no mention of any expectation that troops could be reassigned to the A Shau once the monsoon season had passed.

As the failure to contest the A Shau Valley during the Westmoreland years (March 1966–January 1968) did much to enable the enemy assault on Hue during Tet 1968, so too the abandonment of the allied presence in the valley in late September 1969 was recognized and exploited by the NVA. General Wright's successor, MG John J. Hennessey (CG, 101 Abn Div, May 1970–Jan 1971), quoted from his Senior Officer Debriefing Report:

Early in 1970, the cadre of MRTTH [NVA Military Region Thua Thien Hue] concluded that if allied units did not conduct operations in the upper A Shau Valley before June, the NVA would be able to improve its logistical position to the extent that it would be able to force allied units from the canopy in western Thua Thien Province. Their objectives included:

(a) Extending their cache system closer to the lowlands of Thua Thien Province and the piedmont of Phong Dien District.
(b) Increasing control of the piedmont in Phong Dien District.
(c) Strengthening the VCI in the area to the point that a VC Government could be established.
(d) Disrupting the Pacification Program.

Encouraged by the fact that allied forces conducted only limited operations in the A Shau Valley, MRTTH ordered the offensive against FB Ripcord.[33]

On 23 July, RIPCORD was closed. Thus, while the 101st Airborne Division marshalled its forces in the lowlands and piedmont of Thua Thien to protect pacification, the enemy attacked that same structure indirectly, capitalizing on the failure to maintain allied ground forces in the A Shau Valley and slowly but surely reestablishing base areas proximate to the population centers.

3RD FORCE RECONNAISSANCE COMPANY AND OPERATION RANDOLPH GLEN

October–November 1969

On 2 October, two ARC LIGHT strikes were directed against the former DEWEY CANYON AO. One hit three kilometers south of Tiger Mountain, and the other was centered on Lang Ha, where Route 922 enters South Vietnam from Base Area 611.[1] Throughout the month, several radar-directed air strikes (Sky Spot) were also employed against likely enemy concentrations.

Ranger teams from L/75th Infantry were active elsewhere in the 101st Airborne Division AO, but there were no reported insertions in or near the A Shau Valley. Allied intelligence was once again reliant on aerial surveillance—never a dependable source about a well-hidden enemy who usually moved at night—and signals intelligence. Hence, most of what was "known" was guesswork or outdated by the time it was processed.

With the close of all allied FSBs in the valley and the onset of unfavorable weather which hindered Allied aircraft and surveillance capabilities, the enemy is now being afforded the opportunity to reopen the valley supply lines. It must be noted however, that to date Allied surveillance has not detected any large-scale enemy movement, either into or across the valley. The absence of this anticipated movement may be attributable to limited surveillance due to restrictive weather conditions during early October or the enemy may have been successful in establishing a satisfactory bypass circumventing the A SHAU."[2]

So read the "Quarterly Operational Report – Lessons Learned" of XXIV Corps at the end of the month.

In November, L Company, 75th Rangers sent in a number of patrols to the Rao Nai Valley, east of A Shau, with no enemy contact. No ground reconnaissance took place in the valley itself, but it was soon evident that the North Vietnamese Army was beginning to reestablish itself in western Thua Thien. "Visual reconnaissance sighted new .50 cal position, bunkers, mortar pits and trail activity in the A SHAU Valley to further support the imminent infiltration of troops into the province. It was speculated that the 29th and 803d Regts supported by BINH TRAM 42 would move back into the A SHAU Valley from their positions in BASE AREA 611."[3]

Ominously, late in the month, pilots conducting air strikes in the vicinity of Tiger Mountain reported anti-aircraft air bursts at 5,000–10,000 feet, indicating that heavier 37mm or 57mm weapons were being reintroduced into the area.[4]

3rd Force Reconnaissance Company in the Valley – December 1969

In May 1969, when President Nixon's intention to withdraw American ground forces from Vietnam was first announced, CG III MAF (LTG Herman Nickerson, Jr.) realized that once the U.S. maneuver battalions began to pull out, his forces would be spread thinner and thinner, and maximum economy of force would be critical to the continued performance of his mission. General Nickerson commissioned a staff study to analyze how the Marines might optimize their assets in order to continue to accomplish this goal. One of the key points to come out of that study, published 22 September 1969, was the recognition that intelligence about the enemy's intentions and capabilities would become increasingly critical in order to avoid surprise attacks on thinning American lines. Timely intel was a necessity to enable the best use of the air and artillery assets to disrupt the enemy plans, supply lines, and caches, and to destroy its forces when(ever) they massed to attack. "Of particular interest was the study's finding that intelligence assets in the region lacked coordination – some were controlled by the G-2 or G-3 of the divisions, some by the Army's tactical units or Special Forces, and still others by MAF itself. Nothing was being done to coordinate reconnaissance efforts to prevent duplication or to eliminate 'blind areas' that were not covered at all."[5]

Earlier in September, Nickerson had observed in a message to LTG Zais, CG, XXIV Corps, "that the number of long-range reconnaissance patrols maintained in the field by the 101st Airborne Division has been minimal." The Corps Commander's response took into consideration the imminent closure of all allied bases in the valley: "Employment of [patrols] is considered hazardous because artillery support will be limited to nonexistent, communications will be difficult, and weather in the coming five months [monsoon season] will impose severe restrictions on insertion, support, and extraction of teams."[6]

With that in mind and concerned about the need to keep tabs on the enemy as the NVA reasserted control of the A Shau, CG III MAF had employed the 3rd Force Reconnaissance Company in a temporary role near the Hai Van Pass, while nurturing the hope that the elite Marine unit could soon be more fruitfully employed along Route 548. On 1 October, Nickerson made a formal offer to Zais to employ the Force Recon Marines in the A Shau, an offer that was well received: "I would be delighted to employ the Marine Force Recon Company in the XXIV Corps AO. I have seen it in operation along with the recon teams of the 3rd Marine Division Recon Battalion and I know these teams to be as effective as they are gallant."[7] Thus, the Force Recon company would be the next in a long series of elite units, army and marine, to play a role in the saga of the A Shau Valley.

It would take some weeks to iron out the wrinkles in the planned employment of the Marines, but by mid-December they would begin their insertions, supported by the 2/17th Air Cavalry in the same manner as L/75th Infantry. In the meantime, the Force Recon Marines engaged in an intensive (re)training program to prepare them to work in the A Shau environment and with army and air force support. At the top of the list was rappelling techniques, as the Marines were used to walking to and from their AOs. Also covered were advanced patrolling techniques and the use of air support.[8]

In the week before the Marines began their insertions, three final patrols from L/75th Infantry (Teams Stingray, Volkswagen, and Ferrari) were sent into the central valley area. Volkswagen (YC447955) went in on 6 December to the same LZ (Hill 859, 2 km east of the abandoned Ta Bat airstrip) designated to be used as a radio relay site for the Marines. The other two teams inserted on the west wall of the valley, 5 and 10 km south of Hill 859. The patrols remained in the area for 3–4 days, their extraction delayed by weather, and made no contact with the enemy.[9]

As if to validate General Zais's observation that "communications will be difficult," the initial attempt to insert a marine two-team communications relay unit into the same location that had been used by the last Rangers patrolling the area, and abandoned just a week earlier, resulted in the loss of a slick carrying one of the teams to a 12.7mm heavy machine gun. There were no deaths, but five Marines and one helicopter crewman were injured and evacuated successfully. The relay site would finally be occupied another week later.[10]

On 20 December, after a week of weather delays, the Marines began their patrols in the A Shau Valley, a mission that would occupy them for the next two months.*[11] The nearest allied firebase was BASTOGNE, and the 175mm guns based there were the only artillery support available. The first group of insertions took place on 20–21 December and included six teams at five locations. The Force Recon teams each consisted of six men. Delays in insertions and extractions were not uncommon throughout the month, although no units were left on the ground so long as to compromise health or safety.

---

* Although Lanning (page 181) states that the 3rd Company would patrol the A Shau "over the next five and a half months," and Norton (1969, page 289) says "three and a half months," both are incorrect. The Company would be an early casualty of "redeployment." "The 3d Force Reconnaissance Company ceased patrolling in February [1970], although the unit, almost at zero strength, remained in Vietnam until July."

The first day, Team Boxhill (2-1) inserted near the junction of Route 547A, and Team Atlas (3-1) went in about 1 km west of abandoned FSB CURRAHEE. Teams Tinny (1-1) and Savvy (2-3) occupied ZULU RELAY site on Hill 859 on 21 December, and Team Garlic (1-3) inserted on the east wall, 3 km north of Dong So Ridge near the abandoned village of La Dut. Each of the three actively patrolling teams was responsible for a reconnaissance zone (RZ) two or three kilometers on a side.

Team Boxhill, Recon Zone YC4099/4297, observed enemy movement to the west as soon as it landed, and on the morning of 22 December spotted 10 NVA in a tree line across the valley floor. It called on 2/17th Cavalry gunships and the 175mm guns at BASTOGNE with unknown results. At 2000H on the night of 23–24 December, the Marines ambushed a small NVA platoon at short range, then quickly moved to an alternate location. Within 30 minutes, two gunships and a command and control (C&C) helicopter were on station. The recon team then used the C&C ship to coordinate gunship attacks on the enemy location. On 24 December, the aerorifle platoon of A/2/17th Cavalry was flown in, and together with Team Boxhill, searched the area of the previous day's firefight. There was no report of enemy bodies or captured weapons, and both the Blues and the Marines were extracted at 1625H on Christmas Eve.[12]

Team Garlic, RZ YD3307/3505, also observed an NVA troop movement early on the 22nd, but in much larger numbers. Ninety to one hundred enemy soldiers, heavily armed and with an 82mm mortar, were seen on the valley floor at 0915H, moving quickly eastward with a large interval (10–15 meters) between men. Cobra gunships arrived and attacked throughout the day, but the results of these efforts were not known. The area was one of heavy growth, elephant grass, and small gullies, making observation from both air and ground difficult. During the action, an LOH from B/2/17th Cavalry was shot down by 12.7mm antiaircraft fire, crashing on the valley floor, but both crewmen were rescued. The Blues of Bravo Troop arrived to prepare the downed helicopter for extraction, which was completed by mid-afternoon. During the day, the pilots of the Air Cavalry also discovered a complex of about 100 bunkers near the center of the team's RZ. For the day, the Cobras claimed two confirmed enemy KIA and 10 bunkers and 15 structures destroyed. Forty-four sorties were flown by the 2/17th Cavalry. An air strike late in the day was unable to deliver its ordnance due to increasing clouds. Early on the morning of 24 December, Team Garlic was near the bunker complex discovered two days prior, when they heard numerous sounds from the enemy camp. Air strikes were unavailable as the weather had "socked in" Da Nang airbase, but late in the day the ubiquitous Cobras of the 2/17th Cavalry arrived to attack. Only one enemy KIA was confirmed, but the camp was saturated with rockets and machine gun fire. On 26 December, another enemy column, this time of 18 men with a mortar, was spotted near the bunkers, but morning fog prevented air support. The team was finally extracted at 1130H on the 26th.[13]

Team Atlas, RZ YC3596/3894, sighted a single enemy soldier on the 22nd. One team member with an eye injury was evacuated on 25 December under "arduous flight conditions, but Atlas otherwise had an uneventful patrol in the area between Hill 996 and the former CURRAHEE location, until extraction on 26 December.[14]

The next round of patrols began on Christmas Day, with the insertion of Team Roach (3-2) (RZ YC3605/3803) and Team Snaky (3-3) (YC 3803/4001) into areas on the east wall of the valley between Team Garlic and Team Boxcar. One member of Snaky had to be medevaced by sling hoist on the 26th after suffering a severely sprained ankle. Both teams would remain on the ground until 30 December, making no contact with the enemy.

Teams Pony (1-2) (RZ YC4693/4891) and Spinner (4-2) (RZ 3102/3400) were inserted on 26 December, along with a new radio relay unit consisting of Teams Bitter (4-1) and Coffee (4-3) to replace Tinny and Savvy at site ZULU (Hill 859). Pony's stay on the east side of the valley three kilometers south of the radio relay site was uneventful, and the team was extracted on the last day of the year.[15]

Team Spinner went into the hills west of the Punchbowl, where the advance of the 1st Cavalry Division had come to a halt in May 1968. On the evening of insertion at 1900H, the patrol came across a sleeping area astride a large trail (YD340004) with accommodations for 200–300 men. On 28 December, just a few hundred meters away (YD330012), another camp for 100 men was discovered. The area covered by this patrol would be revisited by B/2/17th Air Cavalry during Operation RANDOLPH GLEN on

25 January 1970, when gunship and air strikes would result in 16 NVA KIA.[16] On the afternoon of the 29th, the team members began to hear enemy signal shots, indicating that they were probably being tracked by NVA counter-recon elements. The team was extracted without opposition on 31 December.[17]

The XXIV Corps intelligence summary for December 1969 cites the evidence produced by the 3rd Force Reconnaissance teams as "confirming the eastward movement of enemy forces out of sanctuaries in Laos." This intelligence was especially important, as "[p]oor weather throughout XXIV Corps AO in early December prevented surveillance of enemy activities in the western areas."[18]

3rd Force Reconnaissance Company in the Valley – January 1970

During January 1970, the Force Recon Marines continued to patrol on both sides of the A Shau Valley. The radio relay site at Hill 859 was manned continuously. On 31 December 1969, Teams Garlic and Savvy took over. On 9 January 1970, Teams Pony and Spinner; on 14 January, Teams Tinny and Atlas; on 19 January Teams Roach and Bitter; and on 24 January, Teams Artic (sic) and Snaky. The radio outpost was attacked a single time, when the NVA used mortars, RPGs, and automatic weapons to hit on 10 January. Teams Pony and Spinner had three men WIA, only one requiring evacuation, and helicopter gunships arrived to chase the enemy away.[19]

Including patrols that were inserted on 30 December, there were a total of 11 recon teams prowling the AO in 29 patrols at various times during the month. There were eight contacts with the enemy, resulting in 29 NVA KIA, 1 Marine KIA, and 14 WIA. There were also eight sightings of the enemy in platoon or greater strength and two truck sightings. The following is a recap of these events.[†] On 5 January, Team Bitter (4-1) encountered and killed a single NVA soldier, about 2 km northwest of Hill 859, near a large new enemy bunker overlooking Route 547A.

---

† The information here includes the enemy contacts resulting in casualties. There were daily reports of the discovery of enemy installations and frequent incidences of gunship or air strikes called against observed or suspected enemy troops, trails, bunkers, and structures. Those are not included.

On 6 January, in the hills about 7 km east of Tiger Mountain, Team Snaky (3-3) twice engaged pairs of NVA troops, killing at least one of the enemy. Operating back in the area where Team Spinner had located large enemy camps at the end of December, Team Artic (sic) (2-2) spotted 15–20 enemy troops but did not engage. An even larger group of NVA was sighted on 17 January when Team Coffee (4-3), landing in the foothills of the east wall across from Dong So Ridge, observed 82 enemy soldiers with three light machine guns about 300 meters away. "Pink Teams" (LOH plus gunship) were called, and in a repeat of the battle just to the north on 22 December, a LOH was downed by intense 12.7mm antiaircraft fire. An F-4 fighter-bomber that arrived to assist was unable to do so due to low cloud cover. Coffee, with its position exposed and the large enemy unit nigh, was extracted under small-arms fire just hours after insertion. After the Marines were lifted out, 50 rounds of 175mm artillery fire were directed at the area. No enemy body count was reported.[20]

A UH-1H slick from A/101st Aviation Battalion heading for ZULU RELAY was shot down less than 1 kilometer northeast of Hill 859 on 14 January, with the loss of three crew members (WO1 John M. Rizzo, Jr., SP5 Clarence S Turner, III, and SP4 Gerald L. Schwuchow) and a passenger, CPL William G. Little, USMC, from the 3rd Force Reconnaissance Company.[21] At the time of its loss, the helicopter was on final approach to ZULU RELAY. The fourth crew member and five Marines, all believed to have been from Recon Team Artic, were wounded and evacuated. Two other helicopters were hit with enemy ground fire in this vicinity on the 14th. CPL Little was the only 3rd Force Recon soldier lost during the month, an outstanding testimony to the skill and professionalism of the Recon Marines under the very challenging conditions in the A Shau Valley AO.[22]

On 19 January, Recon Team Spinner (4-2) inserted near the valley floor, just 1.5 km east of the old Ta Bat airstrip. Moving uphill to the north over jungle-covered mountains, the patrol made contact on the 22nd with 10 NVA, killing 3. As the 2/17th Cavalry helicopters provided support, an LOH was shot down by a rocket-propelled grenade, and two recon team members were wounded. The Blue platoon from B/2/17th Cavalry was brought in to reinforce, and the combined force was extracted later that day after medevac of the helicopter crew and Marines was completed.[23]

Team Boxhill (2-1) was inserted on the 8th into the middle of a brand-new enemy bunker complex (YDC386956). The Marines engaged an estimated platoon of enemy soldiers, killing six, with the door gunners of the Air Cavalry accounting for two more, and was extracted under fire just 40 minutes after the firefight began. "Bad Luck" Boxhill returned to the valley on 11 January, about 1 km southeast of the fight three days previously. Once again, the enemy was close and the team was soon engaged in battle with an estimated 15 enemies. And once again, the final score was Marines 6, NVA 0, before Boxhill moved to safer ground and was extracted the next day. The gunships of the 2/17th Cavalry were in continuous support during this time.[24]

Boxhill, whose members participated in six patrols during the month (twice as many as any other team), was inserted for the third time in the month on 16 January. The LZ was in the Punchbowl, on the southern flank of Dong So Ridge, and the insertion was by C-46 from HMM-161.‡ The lift helicopter experienced enemy fire as it departed, and the patrol "immediately sighted five NVA five hundred meters North of their position. Team had enemy movement to the East, North and the South, with small arms being fired at their cover aircraft." Cobra gunships, initially Marine from HML-367 and then Army from F/2/17th Air Cavalry, provided heavy supporting fire. The helicopters observed seven trucks in two groups hidden under trees at the base of Dong So, as well as many bunkers and caves. Marine A-6 Intruders were called in to hit the truck park. The team was extracted less than one hour after landing, with no casualties.[25]

Boxhill finally cut a break (sort of) when next inserted to the south, on the western side of Route 548 about 7 km south of Hill 859, on 19 January. The team spotted a truck on Route 548 to the east. The bad news was that the weather was socked in, and by the time an aerial observer arrived the next morning, the truck was gone. The team was extracted on schedule on 23 January but had time for one more patrol before the month was out. Inserted on the 26th about 1 km northwest of the junction of Routes 547A and 548, the Marines were engaged two days later with 17

---

‡ Although helicopter support for the Force Recon Marines in the A Shau usually came from the 2/17th Cavalry, Marine helicopters were occasionally employed.

NVA troops, killing 6 and possibly 10 before calling in the Air Cavalry and Aerial Rocket Artillery and extracting under fire.[26]

After dark on 24 January, three teams and the radio relay unit on Hill 859 all reported headlights at multiple locations moving down Route 548 between Ta Bat and A Shau. All available air support was requested, but ARA gunships had to leave due to low fuel levels before a target could be identified, and overcast conditions prevented flares dropped from "Spooky" from revealing the enemy locations. Two and a half hours after the sightings, 175mm guns at Bastogne began firing H&I (harassment and interdiction) salvos at the last known targets, but no results were recorded.[27]

The final action during the month occurred far to the south, about 2 km northeast of the old A Shau SF camp. Recon team Roach (3-2) was inserted on 27 January and soon found a heavily used trail. The next day, a series of rifle shots alerted the Marines to the presence of enemy searchers nearby. The team ran into four North Vietnamese, killing all but having two of their own wounded. The team was then extracted on the evening of 28 January.[28]

The purpose of the recon teams, of course, was not combat but intelligence. Two things Force Recon Marines could do that no air unit could was acquire enemy documents and identify specific NVA units in the valley. The XXIV Corps intelligence summary at the end of January shows the importance of the Force Recon effort to allied knowledge about the NVA in the A Shau Valley:

On 8 January a reconnaissance team operating in the A SHAU Valley discovered a battalion size base camp (vicinity YC 3793) with "E8" inscribed on a support to a flag in the center of the area. "E8" is confirmed AKA for the 29th Regt. [This was Team Tinny and the location is just south of the former CURRAHEE site.] Other sightings of large base camps and heavy eastward movement of enemy forces and contacts by recon teams in the A Shau continued. This heavy trail activity and sightings of several trucks in the A Shau area indicate possibly a transportation unit (BT42) is supporting movement into the Valley area.[29]

3rd Force Reconnaissance Company in the Valley – February 1970

On 5 February, Recon Team Garlic (1-3) inserted about 2.5 km north of Dong Ap Bia at 0917H. At 1130, the team was ambushed by an estimated 50 enemy soldiers. CPL Adam Cantu, CPL Allen M. Hutchinson, and LCPL Daniel Savage were killed in the opening minutes of the firefight. CPL Charles T. Sexton took charge of the surviving team members and used his radio to call in helicopter gunship and air strike support. He would receive the Navy Cross for his actions on this day:

On 5 February 1970, Corporal Sexton was a member of a six-man reconnaissance team which was patrolling deep in the A Shau Valley when it came under a heavy volume of small arms and automatic weapons fire from approximately fifty enemy soldiers occupying well-concealed emplacements in the dense elephant grass. During the initial moments of the attack, three Marines were mortally wounded and two were seriously wounded. After a rapid assessment of the precarious situation, Corporal Sexton directed the fire of his two wounded companions and moved about the fire-swept area to collect hand grenades and ammunition from his fallen comrades. Utilizing his radio, he then reported the situation to his commanding officer and requested assistance. For the next several hours, while the enemy attempted to encircle and overrun his position, Corporal Sexton repeatedly adjusted helicopter and fixed-wing air strikes on the hostile unit, hurled hand grenades, shouted encouragement to his wounded companions, and simultaneously furnished a running commentary to his company commander until a reaction force arrived to lend support. His heroic and determined actions were an inspiration to all who served with him and undoubtedly saved his fellow Marines from further serious injury or even death.[30]

In response to CPL Sexton's call for help, the Blue Platoon of D/2/17th Air Cavalry was inserted at the ambush site at 1219H. The enemy was driven off by air support and the Blues, and the wounded and dead of Team Garlic were medevaced at 1300H, followed by extraction of the survivors and the Blues at 1400. SGT Chester A. Mollett is not mentioned in any of the reports on the events of 5 February, but he died on 26 February of wounds received on 5 February, almost certainly in this action. SGT Mollett had previously received the Silver Star for action on 7 August 1969 with F company, 2/3 Marines.[§31] A second team inserted on 5 February also ran into trouble soon after landing, possibly incurring "friendly fire" casualties. Team Artic (sic) (2-2) "rec'd fire from birds [helicopters]" and was extracted an hour and a half after insertion, with two men wounded.[32]

On 7 February, disaster again struck the 3rd Force Reconnaissance Company. Team Snaky was inserted at 1000H one kilometer north of the Route 547A/548 junction. Just five minutes later, the team was ambushed, losing two men KIA and two WIA, one of whom would soon succumb to his wounds. Team leader CPL Ted J. Bishop, LCPL James M. Fuhrman, and platoon guide and pointman SGT Arthur M. Garcia, Jr., were killed. The remaining team members fought for their lives, led by LCPL Paul S. Keaveney, who was awarded the Silver Star for his valorous actions in spite of serious wounds. The aero-rifle platoon of B/2/17th Cavalry was inserted. The surviving wounded member of the team was medevaced at 1312H and the survivors and aero-riflemen were extracted from a cold LZ at 1415H.[**33]

There were at least seven additional recon teams from the 3rd Force Recon Company active in the A Shau Valley in the first half of February, and the radio relay site (renamed XRAY) continued to be manned. Team Boxhill's reputation for attracting enemy attention appears to have been confirmed once again during its final insertion. On 14 February it landed at 1320H but had to be extracted at 1540H, almost certainly indicating stronger enemy forces in the area. No contact or casualties were reported.

----

§ The patrols of the 3rd Force Reconnaissance Company during December 1969 and January 1970 are well documented in monthly Command Chronologies. For February 1970, however, no similar report has been located. As the company was withdrawn from the field and reduced to cadre size in late February, it is likely that no report for that month was prepared. The Daily Staff Journals of the 2/17th Cavalry and secondary works are thus the only existing written sources.

** The Keaveney Silver Star Citation may be found in Norton's book.

On 15 February, with the extraction of Team Atlas and the XRAY site team, the 2/17th Cavalry journal reported "ALL MARINES out of AO."[34]

At 0800H on 16 February, the 2/17th Cavalry slicks inserted Recon Teams Roach and Pony into landing zones in the Da Krong watershed, far to the northwest of the A Shau (YD114190 and YD126161). These teams likely were the beginning of an operation to track a single NVA regiment as it entered and moved across western Quang Tri Province.[35] This is the last reference to Force Recon team insertions by the Air Cavalry of the 101st Airborne Division. The final note in the Cavalry journals is on 25 February: "3d Force AO is no longer in effect."[36]

The 3rd Force Reconnaissance Company was one of only two USMC units during the Vietnam War to receive the Army's Valorous Unit Commendation. It was awarded for the Force Recon Marines' actions from 7 December 1969 to 16 February 1970.[37] With the withdrawal of the Marines from the A Shau in late February, yet another nail was driven into the coffin containing American efforts to deny the enemy unfettered use of that critical valley. Following in the wake of the removal of maneuver battalions and firebases in September, the most intense and effective ground reconnaissance operation was now also a victim of the de-Americanization of the war.

The Rest of the Story – Early 1970

The ground reconnaissance teams of Company L, 75th Rangers and, early in 1970, of the USMC 3rd Force Reconnaissance Company, were of critical value to the allied harassment of the North Vietnamese Army's attempts to move freely about the A Shau Valley and to identify and destroy new enemy encampments. Those small and vulnerable teams were not, however, by any means the only American units involved in this effort. Almost all of the other components of this campaign were air units.

Signals intelligence, an "ether-based" component, if you will, of the battle for the A Shau, provided top secret grid and unit identification information on the suspected sources of enemy radio transmissions. This intel was referred to as URS, "usually reliable source," in the targeting reports of the 2/17th Cavalry.[38]

The Huey Cobra gunships of the Air Cavalry, whether patrolling independently in certain designated free fire zones or operating together with the OH-6A "Loach" light observation helicopters, generated their own sightings of enemy troops and installations and provided the firepower to attack targets of their own or other intelligence identification. Together with the LOHs, the gunships formed "Pink" teams of one gunship and one LOH, "Heavy Pink" teams of two gunships and one LOH, or "Red" teams of two gunships. The OH-6A crews flew at ground level, constantly daring any NVA below to expose themselves by firing. Miraculously, losses among OH-6A crew over the A Shau were incredibly low. During all of 1970, only one LOH pilot was lost. On 2 March, operating over the Warehouse 54 area, WO1 James J. Burgoyne was KIA by ground fire and two crewmen were WIA. The aerial observer was able to fly the aircraft to LZ Sally, where it crash-landed.[39]

Another tool in the bag of the 2/17th Cavalry was the "People Sniffer" Huey. These helicopters, equipped with the XM-3 Airborne Personnel Detector, flew low-level missions over the A Shau on a daily basis, detecting suspected sources of human body odors under the jungle canopy. While enemy countermeasures, such as hanging bags of water buffalo urine in unoccupied areas, were somewhat successful, the "sniffers" provided primary and secondary data on suspected enemy troop locations.

Higher in the sky, the forward air controllers (FACs) directed the Air Force, Marine, and Navy fighter-bombers, as well as the aerial rocket artillery (ARA), which provided the heaviest ordnance for attacking and were critical to the timely and accurate placement of bombs and napalm when recon teams were under attack. They also kept a sharp eye out for signs of the NVA, generating their share of target acquisitions.

The aircraft controlled by the FACs were referred to as "tac air," or tactical airstrikes. Those sorties, critical though they were, were the smaller part of the fixed-wing air activity over the valley, most of which was comprised of preplanned missions of several types.

1) United States Army OV-1 Mohawk surveillance aircraft flew search missions using side-looking airborne radar (SLAR), which could potentially identify moving trucks at a distance

of up to 45 kilometers. The disadvantage of the Mohawks was that they were unarmed, and this often resulted in delay between target spotting and engagement. Prior to 1 January 1967, they had been equipped with rockets and machine guns, but when the Army and the Air Force sorted out who would control which aviation roles, the Army had to defang its armed fixed-wing aircraft, and the Caribou light transport units were transferred to the Air Force. Later models of the Mohawk also used an infrared sensor to find the enemy.[40]

2) "Sky Spot" radar-directed air strikes by fighter-bombers against predesignated targets at specified times. These targets were located using the full range of intelligence sources, including strings of acoustic sensors that became an ever more important source as the allied ground presence in the valley waned. During February 1970, for example, over 90 such sorties were flown in or near the valley, of which about 90% were directed against the area between A Luoi / Ta Bat and the area where Route 922 enters Vietnam west of Tiger Mountain.[41] BDA (bomb damage assessment) often did not occur at all, but overflight by FACs or LOHs might be used, especially if secondary explosions were observed.

3) ARC LIGHT B-52 strikes were similar in use to the Sky Spot missions but involved the saturation bombing of 1 km x 2 km "boxes" and required a longer lead time.

4) From time to time, 12–18 hour delay-fuse bombs were dropped on targets where the enemy might be expected to concentrate. This type of mission was uncommon, as the fuses did not always work as planned and might remain active so long that allied ground units would be endangered or that the enemy might defuse and use for its own purposes.

5) At night, C-119G gunships (4 x 7.62mm miniguns) flew over the valley on truck-hunting patrol, equipped with night vision devices and capable of putting out 6,000 rounds per minute. In February 1970, the upgraded C-119K (3 x 7.62mm miniguns and 2 x 20mm miniguns) came into service, with explosive shells capable of causing much greater damage.[42] A report of one gunship strike on 24 February: "At 240140 hr Air Force C119 eng[aged] an unidentified type vehicle, res[ult] 5 secondary explosions and 2 secondary fires Vic[inity] YC393983" (on Route 548 between A Luoi and the junction with Route 547A). A visual reconnaissance was requested for first light the next day, but there is no record of this occurring.[43]

6) CS gas missions involved bombing with 350-lb packets of persistent (30 days) riot gas powder to deny the enemy the use of areas such as likely sites of bridge or road construction.

7) "TRAIL DUST" missions were flown by C-123 aircraft to spray Agent Orange or other defoliants.

Each of the air missions required advance coordination with the low-flying helicopters and the recon teams on the ground to ensure that no "friendlies" became victims. The number of each type of mission would vary from month to month, but the overall trend was down, as the "redeployment" affected air units, too.

Wrapping Up Operation RANDOLPH GLEN – to 31 March 1970

After the departure of the Marine recon teams in February 1970, it would be some time before another ground recon effort was made in the valley. The controlling headquarters, 101st Airborne Division, did not send any teams in during the rest of Operation RANDOLPH GLEN. That division-wide effort (7 December 1969–31 March 1970) was assigned "to establish, in coordination with 1st Inf Div ARVN, a belt of security on the periphery of the populated area of Thua Thien Province."[44] The A Shau Valley thus came within the reconnaissance zone of the 2/17th Air Cavalry with attached L/75th Infantry (Ranger), but not within the operational zone of any maneuver battalions, and aside from a few "artillery raids" conducted by batteries from reestablished FSBs BLAZE and VEGHEL along Route 547, there were no units other than the divisional air cavalry operating in the valley.

There was one notable exception to the "no ground units" history during March. On the last day of the month, most of the Ranger company was committed to a traditional commando-type raid, when it was inserted in three groups to destroy three enemy bridges along Route 548 in the central valley. Frank Johnson, who led one of the three groups, had this to say about his mission: "What a farce! We were broken into three forces of four teams each to attack and blow up three bridges in the A Shau Valley. . . . We went into the A Shau at 0700. As the first slick went in, the second slick's rotor struck the first slick's tail rotor and the first slick crashed." One man received a severe concussion in the incident. When one of the teams went to investigate the demolition target, "[It] discovered the bridge to be two slabs of PSP (perforated steel planking) . . . across a creek three feet wide!"[45] The bridges were all blown and there were no other casualties. Just five days later, the Air Cavalry Daily Staff Journal reported: "msn to VR [visual recon] bridges blown last wk – from General Hennesy [sic]: The bridges have been rebuilt. Unconfirmed rpts of machine guns from unk source at location."[46]

While RANDOLPH GLEN did a credible job of keeping the NVA from the coastal area, the North Vietnamese Army units oozed out of the A Shau like lava from a volcano, slowly but very threateningly moving eastward. By the end of March, intelligence "located sizeable elements of the 29th NVA Regiment east of the A Shau Valley. There was substantial evidence that elements of the 803rd NVA Regiment also moved eastward into the A Shau Valley to establish lines of communication and anti–aircraft sites."[47] Among the NVA units identified by XXIV Corps intelligence at the end of March, several were in or near the A Shau Valley:

HQ, Military Region Tri-Thien-Hue (YC2199) in Laos, west of Dong Ap Bia

HQ, 324B Division (YC2295) in Laos, west of Dong Ap Bia

675th Artillery Regiment (BA 611)

29th Infantry Regiment (YD2503) in Tam Boi area, on the Laotian border

803rd Infantry Regiment (YD3106) on Dong Tien Cong Ridge between Tiger Mountain and Dong So Ridge

802nd Battalion, 6th Infantry Regiment (YD2010), 5 km west of Tiger Mountain

806th Battalion, 6th Infantry Regiment (YD2611), near the valley floor, 1 km north of Tiger Mountain

K35 Artillery Battalion, 6th Infantry Regiment (YD3807), in Warehouse 54 area[48]

23

OPERATION TEXAS STAR

Redeployment and Reality

The idea that the armed forces of the Republic of Vietnam could stand alone against the North Vietnamese Army and the Viet Cong was ludicrous. The beefing up of the numbers of the South Vietnamese forces through mass enrollment in the only partly armed Peoples Self Defense Forces (PSDF) and other local militia (Popular and Regional Forces) could not mask the truth. The withdrawal of American ground forces, which totaled over 11 divisions* at full strength, reinforced by massive numbers of artillery, engineer, and helicopter battalions, was only part of the allied disengagement. Gone, too, would be two and one-third divisions of South Koreans, a division of Thai troops, and a brigade of the Royal Australian Army.

In their *Report on the War in Vietnam,* as of 13 March 1968, ADM Sharp and GEN Westmoreland reported that 288 allied maneuver battalions faced 190 NVA and VC battalions.[1] Only 155 of the allied battalions were in the RVNAF, and the disparity between enemy and RVN battalions was most pronounced in I Corps (later known as Military Region One, or MR1), where 76 NVA/VC units opposed 34 RVNAF. The addition of a new, untested ARVN division (the 3rd) and other organizational sleights of hand (e.g., conversion of CIDG units into "ranger border

---

* 1st and 3rd Marine Divisions, 1st Cavalry and 101st Airborne airmobile divisions, 1st, 4th, 9th, 25th and American (23rd) Infantry Divisions, plus two Marine regiments (26th and 27th), 11th Armored Cavalry Regiment, 3rd Brigade, 82nd Airborne Division, 1st Brigade, 5th Infantry Division, 199th Light Infantry Brigade, and the 173rd Airborne Brigade.

defense battalions") could not disguise the fundamental truth that a post–Vietnamization force reduced by 133 battalions could not possibly prevail where the larger, more heavily supported allied army had been unable to do so. All in all, the regular forces opposing the North Vietnamese and their southern comrades were reduced by over 50% in "foxhole" strength (U.S. and "Free World" battalions fielded more infantry troops than comparable ARVN formations), artillery, and aircraft (fixed- and rotary-wing), and that says nothing of the endemic desertion and corruption that plagued the South Vietnamese Army (ARVN) and the weakness of the South Vietnamese logistical system. The NVA's major defeat at Tet in 1968 was essentially self-inflicted, and while they would suffer additional setbacks, as during the 1970 Cambodian incursion and the 1972 "Easter Offensive," their ability to control the level of combat by avoiding allied search and destroy operations and, if need be, by retreating to their Laotian and Cambodian sanctuaries, ensured that they could pursue their "protracted war" strategy indefinitely.

Equally absurd was the notion that American air power was a trump card that could be played to win every big battle, serving in essence as a strategic and even tactical reserve. The U.S. had long pursued that belief during the "American" war. While massive (and expensive) air and artillery support could savage imprudently massed assault formations, it was far from cost- effective against the enemy when it remained hidden in its jungle base areas and/or sheltered in well-constructed bunker and tunnel complexes. One result of the reliance on air power was the consistent failure of American ground forces to trap enemy units, since air replaced ground force reserves in numbers necessary to defeat the enemy through maneuver, allowing the NVA and VC to consistently avoid or escape from "cordon" and "piling on" tactics. Also overlooked or discounted was the collateral deleterious effect on the civilian population caused by the bombing destruction of countless South Vietnamese villages, the corresponding tens of thousands of civilian casualties, and the creation of hundreds of thousands of refugees. While air power could do much to keep the "kill ratio" high and the number of allied casualties low, it had not and would not supplant the need for infantry. Indeed, its use as a "force multiplier" throughout the American war may have encouraged the illusion that we were "winning," while contributing to the misperception that Westy or Abe ever had enough "grunts" (infantrymen) to accomplish more than a stalemate against the elusive and omnipresent enemy.

Nowhere in Vietnam would the shortage of infantry and the limitations of air power be more evident than in the A Shau Valley, as the war gradually transitioned to its post-American era. As the redeployment continued in I Corps, the 101st Airborne Division (Airmobile) tried to maintain the initiative and to keep the enemy off balance and away from the populated lowlands and major American bases. This would prove to be a most challenging mission.

April 1970: In the Valley

Operation TEXAS STAR (1 April–5 September 1970) was one of the last major United States ground operations of the Vietnam War, carried out during a time when the Vietnamization was rapidly—perhaps too rapidly—progressing.

> The only offensive campaign for all of Vietnam planned by the U.S. Military Assistance Command Vietnam (USMACV) for the summer of 1970 called for the 101st Airborne Division with the support of the Army of the Republic of Vietnam (ARVN) 1st Infantry Division, to conduct operations in the A Shau Valley, the North Vietnamese "warehouse area" and against the branches of the Ho Chi Minh Trail coming into South Vietnam in both Quang Tri and Thus Thien Provinces.[2]

The price of allowing the enemy to prevail in the valley in the early 1960s, culminating in the fall of Camp A Shau in March 1966, now continued to be paid by the last full American division in country four years later, during the final months of the "American War."

By the end of September, the USMC would be well on its way out; only the 1st and 5th Marines and one battalion of the 7th Marines remained in the field, operating in Quang Nam and Quang Tin Provinces in central I Corps.[3] The ARVN 1st Infantry Division and the U.S. 1st Brigade, 5th Infantry Division were trying to fill the Marines' boots along the DMZ and in Quang Tri Province. The USMC Combined Action Platoons (CAPs) were gone from Thua Thien, leaving the 101st Airborne Division with an increasing burden in securing the populated lowlands and no maneuver battalions for employment in the A Shau Valley. Henceforth, the most forward fire bases in Thua Thien were to be located along Route 547 between Hue and the A Shau and at isolated hilltops in the jungled mountains between the valley and the fertile coastal plain. TEXAS STAR was fundamentally a delaying attack, "designed to find, fix, and destroy enemy forces, caches, base camps, and interdict enemy movement into the populated lowlands. Extensive airmobile combined operations are to be conducted in the area east of the A Shau Valley and west of the populated lowlands of Thua Thien Province."[4] Unfortunately, with the abandonment of the A Shau in late 1969, the camel's nose was once again under the tent flap; the rest of him was keen to follow. And with its ever-expanding responsibilities, the 101st Airborne Division would have few assets to employ in the hinterlands out toward the Laotian border, aka the "division reconnaissance zone."

In April 1970, Bravo Troop, 2nd of the 17th Air Cavalry, was the unit tasked with primary responsibility for observation and interdiction of the NVA in the A Shau Valley. The Air Cavalry performed its armed reconnaissance role with the usual professionalism, using Pink teams and sometimes the aerorifle Blues to find and attack vulnerable targets, while working with the FACs to identify larger or better defended targets for Sky Spot, tac air, and ARC LIGHT missions, and to conduct subsequent BDA (bomb damage assessment). On 5 April, the new scout platoon leader of Bravo Troop was killed on an orientation mission when his OH–6A LOH was hit by small arms fire over the valley. CPT Olan J. Howe was just beginning his second tour of duty in RVN.[5] The troop responded, hitting two 12.7mm machine gun positions and killing 3 NVA gunners.[6]

On 16 April, acoustic sensors at a number of locations in and around the northern reaches of the valley were activated by suspected enemy vehicular movement, but the weather kept American aircraft from striking. It was evident that the enemy would take advantage of poor daylight weather conditions over the A Shau to move as much or more than in darkness.

17 April would be one of the busier days for B/2/17th Cavalry. At 0900H, an OH-6A was reported shot down at the southern end of the valley (YC493837), apparently by a heavy caliber anti-aircraft weapon that hit it at 1,500–2,000 feet altitude. The aircraft was destroyed, but two crew members were quickly rescued. Several other 2/17th Cavalry helicopters were shot down this day, including another OH-6A about 5 km southwest of FSB RIPCORD, an AH-1G Cobra gunship near Route 547, and a UH-1H slick. There were no deaths in any of the crashes, but most of the day was spent retrieving the crews and destroying the damaged aircraft. In the early afternoon, word came down: "NO MORE A/C [aircraft] WILL BE DESTROYED W/O PERMISSION OF A GEN[eral] OFFICER." The standard operating procedure was to sling out and repair all possible damaged helicopters, but so many were being finished off in place on 17 April that higher headquarters had become concerned. All told, 10 helicopters were hit by enemy fire.

20 April brought another rash of sensor activations. At night, trucks were detected in the north, between Tiger Mountain and Tam Boi on the Laotian border (YC231072), on Route 548 north of Dong So Ridge (YD325060), and also in the valley center, east and southeast of FSB CURRAHEE (abandoned) (YC415955 & YC420970). The targets were engaged with 175mm artillery fire from FSB BASTOGNE and 20mm cannon fire from Air Force gunships, with three secondary explosions, seven secondary fires, and at least six trucks claimed destroyed. The next night, the pattern was repeated, with sensors going off on route 548 east of Tiger Mountain (YD300090), at two locations near the narrowest point in the valley next to Dong So Ridge (YD365030 & YD360033), and again west of Tiger Mountain (YD225060). The C-119s claimed more trucks eliminated. Damage assessment both days was hindered by poor weather over the A Shau. When a heavy Pink team from Charlie Troop was finally able to enter the valley on the afternoon of 21 April, it reported "heavy automatic weapons fire" from the southeastern end of Dong So

and "extremely heavy automatic weapons fire" from the valley floor, two kilometers to the east. No BDA report is recorded.[7]

One of the aforementioned sources of information about the enemy was signals intelligence (SIGINT), a combination of radio direction finding to pinpoint NVA headquarters and translations of intercepted wire or radio messages. Effective triangulation of radio sources required multiple vectors, and a key role in that activity was played by air force crews in EC-47 electronic warfare aircraft. On 22 April 1970, one of those top secret planes was hit by enemy fire over Laos, crashing on the edge of the A Shau Valley floor about one kilometer northwest of the Route 547A/548 junction (YC415973).

> The aircraft was flying at 5,500 feet as the crew attempted to locate and fix enemy radio broadcasts when it was hit by automatic weapons fire or 37mm AAA. With the aircraft losing oil and power, 1LT Wall flew east in the hope of reaching an airfield but he was forced to crash land the aircraft.... 1LT [George M.] Wall and SSgt [Michael R.] Conner were killed in the crash landing but the others were rescued by USAF SAR helicopters.[†8]

A highest priority mission was soon underway to rescue the crew (including technical personnel highly trained in top secret operations) and salvage or destroy the secret communications intelligence equipment on board. The crash was first reported at 1045H. By 1100, a heavy Pink team of B Troop was en route to the crash site, soon followed by aerial rocket artillery gunships and an OH-6A with an artillery forward observer on board. Just six minutes later, the aerorifle platoon of Bravo Troop was also on its way. By 1120, the Pink team was over the crash location with the crew in sight, and shortly thereafter the Blues were on the ground securing the area. There was some confusion over the total number of crewmen on the EC-47, a "false news" report that a parachute was seen before the

† It is not clear from the 2/17th Cavalry journal whether the surviving crew members were evacuated by Air Cavalry helicopters or Air Force SAR Jolly Green Giants.

plane crashed, and some uncertainty about whether the air force wanted any salvageable equipment to be evacuated or destroyed in place. The air cavalrymen reported that the gear was 99% destroyed in the crash and ensuing fire. Everything was eventually sorted out, with two crew members dead, one wounded, and six others with only minor injuries. All were evacuated. The enemy, too, hastened to the downed prize, and by 1235H the scout helicopters were receiving fire from 300 meters to the west. ARA and air strikes were called in, while the EC-47 was prepared for demolition with thermite grenades and C-4 explosives. At 1341H, "permission to burn and blow" was received, and by 1430H it was all over, with the Blues extracted under fire but no one hit. As if to be doubly sure that there was nothing left for NVA scavengers, an air strike was directed at the airplane carcass at 1455H.[9]

Immediately as the drama at the EC-47 crash came to an end, it was decided – since there were already major elements of the 2/17th Cavalry in the area – to put the aerorifle platoon back on the ground just 1500 meters to the south, where one of the enemy bridges blown on 31 March had quickly been reconstructed. At 1500H, the three slicks with the B Troop Blues were joined by a fourth with combat engineers and inserted at the NVA bridge (YC425957). The LZ was green and the bridge, 8 x 20 meters and constructed of logs and captured PSP matting, was blown (again) at 1513H. By 1530H, the extraction was complete with no enemy fire and the B Troop riflemen could finally call it a day.[10]

Throughout the month of April, all of the instruments in the air interdiction orchestra were playing notes at one time or another, in spite of weather that frequently delayed or cancelled scheduled operations. Air cavalry scout, gunship, and People Sniffer helicopters were over the valley daily. Tactical air strikes and frequent radar-directed Sky Spot strikes – the latter most often used in the area west of Tiger Mountain where Route 548 enters from Laos – were called. Acoustic sensor strings dropped by helicopter and SLAR-equipped Mohawks contributed to the target identification process, as did SIGINT from EC-47 and other sources. And Agent Orange (Trail Dust) continued to be put down every 10 days or so at different locations, while persistent CS gas was used where the enemy seemed to be engaged in engineering activities. There were no Ranger reconnaissance teams inserted into the A Shau AO during the month.[11]

May–June 1970: Air Cavalry in the Valley

In May, as during April, it would continue to appear that the strongest NVA activity in the A Shau watershed was in the northern area, and this would all make sense when the siege of FSB RIPCORD was revealed in July. For the second consecutive month, there were no recon teams inserted in the A Shau area. The closest they came was near the end of the month (29–31 May), when teams from L Company, 75th Rangers were inserted into the Rao Nai Valley to the east, where there were several sightings and contacts with small enemy elements.

On 8 May, between Lang Ha and Tam Boi along the Laotian border, a B/2/17th Cavalry team of two Cobras and a C&C Huey responded to a FAC sighting by the 220th Reconnaissance Aircraft Company, spotting trucks in the open. Two trucks and an NVA machine gun were destroyed and three enemy KIA. On the afternoon of 21 May, an OH-6A of B/2/17th Cavalry was shot down within 200 meters of where a similar incident had occurred near the A Shau airstrip on 17 April (YC495839). The weapon was identified as a 12.7mm machine gun. When the command and control helicopter attempted to land to extract the LOH crew, it was driven off by heavy fire. A Pink team and no less than eight Cobra gunships from B and C Troops converged on the downed helicopter, where an entire company of NVA infantry was now trying to maneuver to capture the crew. While the gunships poured on rocket and machine gun fire to keep the enemy's heads down, the Squadron Commander's Huey landed, and incurring only two hits from enemy fire, took aboard the relieved pilot and observer. The LOH was destroyed by ARA. U.S. air activity declined significantly in May, with little fixed-wing activity, except the omnipresent FACs.[12]

June was the slowest month for enemy sightings in the valley in many, many months, and for the third consecutive month, no allied recon teams were inserted. The XXIV Corps quarterly report noted: "During June 1970, the enemy forces in western Thua Thien Province broke contact and then avoided contact with allied units in the area."[13] Rangers patrolling in the adjacent Rao Nai watershed did have some contacts, including Team Wimpy, inserted on 6 June just 3 km east of Route 548 north of A Shau and then immediately extracted after detecting enemy forces all around. On 9 June, the C Troop Blues were inserted at YC545842, 5 km east of the A Shau airstrip. They patrolled north-westward for 1,500 meters. A supporting LOH was hit by enemy fire, losing its radio, and the aeroriflemen found a small camp with recently built bunkers, but there was no contact by the ground element with the enemy and the platoon was extracted without incident five and a half hours after landing. Near the end of the month (27–28 June), eight-inch self-propelled howitzers were temporarily moved down Route 547 to FSB BLAZE for an artillery raid,[14] but air interdiction in the A Shau Valley continued to decline.

March–July 1970: TEXAS STAR in Central and Western Thua Thien Province

Even as the enemy sightings in the A Shau Valley tapered off, it was clear from intelligence elsewhere that the NVA combat regiments were already through the valley, in motion toward the new allied defensive line to the north and east and the heavily populated coastal area beyond.

> During May 1970 enemy units were active in the western areas of [Thua Thien] Province as the 6th Regiment conducted operations in the Fire Support Base Ripcord area, 803d Regiment in the Fire Support Bases Maureen-Kathryn area and the 29th Regiment in the Fire Support Base Veghel area. The enemy's intention was to keep allied forces from discovering their caches.[15]

While the enemy attacks on firebases in the hills between the A Shau and the lowlands did serve to thwart the allied offensive designs against major NVA bases in western Thua Thien, their real purpose was offensive, not defensive. The NVA 324B Division history relates:

> Military Region Headquarters decided to continue the attack and apply strong pressure to the enemy's entire defense line in order to force him to respond in many different locations at the same time [note the correlation between this objective and GEN Hennessey's change of plans, discussed below, due to "overextension!"] Meanwhile we would gradually move our local force units down to operate in the lowland foothills to provide direct support to our grass roots movement aimed at disrupting the enemy's pacification campaign. At the same time the Military Region would use its own main force units to attack and shatter a number of the important links in the enemy defensive network in order to cause it to disintegrate and collapse. Building on this foundation we would open [the] gateway down to the lowlands.[16]

The allied front line, which at times during 1969 had been as far forward as the Laotian Border, with fire bases in and west of the A Shau Valley, had now retreated to a string of FSBs that were at best barely within 155mm howitzer range (14,600 meters) of Route 548. When completed, the new 101st Airborne Division front line extended close to 30 km southeast to northwest, from FSB VEGHEL on Route 547A to FSB RIPCORD, with KATHRYN and MAUREEN in the middle, just 4 km apart. Enemy regiments which previously had been bottled up in Laos in Base Area 611, just inside Vietnam west of the A Shau, or in the "Warehouse" area, were now free to undertake offensive actions in central Quang Tri and Thua Thien Provinces. The chickens of "redeployment" were beginning to come home to roost. The advance of the NVA regiments could not have taken place if the allies had continued to physically block the enemy LOC in the A Shau and to search, disrupt, and destroy its base areas and caches in and around that critical valley.

Although the allies knew that enemy units were maneuvering well to the east of the A Shau Valley in May and June of 1970, they knew next to nothing about the precise composition and disposition of those units, and nothing whatsoever about their objective(s). MG (Ret), then Colonel Benjamin L. Harrison and in command of the 3rd Brigade of the 101st Airborne Division, has this to say about his intelligence reports in early to mid–1970:

> The poor and limited intelligence provided to the 3rd Brigade commander at this critical time is somewhere between disappointing and disgusting. . . . Col (Ret) Lewis Sorley, at my request, reviewed his extensive notes of the MACV Weekly Intelligence Estimate Updates (WIEU) for the period Jan-Aug 1970 and the only mention found of the 324B Div and the 304B Div was during the 11 Jul 70 WIEU which reported: "In the A Shau Valley: Two divisions: 324B [which left South Vietnam in Mar-Apr 1969, returned in Nov 69], and the 304B [which went north in Nov 69 and began coming back in the last few days]." This was two months after these two divisions had been ordered to annihilate Ripcord. . . . The fact that this is all MACV Weekly Intelligence Updates found with a direct connection to the Ripcord battle at the MACV level is pathetic.‡17

TEXAS STAR was a division–sized operation, but the 1st Brigade was committed to action in southern Thua Thien and the 2nd Brigade to pacification near the coast, leaving only the 3rd Brigade (COL Benjamin L. Harrison – 3/187th Infantry, 1/506th Infantry, and 2/506th Infantry) to operate to the west. Moreover, the 3/187th was also committed in the lowlands and the 1/506th was initially tasked with protecting new fire support bases, so that only a single battalion was available for operations around Hill 927, soon to be known as FSB RIPCORD.18

------

‡ While allied intelligence in western Thua Thien was deficient, the North Vietnamese Army ran a signals intelligence operation that was superior to or exceeded that of the allies in product, if not technology. Citing an interview with Major Ho Van Thuoc, Operations Officer for the Viet Cong Thua Thien Province Regional Forces, "they [the VC] listened to our radios all the time. He said their radio intercept squads could understand English. . . . Thuoc told us that they knew 'when and where you move.' Stu Vance of the CIA concluded in 1969 that the enemy had more than 600 listening posts – 200 of them manned by people who knew English."

Beginning in March, as the new U.S. firebases were constructed in the hills of Thua Thien far east of the A Shau, the NVA regiments were already in place to oppose them. FSB GRANITE (YD439188), within 10 kilometers of the lowlands and more than twice that distance from the A Shau, was established in early March to provide artillery support for the building of RIPCORD. The 29th NVA Regiment attacked GRANITE on 20 March, with U.S. losses of 11 KIA and more than 20 WIA.[19] On 30 April, they assaulted once again, this time resulting in 8 U.S. KIA, 1 MIA, and 41 WIA.[20] The enemy ability to mount such attacks so close to the lowlands was a clear sign that the tide of the war in Thua Thien was moving east. The NVA hit FSB MAUREEN (Hill 980) on 5–7 May with mortar and small ground attacks, killing 6 GIs and wounding 12.[§][21] On 23 May, the mortars were aimed at FSB KATHRYN, resulting in 3 U.S. KIAs.[22]

Enemy actions outside the valley compelled the commander of the 101st Airborne Division, MG Hennessey, to rewrite the script for TEXAS STAR. On 6 and 7 May, enemy ground attacks on FSB HENDERSON (a former USMC base in Quang Tri Province converted now to ARVN use) and FSB MAUREEN (see above), the planned eastern anchor for the allied offensive, resulted in a total of 100 U.S. and ARVN casualties and the loss of large amounts of ammunition and fuel just as the ARVN 1st Division was commencing overland movement to reopen FSB BRADLEY on the east wall of the A Shau.[**] As a result of the attacks on HENDERSON and MAUREEN, "Major General Hennessey announced that the Division and particularly the 3rd Brigade, had become overextended and the plans to push further into the A Shau/Warehouse Area have been cancelled. The ARVN forces withdrew back to [FSB] O'Reilly and Barnett."[23] Neither FSB BRADLEY nor AIRBORNE would be reopened,

---

§ MAUREEN was first occupied in 1968. Construction of the new firebase began on 25 July 1970.

** The ARVN lack of air mobility at this stage of "Vietnamization" says as much as anything else about the likelihood of its being able to stand up alone against the combined VC/NVA enemy.

and the Air Cavalry Squadron of the Screaming Eagles would remain the main allied unit operating in the A Shau AO.

Throughout most of Thua Thien in July, "the enemy units in the province were relatively quiet, staying in their base areas in the mountains just east of the A Shau Valley."[24] The exception was the area around FSB RIPCORD. As planned, a key component of TEXAS STAR was to be a multibattalion search and destroy / reconnaissance in force operation into the "Warehouse." The Warehouse 54 designation, originally provided by a POW the previous year, had been shortened in name and expanded in scope to include an ill-defined but large area roughly bordered by the former FSB AIRBORNE (YD354071) to the south, to several kilometers east of former FSB BRADLEY (YD278119) on the west, to several kilometers south of FSB RIPCORD (YD343194) on the north, and to several kilometers west of FSB MAUREEN (YD429122) on the east. Co Pung Mountain (Hill 1615 – YD336107) was the dominant terrain feature and one of the highest mountains in the area. The streams in the jungled hills to the west, south, and southeast of Co Pung drained into the A Shau Valley.

Under the original plan, Hill 927 (FSB RIPCORD) was to be reestablished in March 1970, before the start of Operation TEXAS STAR, to provide artillery support for the reopening of firebases BRADLEY and AIRBORNE (the site of the first contacts in Warehouse 54 the preceding year) and the opening of FSB KATHRYN, and enable the Screaming Eagles and 1st ARVN Division to invade the Warehouse area from multiple directions.†† The battle for RIPCORD is outside the scope of this book, but is described in detail in Keith Nolan's *Ripcord*,[25] John Roberts's *Operation Texas Star*,[26] and Benjamin Harrison's *Hell on a Hill Top*.[27] What happened at RIPCORD, however, was a direct result of the decision to pull allied troops out of the A Shau Valley in September 1969. What is of interest here is how the RIPCORD debacle resonated throughout western Thua Thien, compelling even further constraints on allied plans for the Warehouse operation.

---

†† Hill 927 was the site of a 1st Cavalry Division firebase named CARROLL in 1968.

When the first attempt was made to open RIPCORD in March 1970, the result was a series of "hill battles" not dissimilar, on a smaller scale, to those around Khe Sanh three years earlier. This included Hill 1000 to the west, Hill 902 to the south, and Hill 927 itself. The enemy had beaten the 101st Airborne Division to the punch, reoccupying the Warehouse and moving his infantry forward into the vacuum north and east of A Shau created in late 1969. Abandoning the initial attempt to occupy Hill 927, the 2/506th Infantry (LTC Andre C. Lucas) withdrew, to try again on 1 April as TEXAS STAR commenced. It took 10 days, renewed fighting on the peaks and slopes nearby (Hill 805, 1000, and 1298), and reinforcement by the division reserve – 2/501st Infantry (LTC Bobby Brashear) – before the brigade was able on 11 April to provide enough security at the landing zone atop Hill 927 to bring in the engineers and begin construction of the artillery positions and support structures.[28]

Once construction of RIPCORD was under way, the North Vietnamese began "preparing the battlefield" for the siege of Hill 927. According to former division commander MG Chu Phong Doi, the NVA 324B Division received orders on 19 May to eliminate RIPCORD.[29] Heavy weapons (82mm and 120mm mortars; 57mm and 75mm recoilless rifles) and 12.7mm antiaircraft guns were en route and multiple firing positions were constructed and camouflaged. Six weeks later, on 1 July, the NVA began its siege of RIPCORD with a devastating attack on an American company three kilometers to the southeast on "Re-Up Hill;" 22 days later, after a prolonged series of engagements in the hills around RIPCORD as well as at the firebase, RIPCORD was evacuated.

It was the NVA attack on RIPCORD that finally derailed the allied plan of attack (CHICAGO PEAK) into the Warehouse,[30] earlier sidetracked by the attacks on HENDERSON and MAUREEN (6–7 May), rather than the allied attack into the Warehouse that would upset the enemy's plans. While the enemy might not have reached the lowlands yet, its base area in the Warehouse was intact.

As a result of the heavy enemy resistance and then the abandonment of RIPCORD, it was decided to scrap the plans for reopening AIRBORNE and BRADLEY.[31] CHICAGO PEAK (a suboperation of TEXAS STAR) was redrafted, finally commencing on 25 July, when artillery was lifted to FSB MAUREEN to provide support. On 30 July, elements of the ARVN 1st Infantry Division combat assaulted onto the top of Co Pung Mountain to begin search and destroy in the Warehouse. Just six days later, the NVA again threw the allied plan into disarray, when a siege of ARVN FSB O'REILLY, the next in line to the northwest of Hill 927, was begun. O'REILLY, too, would eventually be abandoned on 7 October, before the monsoon arrived,[32] but the NVA 6th Regiment's attack forced the premature cancellation of CHICAGO PEAK on 12 August, as the ARVN troops were needed to the north. There were, in the end, no significant caches or base camps discovered during the brief ARVN presence on Co Pung.[33] In *Ripcord*, Keith Nolan observed: "Even the best-laid plans tended to come to ruin in connection with the A Shau." TEXAS STAR was not a "best laid" plan, but may have been the best the allies could come up with, given that the A Shau was controlled by the NVA, and, with redeployment well advanced, no other troops available.‡‡ As Nolan also remarked about the siege of FSB RIPCORD, "what was needed was another infantry brigade."[34]

Meanwhile, in the A Shau Valley, the only notable action during July and August took place on 11 July, when an F-4 Skyhawk was shot down by .51-cal fire over the A Shau airstrip. The two-man crew, from Marine Air Group 11, bailed out and reached the ground safely. Their rescue was accomplished by an Air Force SAR helicopter, which sustained 15 hits during the very hot extraction.[35]

Similar to the "commando raid" on NVA bridges on Route 548 at end of RANDOLPH GLEN on March 31, the impending end of TEXAS STAR brought recon teams from Company L, 75th Infantry, the nearest to the valley they had been since that one-day foray five months previously. Teams Philippines, Libya, and Pakistan were inserted between 3–5 September on either side of Route 547A a few kilometers from its terminus at Route 548 in the valley. Only Team Philippines reported any

---

‡‡ The divisional reserve 2/501st Infantry was committed to the battle but could not alter the outcome.

enemy sightings and had one minor contact, and all three patrols were extracted from green LZs several days later.[36]

A Better War?

The mission of TEXAS STAR was to "interdict enemy movement into the populated lowlands." "The basic mission [of Operation CHICAGO PEAK] was to increase the security of the populated lowlands. . . ."[37] Phu Bai, a major allied base south of Hue, was attacked by fire on 2 August and 28 August. Camp Evans, the main 101st Airborne Division base, was hit on 3 and 30 August and on 1 September. Camp Eagle, another major base, received twelve 122mm rockets on 28 August. Hue City was hit by 122mm rockets on 6 August, resulting in 14 ARVN KIA and 63 WIA, and again on 28 September.[38] The mission of TEXAS STAR was not accomplished.

In his provocative work *A Better War*, noted Vietnam War veteran and historian Lewis Sorley presents the conclusion: "There came a time when the war was won. The fighting wasn't over, but the war was won. This achievement can probably best be dated in late 1970, after the Cambodian incursion in the spring of the year."[39] It is difficult to support this hypothesis from the perspective of the war in Thua Thien Province in late 1970. Throughout Operation TEXAS STAR (1 April–5 September), the enemy had consistently attacked allied positions with indirect fire as well as sappers, demonstrating who held the initiative, and compelling the 101st Airborne Division to focus on defensive patrolling of the areas proximate to their bases, rather than running reconnaissance in force operations westward toward the A Shau. Fire Support Base RIPCORD had fallen, the 101st Airborne Division's plans to take the offensive against the Warehouse area were on the shelf, and its battalions were heavily committed to what was supposed to be a South Vietnamese responsibility, pacification. The tide of the allied initiative, which had once lapped up on the Laotian border, had now receded to a (broken) line of isolated firebases midway between the A Shau and the coast, and the hope that the Route 547 corridor could be defended. The enemy was once again rebuilding its A Shau roads and bases, and its regiments were slowly but surely moving eastward.

24

OPERATIONS JEFFERSON GLEN, DEWEY CANYON II, LAM SON 719, AND LAM SON 720

JEFFERSON GLEN/MONSOON PLAN 70:
September 1970–January 1971

When TEXAS STAR came to an end on 5 September 1970, it was immediately replaced with another division-wide operation, JEFFERSON GLEN. The mission of the new operation was "locating and clearing enemy forces, staging areas, and forward cache sites; protecting vital lines of communication; and pacification."[1] Even as the 101st Airborne Division continued to maintain a very high level of activity with three full brigades throughout the northern part of Military Region 1, American units in the rest of the country were packing their bags and heading home. By the time JEFFERSON GLEN drew to a close in early October 1971, the Screaming Eagles would be one of just two full U.S. divisions remaining, and two brigades of the other (Americal Division) departed in early November 1971. As of 15 November 1971, there remained of the combat troops once in Vietnam, in addition to the 101st Airborne, one brigade of the 1st Cavalry Division, the 196th Light Infantry Brigade, and one squadron of the 11th Armored Cavalry Regiment.

The last several months of 1970 were relatively quiet in the valley, as high winds and heavy rains were frequent and the NVA were busy replacing losses and licking their wounds from the RIPCORD and O'REILLY battles. Only the more significant contacts with the enemy in the valley are described in the following paragraphs of this section. Most of the infantry battalions in the 101st Airborne never came close to the A Shau Valley, and the few occasions when the airmobile infantry did so were of brief duration. The long-range patrols of Company L, 75th Infantry (Ranger) also were in the A Shau area only sporadically and for short missions. The men and helicopters of the 2nd Squadron, 17th Cavalry, on the other hand, were over and in the valley and the mountains around it every single day. The story in this chapter is therefore primarily their story.

One innovation introduced to the recon toolkit during 1970 was helicopter-mounted forward-looking infrared radar (FLIR). This new technology could pick up radiation in the infrared spectrum, making it useful at night and through fog, smoke, haze, and even jungle foliage. First introduced into Vietnam on UH-1G Hueys in February 1970,[2] by that summer it was in use in the A Shau by the 2/17th Air Cavalry. On 18–19 September, using FLIR and Cobra gunships, C/2/17th Cavalry knocked out a four-ton truck on the valley floor, and subsequent air strikes eliminated a second truck as well as a 12.7mm anti-aircraft machine gun.[3]

The Rangers of L Company, 75th Infantry (Ranger) were active in the valley again in early September. Several recon teams inserted near the end of TEXAS STAR (see preceding chapter) were withdrawn, but on 7 September a new group of missions commenced. RT Columbia with 12 men went in first on Hill 859 (YC448955), reestablishing the radio relay site used previously. Later in the morning, RT Buffalo was inserted on the south side of the Rao Lao, 5 km west of Route 548 (YC370930). The final team to go in was Albany, with an LZ deep in the hills west of the A Shau, 2 km from the Laotian border (YC424880). All of the landing zones were green. Buffalo heard some signal shots on 8 September and enemy movement to the west the next day, and Columbia reported enemy troops both east and west on 12 September, but the Rangers made no contact with NVA units. Albany and Buffalo were extracted on 11 September, and Columbia came out on the 13th.[4]

Three more teams went into the AO on 12–13 September. Recon Team Atlanta inserted 3 km north of FSB RENDEZVOUS (YC432995) and would patrol to the east for one kilometer until extraction on 16 September. RT Detroit also went in on the 12th, landing on the east wall of the valley 2 km from Route 548 (YC483920). This team moved south about one kilometer before extracting on 17 September. The third team in was Denver, inserted on 13 September on Dong Re Lao (Hill 1487 at YD405038). Denver observed three NVA soldiers on 15 September and called in a Cav gunship, resulting in one confirmed KIA. Its position disclosed, Denver was extracted without taking fire later that day. Neither Atlanta nor Detroit made contact with the enemy; Atlanta was withdrawn on 16 September and Detroit the next day. Although the latter team reported enemy movement around the pickup zone, it took no fire.[5]

On 23 September, a note in the Daily Staff Journal of the 2/17th Cavalry reported: "URS ["usually reliable source," meaning signals intelligence] received from G-2 [Intelligence] obs YD288132 believed to be enemy Div Hq. passed to B 2/17." This location is about 6 km northeast of Tiger Mountain and the same distance northwest of Co Pung Mountain. The elite ARVN Hac Bao (Black Panther) company was inserted 2 km southeast of the reported NVA divisional headquarters (YD308118) on 25 September. One ARVN soldier was killed and a .51-cal machine gun was captured in this one-day operation. By 1630H, the ARVN troops were withdrawn from a green pickup zone.[6]

A final group of Ranger team insertions began on 28 September on the east wall of the valley. RT Eglin (YC488899) and RT Houston (YC503887) were dropped about 6 km north of A Shau and RT Anaheim 2 km farther north (YC482925). Team Jamaica followed on 29 September, inserting 2 km east of A Shau (YC515840). All four teams were extracted mid-day on 3 October without having made contact with the enemy.[7] This was the final set of Ranger patrols in the A Shau Valley during 1970. It was evident from the repeated lack of enemy sightings by Ranger teams in the middle and southern parts of the valley that the enemy's focus was directed to the north, around FSB O'Reilly, in the early fall of the year.

In October, "the Squadron was restricted from significant activity due to poor conditions throughout the month."[8]

On 13 November, troops from D/2/17 Cavalry and a demolition team from the 326th Engineer Battalion were inserted in the central valley and patrolled along Route 548, destroying one large bridge and heavily damaging four smaller ones with demolition charges; there was no enemy contact. Elements of D Troop were back in the valley just two days later, when they were inserted into the very dangerous area of Route 922 south of Tiger Mountain. They destroyed several bunkers and a bridge constructed of logs and steel planking, and captured the barrel and breach assembly from a 37mm anti-aircraft weapon. Also on 15 November, a Pink team from A/2/17 Cavalry spotted a 3/4-ton truck in the valley 6 km east southeast of Tiger Mountain. Although receiving .51-cal antiaircraft fire, the Air Cavalry destroyed the truck and several bunkers at the same location.[9]

On 11 December, the C Troop Blues were inserted and destroyed enemy bridges located by scout helicopters on Route 548 near the A Luoi airstrip. In the southern A Shau Valley on 18 December, a B Troop Pink team received fire from two 12.7mm anti-aircraft guns. The helicopters were not hit and called in an airstrike, destroying both enemy weapons. In late December, D Troop was conducting ground reconnaissance near the A Luoi airstrip when it found and destroyed four more enemy vehicles.[10]

In January 1971, the pace began to quicken once again. On 14 January, near one of the same locations where D Troop had destroyed trucks just days earlier, a B Troop OH-6A was shot down, with the crew safely rescued. Two days later, another Loach went down from .51-cal fire. Again, the crew was extracted, although the aircraft was a total loss. C Troop was also reconnoitering over the valley and observed an estimated enemy company near the former CURRAHEE location. A "Cavalry Cobra Raid" was organized, using an OH-6A, a UH-1H Command helicopter, and five AH-1G Cobra gunships (three from C/2/17 and two from the 4/77 ARA), resulting in three NVA KIA.[11] On 19 January, while patrolling in his LOH west of Tiger Mountain, CPT Wilbur D. Latimer of C/2/17th Cavalry was hit and killed by small arms fire.[12]

Laos and the A Shau Valley: The Plan

On 8 December 1970, General Creighton Abrams held a special meeting at MACV Headquarters to discuss a response to the expanding NVA buildup in Laos. Only seven officers were present, including Abrams, GEN Fred Weyand (DEPCOMUSMACV), and LTG James Sutherland (CG, XXIV Corps). MG Petts, MACV J2 (Intel), provided the latest information. "Following the Cambodian incursion [May–June 1970], the enemy had conducted a massive buildup in the A Shau Valley and although their plans ["an offensive against Hue City"] had been stunted, the foe was still concentrating on efforts to strike at the vulnerable city of Hue."* NVA activity during the preceding August and September was assessed as the prelude to a major effort that had commenced in October, with the passing of the monsoon season in Laos. "As predicted, in the five weeks since the advent of the October dry season, the Communists had greatly increased their activity. . . ." By the first week in December, in spite of an intensified air interdiction campaign, "BA 611, and the A Shau Valley noted increased truck movement." ". . . 90 percent of the material coming down the Trail was being funneled into MR1 [as the I Corps area had been renamed]," leading the MACV J-2 to predict the launching of a four-part enemy offensive as soon as preparations could be completed. Phase I would be "A multi-division invasion of various targets in Quang Tri and Thua Thien Provinces with . . . BA 611 using the adjacent A Shau Valley as a channel pointing at Hue City." Under the guidance of General Abrams, the response developed by the allies was "a spoiling attack toward Tchepone to destroy the stockpiling in BA 604 and 611, and thus stunt the coming NVA offensive."†13

------

* The source of this intelligence is not known. As cited in the preceding chapter, the enemy buildup in western Thua Thien and Quang Tri during the summer of 1970 was mainly north of the A Shau, targeted at FSBs RIPCORD and O'REILLY. The 324B Division history cited in *Hell on a Hill Top* indicates that the enemy's ultimate objective at this time was the allied pacification program in the lowlands and not a new assault on Hue City.

† The genesis of the specific operational plan for LAM SON 719 and General Abrams's real enthusiasm for the operation are not entirely clear. Noted Vietnam War historian John Prados (*Vietnam: History of an Unwinnable War*, pp. 406–410) has done an excellent job of identifying the players involved in the "idea" of the incursion, tracing the initial concept to the summer of 1970 and Henry Kissinger's contemplation of a sequel to the Cambodian incursion. However, still murky is our understanding of exactly how or why the Allied military commanders eventually "bought in" to the plan that would send the cream of the South Vietnamese Armed Forces (including marines and rangers) against an enemy of equal or greater strength (including armor, artillery, and heavy antiaircraft artillery), with absolutely no level of surprise, without American advisors, and with no prior experience in corps- or even division-level operations. Certainly, an overreliance on American air power – including the untried extensive employment of helicopters in a "mid-intensity" battle – played a key part in that process. According to Prados (p. 409), in late January 1971 "General Abrams recommended cancellation after completing Dewey Canyon [II]." But Nixon said "go."

While the A Shau Valley and adjacent Base Area 611 in Laos would not be a target during the assault phase of LAM SON 719 (2 February–6 April 1971), the Communist buildup there was very much on the minds of the allied planners. As the concept of the invasion of Laos evolved, it was envisioned that operations directly related to the attack toward Tchepone would take place in the A Shau Valley at three points during the campaign. First, the 101st Airborne Division would conduct diversionary attacks (i.e., artillery raids) during Operation DEWEY CANYON II prior to the start of LAM SON 719 to try to prevent the enemy regiments in the A Shau / BA 611 area from reinforcing the NVA counterattack against the drive into Laos. Second, in the original plan, after successfully disrupting the enemy base along Route 9 (from Khe Sanh to Tchepone, through BA 604), part of the ARVN forces would withdraw southward inside Laos, down Route 92 to Route 922, and then pass eastward through Base Area 611 and the A Shau on their way back into South Vietnam. Third, in a follow-up operation, LAM SON 720, ARVN troops would return to the Warehouse area several weeks later to wreak yet more havoc on the enemy supply infrastructure. The first and third of these plans were carried out, although LAM SON 720 was much reduced in scale due to losses suffered in Laos. The second never happened, after the NVA rout of the ARVN along Route 9.‡14

DEWEY CANYON II – Late January 1971

The immediate precursor to LAM SON 719, Operation DEWEY CANYON II involved widespread diversionary operations designed to keep the enemy guessing about the direction and timing of the next allied offensive. The major ground actions occurred along Route 9 inside South Vietnam, as American forces moved westward to reopen

---

‡ The withdrawal through BA 611 was the preferred option of COMUSMACV. A withdrawal straight back along Route 9 was a second option in the planning.

the Khe Sanh base and to improve roads, helicopter support, and logistical facilities for the coming air assaults across the border. There were even diversionary maritime exercises to simulate early-stage preparations for an amphibious assault against North Vietnam.[15] The focus here, however, is limited to what was going on in and near the A Shau Valley.

In the first instance, in the certain knowledge that spies in the leaky ARVN High Command would pass along much information about the allied plans to the enemy, a deception "name game" was played in the planning documents, designed to use against it the foe's known sensitivity to allied actions against its burgeoning presence in the A Shau Valley. "DEWEY CANYON II" itself was selected as the name for the operation in the hope that the NVA would assume that this was a follow-up to DEWEY CANYON (I) carried out by the 9th Marines along Route 922 and in the Da Krong and northern A Shau valleys in early 1969. Moreover, two key ARVN landing zones during the assault into Laos were code-named "Tabat" and "Aluoi," to direct attention toward the A Shau and away from the point of the attack down Route 9.[16]

The Screaming Eagles, although given the mission of providing a diversion in the A Shau just prior to the assault into Laos, did not have much to work with. The division was tasked with taking over the area of operations of the ARVN 1st Infantry Division, as that unit was headed west, so no maneuver battalions would be available to use in a feint into the valley. The divisional air cavalry squadron was also assigned to provide support in the area of Route 9 and Khe Sanh, leaving meager forces to deceive the enemy in the A Shau. Intelligence at this time "indicated that the 803rd and 29th Regts / 324B Division were assuming support and security roles for base areas, cache site, and infiltration routes in the A SHAU VALLEY and eastern Base Area 611 areas."[17]

On 28 January, B Battery 2/320th Artillery (105mm howitzers) moved from CAMP EAGLE to FSB VEGHEL for an artillery raid. The

155mm howitzers of B Battery 2/11th Artillery were also transported to a forward base to carry out an artillery raid, although the 101st Division report is confusing, as it records the movement of this unit from FSB BASTOGNE to FSB BLAZE on 29 January, then from FSB VEGHEL to FSB ZON on 30 January. The 1/327th Infantry provided security for these temporary artillery bases.[18] The "artillery raids" did not fool the North Vietnamese.

By 20 March 1971, "the 29th and 803rd Regiments of the NVA 324th Division surrounded the 147th VNMC Brigade Headquarters, the 7th Marine Battalion, and the brigade's artillery at FSB Delta."[19] FSB DELTA was located in Laos, south of Route 9 and about equidistant from Route 92 and the border with Vietnam.§

"According to PWs taken in late March, the 803rd and 812th Regts are moving back into their traditional areas of operation after LAM SON 719. An agent report indicated the 29th Regt is also returning. Other intelligence sources confirm the eastward movement of the 29th and 812th Regts. However, recent analysis indicates that the 803rd Regt remains in LAOS south of QL-9."[20] The 101st Airborne Division's diversion in the A Shau had failed to convince the enemy that the 324th Division needed to remain in the A Shau or Base Area 611 to protect bases and caches there; the NVA High Command certainly knew that there were no battalions available to XXIV Corps to use in search and destroy missions in western Thua Thien. While the allies throughout the war made much of the "superior" mobility provided by the helicopter, the Ho Chi Minh sandals of the NVA infantry served them well, enabling the regiments in western Thua Thien to move north to take part in the defeat of LAM SON 719, and to return south in time to oppose LAM SON 720 just a few weeks later.

---

§ Benjamin Harrison's conclusion that "the crippled 324B Division played no significant role in the critical and strategically important PAVN 1971 Counter Offensive to Lam Son 719" (*Hell on a Hill Top*, p. 215) appears to be incorrect. By the time of the Laotian campaign, the 324B had recovered from the battles of the previous summer and fall and was able to slip out of South Vietnam and into Laos to participate in the NVA counterattacks.

JEFFERSON GLEN / MONSOON PLAN 70: *February–March 1971*

With Operation LAM SON 719 monopolizing the attention of the Screaming Eagles and the resulting displacement of 101st Airborne Division assets, February and March would be months of low-key activity in the A Shau. The movement of the NVA 324B Division to Laos was also, of course, a major contributing factor. On 10 February, a B Troop LOH was shot down by a 12.7mm antiaircraft gun just southeast of Tiger Mountain, with the loss of pilot WO1 Mark J. Robertson and observer SGT Joseph R. Pietrzak.[21] From 11–14 February, B troop gunships destroyed an NVA bulldozer and truck in the valley, but with the other troops of the divisional reconnaissance squadron operating in support of LAM SON 719, no other contacts took place in the A Shau.[22]

One of the more peculiar operational twists to the planning for LAM SON 719 was the prohibition of any U.S. personnel from participating in cross-border operations in Laos. Even as the ARVN were preparing to cross the border en masse with strong helicopter and fixed-wing air support from the Americans, the MACVSOG Operation PRAIRIE patrols in Laos were forbidden after 7 February 1971, as a result of the Cooper-Church Amendment, to include any U.S. Special Forces personnel.[23] Thus, the raid into Laos would be deprived of an important American reconnaissance component. Not to be entirely excluded from LAM SON 719, MACVSOG did conduct preinvasion recon and raids and diversionary actions "over the fence." But starting on 8 February, the main contribution of the SOG warriors to LAM SON 719 would be in the A Shau Valley area.

Seventeen kilometers due east of the abandoned A Shau airstrip lies Hill 1084 (Ha Te), where during earlier operations a fire support base called THOR had been occupied temporarily. On 18 February 1971, a 14-man SOG team led by CPT Jim Butler and including 4 Americans and 10 Montagnard tribesmen was inserted here. In the ensuing action (occurring on 18–20 February per most sources, but 16–18 February per SSG Leslie Chapman's Distinguished Service Cross citation awarded for this battle), almost 400 enemy soldiers were killed by air strikes and the ferocious defense of the recon team's perimeter. This location, however, is not within or adjacent to the A Shau Valley.**

RT Python was one of six MACVSOG teams inserted in mid-February to conduct reconnaissance and to use air power to interdict enemy movements in and around the A Shau, which at this time was still proposed as a primary withdrawal route for ARVN forces invading Laos to the north.[24] No information on four of those teams has been located, but it would appear that they were without American advisors and may have

---

** In one of the more persistent but geographically impossible tales of "A Shau" actions, the THOR site is invariably referred to as overlooking Route 548 in the A Shau Valley. This never happened. Some examples:

- "RT Python occupied abandoned Firebase Thor in order to block PAVN access to Highway 548" (Gillespie, *Black Ops Vietnam*, p. 229).
- "When Recon Team Python Ruled the Valley of Death [i.e., the A Shau Valley]" – title of a story by Mike Perry at the specialoperations.com website. Includes repeated references such as "teams would land in the A Shau Valley," placing RT Python in or next to the A Shau.
- "RT Python, on the Ashau's east wall . . ." (Plaster, *SOG: The Secret Wars of America's Commandos in Vietnam*, p. 322).
- "The remnants of old Firebase Thor on the east wall of the A Shau approximately six kilometers due east of the insert LZ for RT Intruder." "Situated on the barren top of Hill 1084. . . ." "That first night, RT Python watched the valley floor to their west as a procession of truck headlights moved unimpeded down Route 548, accompanied by long lines of troops carrying flashlights" (Yarborough, *A Shau Valor*, pp. 254–256).

Fire Support Base Thor was indeed atop Hill 1084 (aka Ha Te Mountain, YC665835) (Michael P. Kelley, *Where We Were in Vietnam* (Central Point, OR: Hellgate Press, 2002), pp. 500–506.) Hill 1084, however, is not "six kilometers due east" of RT Intruder, but rather 17 kilometers due east of the A Shau airstrip and about 20 kilometers east-northeast of RT Intruder. It does not overlook the A Shau – from which it is separated by several ranges of hills, including the actual "east wall" of the valley, as well as the Dong Tre Gong ridgeline, which at 800–1,000 meters in height would have blocked a sightline even into the Rao Nai (Be Luong) Valley between the outpost and the A Shau Valley. What could be seen from THOR to the west was the valley two to the east of the A Shau where a tributary of the Song Huu Trach flows, or, at the foot of the mountain on the east side, the main branch of the Huu Trach.

There is one other possibility. All information about the location of RT Python is from secondary sources. Is it possible that "Firebase THOR" was not the actual site of RT Python's battle? Could it have been confused in the telling/retelling with Firebase FURY? FURY, at YC534846, *is* on the east wall of the A Shau, *does* overlook Route 548, and is just eight kilometers northeast of RT Intruder's LZ.

been inserted either in the valley itself, or more likely in Base Area 611 to the west. (Coordination with the 2/17th Cavalry would have been necessary if operating in the A Shau, and that has not been documented.) The sixth team was RT Intruder.

Recon Team Intruder's designated landing zone was almost directly on the Vietnam-Laos border, on Hill 1528, about 6 km south southwest of the A Shau airstrip. The hill is part of a long ridgeline that runs along the border at its closest point in the southern valley to Route 548, less than 1 km from one of the streams forming the headwaters of the Rao Lao (A Shau). While the ridge slopes very steeply down to the valley floor, it is much more accessible from the Laotian side. RT Intruder was inserted on 18 February and soon after arrival came across a major trail, large enough for trucks, and even with multistrand communications wire alongside, and then engaged in a firefight with a small enemy unit. Immediate extraction was requested, as NVA signal shots indicated that the enemy was fast closing in. Three slicks from A/101st Aviation Battalion arrived, but as the first attempted to lift out four men and their equipment on STABO rigs (basically, long ropes dangling freely below the aircraft), it did not have enough power and the men all cut loose before the Huey could be dragged down into the jungle. The other two Hueys were able to lift out the five Bru and two of the Americans, but CPT Ronald L. "Doc" Watson, SGT Allan R. "Baby Jesus" Lloyd and SFC Samuel D. Hernandez remained on Hill 1528.

The men still on the ground requested that any further rescue attempt be delayed until the next morning, hoping to convince the enemy that the entire team had already been extracted so that they would leave the area. But another slick with orders to extract arrived less than an hour later. The three men all hooked up to the STABO lines and the helicopter lifted off, to be hit by ground fire and nose-dive down the ridge wall. At the last minute, SFC Hernandez's line either was cut by the helicopter crewman or snagged and broke, and he fell onto the mountain top. The helicopter crashed inverted down the cliff, killing CPT Watson, SGT Lloyd, and the four-man crew from A/101 AVN: aircraft commander WO George P. Berg, pilot WO1 Gerald E. Woods, crew chief SP4 Walter E. Demsey, Jr., and door gunner SP4 Gary L. Johnson.[25] SFC Hernandez spent the night alone on the ridge and was safely extracted the next morning. This was not the end of the story of RT Intruder.

A "BRIGHT LIGHT" (POW/MIA) mission was launched on 19 February to try to retrieve the bodies from the downed helicopter. RT Habu was inserted and able to get to the crash site by rappelling down the cliff, but a ground assault by an enemy platoon, later reinforced to company strength, caused heavy casualties, and an O-2A Covey FAC from the 20th Tactical Air Surveillance Squadron was shot down, killing pilot 1LT James L. Hull and MACVSOG aerial observer SGT William M. Fernandez.[26] That night, the members of Habu, including several wounded, were able to escape the enemy assault only by jumping off the cliff à la Robert Redford and Paul Newman in *Butch Cassidy*. Amazingly, all survived and were picked up the next day. The bodies from the crash site could not be recovered.[27] The rest of the story: because the ARVN did not withdraw from Laos down Route 922, the reconnaissance mission of the MACVSOG teams was irrelevant to LAM SON 719.

In March, the Ranger patrols from the 101st Airborne Division returned to the valley. For the entire month, 21 teams were inserted on the east side of Route 548, within 5 km of the valley center and spread from northeast of Tiger Mountain to about 5 km north of A Shau. There was a concentration near the southern end of Route 547A and another between Ta Bat and A Shau.[28] Only six teams made contact with the foe, killing eight NVA and capturing two,[29] as the 324B Division was still operating in the LAM SON 719 AO. On several occasions, the Rangers called in artillery and/or gunship strikes on suspected enemy concentrations.

On 22 March, RT Fayetteville was operating near Route 547A about 2 km east of RENDEZVOUS (YC450975) when it made contact with the enemy. SP4 David R. Hayward was hit in the chest with small arms fire and was picked up by medevac 30 minutes later but did not survive.[30] The rest of the team was extracted. On 23 March, RT Indianapolis went in 500 meters west of Route 548 near a small stream (YC476870). After observing two NVA truck convoys on Route 548 the night of 25–26 March, the team was engaged in sudden heavy firefight with five NVA the next morning (YC484875), killing one and probably a second. PFC

Joel R. Hankins, the newest man on the team, was hit by enemy fire and died shortly thereafter.[31]

As March expired, so too was LAM SON 719 coming to an ignominious end along Route 9 west of Khe Sanh.

LAM SON 720 and JEFFERSON GLEN: April 1971

In the master plan prepared before LAM SON 719, the follow-up operation, LAM SON 720, was to involve a corps-sized force. "According to Lam's [GEN Lam, ARVN I Corps Commander, later relieved] plan, [LAM SON 719] was to be completed by 31 March. Then, after resting and reorganizing for about two weeks, the 1st ARVN Division, two Marine brigades, and one airborne brigade would conduct operations into the eastern sector of enemy Base Area 611, the A Shau Valley, and the Laotian Salient [which lies west of the Quang Tri – Laos border, south of Khe Sanh and west of the Da Krong River Valley]."[32]

The necessity for a *ground* operation in the northern A Shau Valley was emphasized by USAF General Lucius Clay, Deputy COMUSMACV for Air Operations, at the WIEU on 27 March 1971:

CLAY (re Route 922): "I'd give my left arm if somebody could put a marine battalion down there sitting on that thing for the rest of the dry season [A Shau area, Tiger Mountain]. It's been the principal route the whole year. It's a very difficult area to interdict – flat terrain, it's relatively open, and multiple options in terms of where you want to move trucks. We have put navy cutting [?] and seeding, we have bombed in there, we've had gunships in there, and there [are] just no really good natural interdiction points that aren't easily bypassed, repaired, or moved around. It's just real tough."[33]

While the ARVN caught their breath after the precipitous pull-out from Laos, the 2/17th Cavalry continued to disrupt the NVA logistical system in the A Shau Valley. On 5 April, a Pink team had caught NVA troops in the open about 2.5 km east of the abandoned Ta Bat airstrip (YC451950), killing several. The squadron commander decided that a

mission to recover the enemy bodies and equipment should be undertaken, and the following day a Ranger team rappelled into the site after
preliminary reconnaissance reported no enemy in the area. As the slick
from B/2/17 Cavalry hovered over the landing zone and four Rangers
descended down ropes, the enemy sprang an ambush. Heavy automatic
weapons fire came from a trench full of NVA soldiers, and the Huey
received over 60 hits. SSGT Leonard J. Tremblay, the Ranger serving
on the slick as "belly man" to control the lines of the men descending,
was killed, and the pilot, copilot, crew chief and Ranger Team Leader
Billy Nix received multiple gunshot wounds in the legs. Somehow, the
pilots nursed the crippled aircraft, never to be flown again, back to Phu
Bai.[34] On 10 April, another incident resulted in the death of a hero from
an earlier time. MAJ Robert L. Grof, B Troop Commander, who had
received the Distinguished Service Cross for action on 19 June 1967 while
flying with the 9th Infantry Division, was struck and killed by 37mm
anti-aircraft fire while flying in his C&C helicopter about 8 km west of
Tiger Mountain.[35]

When the recon teams of L Company, 75th Infantry (Ranger), were
inserted in March, they had observed an increasing level of enemy vehicular traffic, especially in the southern part of the valley. On 7 April, an Air
Force Stinger gunship reported the destruction of seven enemy trucks just
north of the junction of Routes 547A and 548 (YC430965). The air cavalry squadron decided to try a new tactic. Instead of inserting recon teams
and then trying to engage the enemy, if spotted, with air and artillery, the
ground troop (D Troop) of the cavalry squadron would be inserted on
the valley floor and would conduct brief sweeps before being extracted
the same day. The tactical trick was that Rangers of L/75 INF, in platoon strength (five six-man teams), would insert with D Troop, but would
then conceal themselves and remain overnight in ambush positions in the
valley.

This plan was executed three times during the month. On 15 April, D
Troop was inserted on the valley floor, 3 km south of the ambush nine
days earlier (YC451920). It was joined by a Ranger platoon led by 1LT
Paul C. Sawtelle. The unit patrolled to the northwest about 1,500 meters

(YC442938), finding signs of truck traffic and a few bunkers, before the air cavalrymen were extracted in late afternoon. Meanwhile, the Rangers had dropped off short of the pickup zone. Before dark, the platoon established an ambush position overlooking Route 548, but no trucks came down the road that night. The next morning, the Rangers moved to better daylight cover and settled in to await their prey. A 10-man NVA patrol appeared, cautiously sweeping the road from the north. The enemy detected the ambush and opened fire first, instantly killing LT Sawtelle and SGT James B. McLaughlin. Helicopter gunships were called in and the enemy withdrew after a brief firefight, leaving blood trails behind. The platoon was then extracted.[36]

The second stay-behind ambush mission was conducted on 20 April, with insertion about 1,500 meters farther south along Route 548 (YC456903). Moving several hundred meters to the northeast, the patrol found lots of trails but no signs of a recent enemy presence, before extraction of D Troop in the afternoon. Now on their own, the Rangers under LT Jim Montano made contact with an enemy force at 1630H and had one man severely wounded. ARA support arrived and the enemy fire was suppressed, allowing medevac of the wounded Ranger. Their position exposed, the remaining Rangers were also extracted by 1800H.[37]

While the air cavalry and Rangers tried unsuccessfully to ambush an enemy convoy on Route 548, the 1st Brigade of the 101st Airborne Division was moving into position to launch LAM SON 720. On 14 April, the brigade set up a forward CP (command post) at reestablished FSB FURY (Hill 825 – YC534846), on the east wall 4 km east of the A Shau airstrip. Troops from the 1/327th Infantry secured the FSB, and three 105mm howitzers from C/2/320th Artillery were flown in to provide support.[38] The 1st Brigade infantrymen commenced patrols in the vicinity of FSB Fury and farther west into the valley.

On 21 April, C/1/327 was on patrol along Route 548, 2 km southeast of the Ranger action two days earlier (YC473890), when it ran into an enemy unit of unknown size. Using its own small arms and mortar fire, the company accounted for five enemy KIA and captured an 82mm mortar.[39] For 0500H the following morning, the 2/17th Cavalry Daily Staff

Journal includes the following entry: "informed by 1st Bde that there [sic] unit 1-327 at YC474888 reports trucks coming up to the E & W of their pos[ition], stopping, and people unloading. Bde requests P[in]k Tm to go out for coverage of unit at 0600H." A heavy Pink team (2 Cobras and 1 LOH) was sent out, but there is no further report on the results of this contact. Most interesting is the notation that enemy trucks were approaching from both directions. The enemy must have had substantial bases not far away (it was 5 a.m. and they would have to get the trucks under cover by daylight), both east and west.

On 23 April, the attempt to insert a radio relay team to facilitate communications with patrols in the valley turned into a tar pit, sucking helicopters and men into a vortex of death. RT Cubs of L/75 Infantry was due to be inserted to establish a radio relay outpost on the west side of the valley. When their helicopter was assailed by ground fire on final approach to the LZ, the team leader opted to land instead at the secondary LZ, about 3 km to the east of where the two stay-behind ambushes had been inserted (YC484920). The mission was to ensure commo with the third ambush, launched the same day. Almost immediately after landing, the team was in contact and the team leader, SGT Marvin Duren, was severely wounded before he could even get to cover. A Huey from B/2/17th Cavalry was sent in to drop off Ranger SSG William Vodden, who had volunteered to take over command. As the helicopter lifted off, it was hit by heavy fire and went down into the jungle to the east. Both pilots, CPT William E. Cullen and CPT Louis J. Speidel, were badly injured and trapped in the wreck; both died there. A medevac helicopter from the 326th Medical Battalion arrived and was able to extract the wounded door gunner and SGT Duren. When the dustoff chopper returned to pick up more wounded, it too was downed by the NVA in the landing zone. Crew Chief SP5 Michael L. Brummer died in this crash. The crew medic, SP5 Robert F. Speer, recipient of the Distinguished Flying Cross for Valor and the Bronze Star for Valor died in ground combat after the crash, and a second crew chief, SP4 David P. Medina, died on 30 April of wounds he received in this action. The surviving Rangers and helicopter crewmen had no communications and hunkered down for the night. Fortunately, the enemy, who occupied strongly fortified bunkers, elected to remain under cover and did not attack.[40]

On the morning of 24 April, recon team member SP4 Johnnie R. Sly attempted to recover a team radio from the landing zone. He was able to do so, thus reestablishing a critical link to help, but was severely wounded and died shortly thereafter. D Troop was called upon to reinforce the beleaguered force at the relay site, inserting several hundred meters to the north. The official record picks up the story:

> Approximately 150 to 200 meters from the [downed] aircraft the lead element of the troop came under heavy RPG, small arms and Chicom grenades fire. The platoons were maneuvered to react to the threat and pick up survivors of the Ranger team and aircraft. The contact on the 24th lasted approximately six hours and resulted in 5 friendly KIA and 14 WIA, one known enemy KIA by small arms fire, 11 enemy KIA by aircraft and an unknown enemy WIA. After the extraction of three US from the downed medivac helicopters the troops withdrew approximately 400 meters north to NDP, medevac our wounded and be reinforced by another company.

The five men killed in D Troop on 24 April were the troop commander, CPT Thomas D. Chenault, SGT Jose A. Soto-Figueroa, SP5 John W. Wilson, SP4 Christopher L. Vollmar, and PFC Thomas H. Taft.[41] That same day, additional troops from 2/502 ABN (Reinforced) and A/1/327 ABN were brought in to reinforce D Troop, landing on the valley floor and approaching from the south and west.[42] Intelligence reported an NVA battalion also marching to the sound of the guns, although it would not arrive in time to participate. After the crash of the heavily loaded medevac Huey, the surviving Rangers and crewmen became fragmented into several small groups. At some point, Ranger PFC James A. Champion left in a search for water; he was never seen alive again. Another Ranger, PFC Issako F. Malo, was taken prisoner and released two years later.[43]

On the 25th, against only very light opposition, the locations of the downed helicopters and the radio relay team site were reached. The medevac helicopter was lifted out, and just before 1900H, the last men from D Troop and the Rangers were extracted from a green pickup zone. A total of 13 Americans were lost in this action: 5 from D/2/17th Cav, 3 from the 326th Med Bn, 3 from L/75 Rgr (1 KIA, 1 MIA, and 1 POW), and 2 from B/2/17th Cav. Twenty additional men were wounded and evacuated. On 27 September 1971, *The Screaming Eagle* in-country newspaper reported that three Rangers had been awarded Silver Stars for their actions "establishing a radio relay." While not identified by names or date, this certainly refers to members of Recon Team Cubs.

The third and final attempt to ambush an enemy convoy on Route 548 came to naught on 23–26 April. Two platoons of L/75 INF were inserted in the valley (YC458906) just before 1300H. They blew up an enemy culvert on Route 548 (not the one they were assigned to blow, but close enough for A Shau Valley work), and one platoon was extracted just over an hour later.[44] The intent was to fool the enemy into thinking that the culvert demolition mission was completed and all troops had been pulled out. But once again, LT Montano's platoon remained hidden. For the next three days, as the battle at the radio relay site waxed and then waned, the platoon was on its own. Although infrared and "sniffer" imaging by aircraft reported enemy movement all around the ambush location, nary a truck nor an enemy soldier was seen, and the final hidden ambush trick was concluded, ineffective, but at least unscarred.[45]

While the 1st Brigade of the 101st Airborne Division worked the southern end of the valley during LAM SON 720, the ARVN were tasked with conducting operations in the Co Pung/Warehouse area and farther west, in the Da Krong Valley and newly designated Base Area 129. Before they could take over that responsibility, the Screaming Eagles would need to secure a base for their use. In the first attempt, on 15 April, the Recon Platoon ("STRIKE Force") of E/1/502 ABN was assigned to establish a landing zone atop Co Pung Mountain (Hill 1615 at YD336107). This ended in failure when two of the slicks crashed on the LZ, with five STRIKE Force troops KIA and 10 WIA. The dead were SSG George A, Pacheco, SGT Jeffry E. Cowley, SGT Thomas J. Gettelfinger, PFC Robert J. Dutkiewicz, and SP4 John L. Wilson, Jr., a medic from HHC/1/502 ABN.[46]

On 22 April, A/2/502 ABN conducted a night patrol in the valley east of the Punchbowl. E/2/502 supported this by establishing a radio relay site atop the abandoned EAGLE'S NEST. On 30 April, with the weather clearing over Co Pung, a second attempt was made to occupy the mountain top, this time by the 2/502 Infantry (LTC Lloyd N. Cosby), B and C Companies having just completed the operation at the radio relay site to the south. The mountain had been hit with B-52s, tactical air strikes, and artillery since the earlier aborted insertion on 15 April. Bravo and Echo Companies were inserted on Co Pung, and in spite of the loss of five men when their helicopter was shot down, the base was established. Lost in this incident were the aircraft commander, WO1 Charles T. Kallaher, the crew chief, SP4 Harold K. Armstrong, and the door gunner, Cantrell M. Daniel, III, from B/101st Aviation Battalion, and two combat engineer passengers from A/326th Engineer Battalion, SP4 James E. Shanks and SP4 Larry T. Yielding.[47]

LAM SON 720 and JEFFERSON GLEN: May – October 1971

For the first eight days in May, the 2/502 Infantry continued to occupy the top of Co Pung Mountain and to patrol the area nearby, as the weather deteriorated with the approaching monsoon season. On 2 May, the NVA hit the hilltop with the first of 18 (11 mortar and 7 recoilless rifle) attacks-by-fire, an average of more than 2 per day. The troops were well dug-in and not a single casualty resulted, while timely and effective counter-battery fire and active patrolling kept the enemy from tarrying in its attack positions. Charlie Company worked the eastern side of the mountain and Delta Company patrolled to the south. On 8 May, the ARVN arrived to assume responsibility for the AO.[48]

The Loaches and Cobras of the 2/17th Cavalry supported the 2/502 Infantry and continued to carry out reconnaissance flights throughout the valley. On 1 May, the C Troop Blues were inserted at abandoned FSB Airborne to extract the crew of a downed helicopter. Heavy enemy fire forced the cancellation of the recovery of the aircraft, but the crew and Blues were extracted without loss.[49] On 8 May, two Cobras and the C&C

ship from Bravo Troop were sent "over the fence" into Laos to check out a reported NVA pipeline.[50] The B Troop Aerorifle Platoon was inserted to destroy an NVA communications site 3 km north of A Shau (YC491865) on 9 May and got into a firefight with the enemy. D Troop was inserted to reinforce, and by the time contact was broken, the squadron had two KIA (1LT John J. Woodrum and PFC James D. Jackson, both of B/2/17) and several wounded.[51] An NVA 2½-ton truck was sighted and destroyed by a B troop Pink team on 12 May, north of A Luoi.[52] On 28 May, an OH-6A was hit with an RPG over the center of the valley, 4 km east of Tiger Mountain (YD284087). The aircraft exploded and the pilot, CPT Paul D. Urquhart, and the observer, SP4 Stephen Chavira, both of B/2/17 Cavalry, were listed as missing in action. The LOH was unrecoverable. A ground reconnaissance on 7 June found no signs of survivors or remains.[53]

During the month of May, ARVN Marines on LAM SON 720, operating in the Warehouse area, engaged in one of the largest battles ever in the A Shau AO. "A Vietnamese marine battalion, supported by tactical air strikes, killed 177 Reds Wednesday afternoon, six miles east of A Luoi, after killing 23 before dawn in the same area. . . . The Marines, airlifted into the A Luoi area, 25 miles southwest of Hue Tuesday, lost 36 men killed and wounded in the afternoon battle, and six killed and 48 wounded in the earlier clash." The Marines weren't the only ARVN troops finding the enemy in the A Shau. The Hac Bao (Black Panther) Company from the ARVN 1st Division got into two small firefights closer to A Luoi and suffered additional losses when they were hit with mortar fire later that afternoon while operating to the southwest of A Luoi.[54] A Ranger team from L/75 Infantry working with the Hac Bao was the only element of the LRP company in the A Shau, having been sent to work with the Black Panthers on 1 May: "This team was instrumental to an ARVN victory over the enemy forces they engaged." Although Company L, 75th Infantry (Ranger), would actively patrol elsewhere in the 101st Airborne Division AO, the team working with the Hac Bao Company was the last to enter the A Shau Valley. The Rangers would continue to fight along with the elite ARVN unit until extracted on 22 July.[55] According to Tom Yarborough, three U.S. helicopters and one 0-2 FAC were downed while supporting the ARVN in May, but the VHPA website lists no KIAs among 101st Airborne Division helicopter units during this time period, so all crews must have been rescued.[56] The only O-2A lost was in a crash-landing at Da Nang on 20 May; both crew members survived.[57]

According to Andrew Wiest, two battalions of the 54th ARVN Infantry Regiment (1/54 and 4/54) conducted operations

> ... into the northern reaches of the A Shau valley, resulting in sharp clashes that indicated that the communist recovery [from LAM SON 719] was well and truly under way. . . . The operations of the 54th around Khe Sanh and in the Da Krong and A Shau valleys [in LAM SON 720] were of great symbolic importance; they were the final allied operations ever in those areas, the forbidding and legendary scenes of some of the hardest fighting of the entire Vietnam War.[58]

"Symbolic" of the impending defeat of the allies, perhaps? If this information is correct, then the ARVN never went back into the A Shau Valley after the departure of the 101st Airborne Division!

On 1 June, B/2/17 Cavalry provided cover for the extraction of the 54th ARVN Regiment from its base on Co Pung Mountain. A/158 Aviation provided the lift ships, and a giant CH-54 Tarhe helicopter was employed to haul out heavy equipment. The first week of June was notable for not one, but two incursions by the 2/17th Cavalry into Laos. On 1 June, aircraft from A Troop worked an enemy concentration several kilometers inside Base Area 611 (YD0505) west of Tiger Mountain. In spite of intense 37mm and 57mm antiaircraft fire, no helicopters were lost, and 16 enemy KIA were claimed. C Troop spent the entire first week of the month at the opposite end of the A Shau, conducting reconnaissance for a Hac Bao raid into Base Area 607, and in the process destroying a 5-ton truck at the southern end of the valley (YC508808).[59] One may recall that BA607 was a target that General Westmoreland had wanted to attack four years earlier. Although plumb targets were found, the Hac Bao raid was canceled. While the other troops operated to the west, B Troop prowled the center of the valley, destroying one enemy truck near the A

Shau airstrip on 1 June and then two more near FSB RENDEZVOUS on the 9th.[60]

Picking up where they left off in June, on 3 July, B/2/17 Cavalry KO'd another NVA truck south of the A Shau airstrip. The NVA struck back the next day just north of the airstrip (YC496843), when an LOH from C/2/17 Cavalry was shot down by heavy fire. A UH-1H chase ship went in to pick up the surviving crew and was itself hit several times and forced to land with hydraulic failure near Route 547. All crew on both helicopters survived, but the LOH had to be destroyed in place. A third helicopter, a Cobra also from C Troop, was downed by small arms fire during the rescue operation (YC492855). It, too, had to be destroyed; both crewmen were wounded but rescued. On 7 July, two ARC LIGHT B-52 strikes were targeted at Route 548 3 to 5 km north of the A Shau airstrip and the trail network that ran to the west in this area, which was centered on the site of the 9 May battle.

Six kilometers east of Hill 937 (Hamburger Hill) (YC386996), an OH-6A from B Troop was hit by small arms fire and went down. Two Pink teams and the Troop's Blue platoon were sent in to provide cover and secure the crash site, and both helicopter and crew were successfully extracted. On 16 July, a B Troop LOH was knocked down as it flew over Route 548, north of Dong So Ridge. Like its sister ships, the helicopter crashed, but the crew, including one wounded, was saved. Twelve enemy soldiers were sighted on 20 July, 6 km southeast of FSB RENDEZVOUS. They were engaged with artillery fire, and five were killed. At the other end of the valley 6 km east of Tiger Mountain, during the last part of the month, B/2/17 killed 11 NVA soldiers and destroyed a 5-ton truck (YD307097).[61]

During the last few months of JEFFERSON GLEN (ending 8 October 1971), the northwest monsoon reduced both flight time and enemy movement significantly. In August, the Air Cavalry continued to patrol in the skies over the A Shau and around Co Pung Mountain, weather permitting. On 7 August, B Troop destroyed an NVA .51-caliber position just a few hundred meters northwest of the battle of 9 May (YC488686). On 12 August, just west of where the Stinger had wiped out seven trucks on 7 April, the guns of B/2/17 Cavalry accounted for another KIA. C Troop also performed recon over the valley but had no significant contacts. On 31 August, LAM SON 720 ended.

There were no significant reported contacts by the Second Squadron, Seventeenth Cavalry in the A Shau AO during September or early October, although aerial reconnaissance continued at a reduced pace. During much of this period, there was an intense focus on the area west of the Route 547A/548 junction, including the areas around abandoned FSB CURRAHEE and Hamburger Hill (aka FSB DESTINY). There was also a perpetual concern with Tiger Mountain and the roads and trails to its west and north.[62]

As JEFFERSON GLEN concluded, the end of the American ground war, like the proverbial "light at the end of the tunnel," was beginning to come into sight. C and D Troops of the 2/17th Cavalry began to stand down during October and November, and B Troop was inactivated in December. The Rangers of L/75 Infantry were placed under control of the 3rd Brigade on 15 October. While I cannot be absolutely certain, it is likely that CPT Robert O. Baker, Jr., of A/2/17th Cavalry was the last American solider to be killed in the A Shau Valley. CPT Baker was piloting an OH-6A LOH over the southern A Shau (YC483861) on 2 January 1972 when he was struck by fire from an AK-47 assault rifle. The observer was able to fly the helicopter to Phu Bai, but CPT Baker died in the hospital there.[63] On 7 February 1972, the last elements of the divisional air cavalry squadron departed the Republic of Vietnam.[64]

25

1972–1975

The Growth of the NVA Logistical System: 1969–1972

When Operation DELAWARE brought allied forces into the A Shau Valley in April 1968, the North Vietnamese Army, in preparation for the Tet Offensive, had just completed major road work linking Route 548 with Route 9 in Laos via Route 922. They had also constructed a new road (547A) between Route 548 near Ta Bat and Route 547 in the vicinity of the Rao Nho / Rao Nai river junction (future location of FSB VEGHEL). Work was ongoing on an ever-expanding network of trails connecting base areas in Laos with the A Shau Valley, a network that included dozens of paths up and down the entire length of the valley, joining route 548 with sanctuaries across the border. For the next year and a half, the allied infantry in the valley made major road work impossible, although the PAVN engineers continued to build a honeycomb of camps and well-hidden trails in the jungles of western Thua Thien Province, especially around Co Pung Mountain (the "Warehouse") in the north and between the A Shau and Rao Nai Valleys to the south.

In the fall of 1969, once allied ground troops were gone and the weather began to improve, the NVA engineers went into high gear. In spite of the efforts of the 3rd Force Recon Company, the air cavalry, and air attacks during RANDOLPH GLEN, by April 1970, CIA satellite reconnaissance photos reported "Route 548 is probably serviceable and in use from the Laos/South Vietnam border . . . to the La Dut (1) stream crossing . . . south of this point the road remains unserviceable."[1] Six weeks later, the

CIA updated: "SVN Route 548 through the A Shau Valley is now serviceable and in use over its entire length. The road is also serviceable into Laos via route 923C to Laos route 614A [at the southern end of the valley in Base Area 607]."[2] Route 9222 in the Da Krong Valley, which paralleled Route 922, 8–10 kilometers to the south, was also extended during the 1969–70 dry season (October 1969–April 1970) from the Laotian border to just 2.5 km northeast of the Route 548 Bypass at Tiger Mountain. All of the NVA roads were "limited all-weather," meaning suitable for vehicles, but of limited use during the monsoon season.

By February 1971, Binh Tram 42, the NVA headquarters for the Route 922/548 area, controlled eight battalions: two antiaircraft, four engineer, one transportation, and one signal.[3] During the 1970–71 dry season (October 1970–May 1971), one focal point for North Vietnamese road construction was the "New Eastern Route." By late 1972, Route 548 – the old trail down the center of the A Shau Valley – would be the key part of an NVA road directly connecting North Vietnam with Da Nang and points south, via the abandoned Khe Sanh Combat Base!

A new route is under construction that will, when completed, connect North Vietnamese Route 103 in the western DMZ with the A Shau Valley without crossing the border into Laos. This route lies on the eastern slope of the Annamitique Mountains and therefore is in the weather cycle of the coastal area of northern South Vietnam, which has its heaviest rains at the beginning of the dry season in the Laotian Panhandle.... When construction of all road gaps is completed, the road distance between the DMZ and the A Shau Valley will be shortened from 125 miles to approximately 65 miles.[4]

In addition to halving the length of the enemy supply line to the A Shau Valley, the new route offered three other distinct advantages.

1) **Since the monsoon weather patterns on the eastern and western sides of the Annamitique mountains were almost opposite, the NVA would have the opportunity to maintain a year-round supply flow by altering routes according to the season.**

2) "Because the new route roughly parallels the coastal tide-
lands of South Vietnam in which the population lives, supplies heading south move progressively closer to key enemy
base areas near these population centers."

3) "The geography of this new route also permits enemy units
and logistical activities to outflank the major ARVN defenses
in northern South Vietnam."[5]

Thus, the threat to Hue and Da Nang would increase, and the advantage of interior lines so long enjoyed by the allies in their coastal enclaves
would now dissipate with the coming of a parallel North Vietnamese line
of communications.

A year later, with all allied troops gone from the A Shau, a memo
from COL Hillman Dickinson, Senior Advisor to the ARVN 1st Division, to the division commander, provides a comprehensive overview of
the expanding NVA logistics network just before the start of the enemy's
1972 offensive, based on intelligence from sensor strings located at key
junctions and choke points:

Heavy sensor activities vic YD0631 on upper Route 616 indicates [sic] possible buildup of storage sites SE in BA 129. Route 616 is not open to Route 548
at this time and engineer activity is endeavoring to link it up with Route
548 in the northern A Shau preparatory to any large scale engagement. . . .
Activity at the intersection of Routes 922 & 548 (vic YD271096) reveal [sic]
steady shuttleing [sic] of supplies across tiger MT., probably into the La Dut
area for trans-shipment up T-7 or down Route 548. Lower T-7 was reportedly
in vic YD 356147, acoustic sensors have confirmed the presence of a tacked
[sic] vehicle on upper T-7, indicating road repair and possible extension
of that route. Along Route 548 in the A-Luoi Area, another trans-shipment
storage point is likely, since the sensor string in that area records more
traffic than does the one located a few kilometers further down the road
at Ta Bat. Because there are no sensor strings along Route 547, the volume
of traffic entering it from Route 548 cannot be determined. In the lower A
Shau Valley, the sensor string on Gorman's road (YC512896) continued to

record traffic at a steady pace. Vehicles are probably supplies to the end of the road (YC546917) for trans-shipment up the Rao Nai River Valley into the current battle area. Gorman's road was interdicted on 23 Mar, the sensor string on lower 548 indicated that 07 vehicles went SE on 22 Mar probably entering Gorman's road to off-load and being trapped there after the interdiction. Two truck parks were reported by VR [visual reconnaissance] in the YD 5591 area and bulldozer activity extending Gorman's road to the East has also been noted here.[6]

This document is quoted here almost in its entirety because it hints at a host of developments of the NVA logistical system with both short- and long-term ramifications, and clearly places the A Shau Valley at the heart of the NVA logistical system in Military Region 1 (formerly I Corps).

First, the completion of the "New Eastern Route" by connecting Routes 616 and 548 was imminent.

Second, when air cavalry and long-range patrol operations by the 101st Airborne Division were phased out over the last several months of 1971, the NVA made haste to establish multiple storage areas and transshipment points up and down the entire valley and to develop new roads eastward to support offensives towards Hue and the coastal lowlands. The roads included "T-7," built around the north side of Cu Pong Mountain, and "Gorman's Road," connecting Route 548 with the Rao Nai Valley, the old route that had been used in the mid-1960s and now was being prepared for vehicular traffic.*[7] Bulldozers were being used on both of these projects, indicating that the NVA believed the trailhead construction sites to be relatively secure. The Binh Tram transshipment locations included La Dut (where the 1/7 Cavalry found two damaged bulldozers on its trek from LZ VICKI to LZ GOODMAN in April 1968), A Luoi and Ta Bat in the central valley, and at the southern terminus of the new Gorman's Road near A Shau.

---

* When completed, Gorman's road would actually be routed into the next valley system east of the Rao Nai, and would join Route 547 east of FSB Bastogne, flanking the main ARVN defensive point along that most fought-over trail.

Third, the memo reveals that intelligence about the enemy in the A Shau Valley and adjoining areas was almost exclusively generated by acoustic sensor strings monitored by the Americans. In a report to his boss, MG Kroesen, on 15 April 1972, COL Dickinson commented at length on sensor operations of Task Force Alpha (TFA), the U.S. sensor monitoring unit at Nakhon Phanom, Thailand, pertaining to the 1st ARVN Division in Thua Thien:

> The total number of sensor strings in the 1st Division AO as of 31 March was 117, [of which] 27 are air-relayed Strings monitored by TFA. . . . Three reseeds of air-relayed strings were performed by the Air Force in the A Shau Valley. . . . Air drops of sensor strings will continue whenever VNAF assets are available and may be extended to the eastern slopes of the mountains bordering the A Shau Valley. . . . Sensor strings located on Gorman's Road and Route 548 have provided valuable intelligence on enemy logistics and troop movements into the Veghel/Kim Quy battle area.[8]

The lack of sensors along Route 547 would soon be remedied. There were no sensors originally, as the ARVN continued to consider Route 547/547A a "friendly" line of communication. That would soon change, and forever. Ironically, as late as early 1971 in the preparations for DEWEY CANYON II, allied engineers had cleared and upgraded the southern part of Route 547A for support of allied "artillery raids." Now, as the NVA began to move on Hue via FSBs VEGHEL and BASTOGNE, "[t]he speed of these communist advances surprised US and RVN leaders alike as they realized that American efforts to open paths with herbicides and bombs now aided PAVN units driving Soviet tanks over widened roads."[9]

The Nguyen Hue Offensive and ARVN Counterattacks: the Role of the A Shau Valley

There is one curiosity about Dickinson's memorandum (26 March 1972), and that is the remark "preparatory to any large-scale engagement."

The North Vietnamese Nguyen Hue Offensive, a.k.a. the Easter Offensive, is usually cited as having commenced on 30 March 1972, with the North Vietnamese invasion of Quang Tri Province.[10] In fact, heavy fighting had been going on in Thua Thien Province along the axis of Route 547 since early March. The 1st ARVN Division launched an offensive, Operation LAM SON 45, on 5 March "to clear areas south of Fire Support Base Rakkasan and west and south of Fire Support Base Bastogne, all in preparation for future division operations towards the A Shau Valley."[11]

Although the battles along the Route 547 corridor were initiated by the ARVN, it is not clear which side, if either, accomplished its mission here during March–May 1972, and the initiative would change hands several times. As early as 19 March, the initiative appeared to have shifted to the NVA. "The enemy is making a determined effort to force the withdrawal of 1st Division units positioned southeast of FB VEGHEL. . . . Recent activity is by far the most significant effort by the 6th [NVA] Regiment in over a year."[12] The fighting was very heavy, particularly around Dong Cu Mong Mountain southeast of FSB VEGHEL, and U.S. air support was critical. On the night of 18–19 March, four ARC LIGHT strikes were used, two in the Dong Cu Mong area and two against storage areas along Gorman's road, just east of the A Shau; for the period 6–31 March, a total of 40 ARC LIGHT strikes supported the 1st Division.[13] The importance of the A Shau base to the enemy attacks toward Hue can once again not be overstated. "There are indications that the NVA has prepared defensive positions along successive ridgelines between VEGHEL and the A Shau Valley. These positions may protect massive storage sites or serve to screen the anticipated deployment of large units into western Thua Thien Province."[14] The defensive lines probably accomplished both objectives, as the 1st ARVN Division never came anywhere near the A Shau and two regiments of the NVA 324B Division would be redirected into the Route 547 battles once the capture of Quang Tri City was assured. Three days later another message from MG Kroesen (Commanding General, First Regional Advisory Command, activated on 19 March 1972 and replacing XXIV Corps as the new HQs commanding U.S. forces and advisors in MR1) to GEN Abrams reemphasized the role of the NVA supply lines originating in the valley:

> Heavy enemy activity continues on Gorman's Road, Route 547, and the T-7 complex providing logistics input to staging areas in the Tennessee Valley / Bastogne area. Within the MR1 allocation [of air strikes], priority for USAF Tac Air will be interdiction of these LOC's on a daily basis.[15]

COL Dickinson, according to the MACV Command History, believed that "the [1st ARVN] division preempted a scheduled 1 April enemy push against Hue from the west and southwest and was the key to the enemy's ultimate failure to take Hue and Thua Thien Province."[16] The ARVN I Corps Commander from 3 May 1972 and former commander of the 1st ARVN Division, LTG Ngo Quang Truong, in his monograph for the Center of Military History, took a slightly different view:

> During the month of April . . . the 1st ARVN Infantry Division fought back several attempts by the 324B Division to gain control of the western and southwestern approaches to Hue City in Thua Thien Province. Hue City was undoubtedly the prime target of enemy efforts in this area. But the seesaw battles that in fact had begun in early March then continued throughout the months of April and May without any solid gains from either side clearly indicated that this was only a secondary front designed to contain the 1st Infantry Division and support the main effort in Quang Tri.[17]

If General Truong is correct, then it must also be considered possible that the battles in March served effectively to tie down the ARVN 1st Division in advance of the NVA assault on Quang Tri, making it impossible for the best division in the Army of the Republic of Vietnam to come to the assistance of the defenders of Quang Tri city. If diversion was the enemy objective, that might explain the tenacity of the 6th NVA

Regiment's commitment in March. The 3rd Regiment of the ARVN 1st Division was held in reserve near Hue, and not sent to Quang Tri Province, where it might have turned the tide.

On 28 April, the 29th and 803rd Regiments of the 324B Division commenced their attack on FSB Bastogne, which fell that day, the ARVN having earlier been forced back from the FSB VEGHEL / Dong Cu Mong Mountain line, and with the fall of Quang Tri City imminent (it happened on 1 May). "At the same time, there was a significant buildup of enemy personnel and supplies in the A Shau Valley." It was at this juncture (3 May), after the fall of the last ARVN Firebase in Quang Tri Province (NANCY – on 2 May) and the loss of FSB BASTOGNE (29 April), that LTG Truong replaced the incompetent LTG Lam as I Corps Commander.[18] Once Truong was in command, additional ARVN strategic reserves (the Airborne Division) were committed, the ARVN lines were stabilized, and counterattacks could begin. U.S. airpower continued to play a major role during this period, including 10 ARC LIGHT strikes on 13 May. BASTOGNE was retaken on 15 May.[19] The ebb and flow on Route 547 would continue for some months. On 26 July, the NVA captured BASTOGNE again, only to be driven out on 2 August by the ARVN.[20] The end of July marked the final turning point in the campaign, and FSB VEGHEL would belong to the ARVN again on 19 September. In Thua Thien, the new status quo was restored, sans American ground troops and with the NVA in firm control of the A Shau Valley. Perhaps both sides were frustrated; the NVA could not take Hue, and the ARVN could not get into the valley. And so it would remain until 1975.

As noted in the preceding chapter, it would appear that no ARVN maneuver battalions entered the A Shau Valley after mid-1971 (LAM SON 720). The interdiction of the enemy line of communications and the destruction of equipment and supplies were therefore entirely dependent on air strikes (with occasional artillery employment, when range permitted), with targeting based primarily on acoustic sensor data and some FAC input. By early April, sensors had been dropped along Route 547 and were located at Route 547 near FSB VEGHEL (YC512029 & YC531008),

Gorman's Road (YC510896), Route 548 near Ta Bat (YC417966), Route 548 north of A Luoi, near the junction of Routes 922 & 548, on upper T-7 (YD357148 and lower T-7 (YD311123), on Route 9222, and at two locations on Route 616 in Quang Tri Province. Collectively, these not only provided targets but also informed allied intelligence about changes in the enemy logistics flow and likely areas of stockpiling. The sensor strings provided both quantitative data and information on the direction of travel; they could also identify tracked vehicles.[21]

At night, heavily armed USAF fixed-wing gunships continued to hunt prey over the valley. Just before midnight on 18 June 1972, a USAF AC-130A Spectre gunship of the 16th Special Operations Squadron at Ubon, Thailand, was making its second attack pass over a target in the A Shau Valley when it was struck in the left engine by a handheld SAM-7 Strela missile, one of the newest weapons in the North Vietnamese arsenal. Three crewmen were instantly blown out of the plane, which broke apart and crashed 1,500 meters east of Hamburger Hill (YC343978). The men blown from the plane, fire control officer CPT Gordon L. Bocher, low light sensor operator 2LT Robert V. Reid, and illuminator operator SSG William B. Patterson, were able to open their parachutes, land safely, and be picked up by a USAF search and rescue helicopter the next morning. The 12 other crewmen were missing in action and were the final American casualties of the Vietnam War in the A Shau Valley. They were pilot 1LT Paul F. Gilbert, co-pilot CPT Robert A. Wilson, navigator MAJ Robert H. Harrison, sensor operator MAJ Gerald F. Ayres, electronic warfare officer CPT Mark G. Danielson, flight engineer TSGT Richard M. Cole, Jr., aerial gunner SSG Leon A. Hunt, aerial gunner SSG Richard E. Nyhof, aerial gunner SSG Larry J. Newman, aerial gunner MSGT Jacob E. Mercer, aerial gunner SSG Stanley L. Lehrke, and illuminator operator SSG Donald H. Klinke. This was also the last U.S. aircraft shot down over the valley.

Once the fighting along Route 547 subsided, the ARVN 1st Division retained responsibility for the defense of Thua Thien Province and Hue City. COL George Millener, Jr., Senior Advisor to the ARVN 1st Infantry Division from 20 May 1972 till 29 January 1973 informs: "During the period September 1972–January 1973 the division assumed a monsoon posture by occupying firebases along key terrain from which company-sized elements conducted offensive clearing operations to deny NVA/VC access to the populated lowlands and to uncover and destroy caches of war material."[22] Try though he might, it is hard to disguise the fact that the company strength "offensive clearing operations" were part of an overall defensive posture, and the ARVN was not strong enough to attempt to disrupt the enemy in the A Shau Valley.

On 27 January 1973, the cease fire was signed in Paris and, in theory, went into effect throughout the Republic of Vietnam.

The Cease Fire in the A Shau Valley: The NVA Logistics Hub

With the end of heavy ground combat for the time being, and absent any ARVN operations, NVA road and base construction could proceed unhindered in the valley. In early May 1973, satellite reconnaissance revealed that

> a new road trace and clearing paralleling 7 miles of Route 548 between A Luoi and Ta Bat airfields may be an effort to shorten the distance along this mountainous segment. To the south, additional construction extending Route 614 is under way. The new construction adds another 4 miles of motorable road south of the A Shau Valley and leaves only an 18-mile gap to road clearing operations on Route 14 [in Quang Nam Province toward DaNang].[23]

Apparently, that information was already outdated, as another CIA report dated 12 May reported that "a gap of only some 10 miles south of A Shau remained uncompleted on this network as of late April."[24]

A week later, relying on communications intelligence, a CIA report noted:

> An element of Binh Tram 42, which has the responsibility of moving supplies into the A Shau Valley region reported on 30 April that it was

receiving 30 cargo vehicles each day. One summary report covering an unspecified period indicated that the Binh Tram had handled more than 6,000 tons of supplies, most of which was received, then dispatched to destinations in the South Vietnam/Laos border area. Most of the cargo appeared to be rice and foodstuffs, but included some ammunition and other ordnance items.[25]

The importance of the A Shau nexus as a key transshipment point continued to grow as the Communist logistical network in western South Vietnam came into full bloom.

What the CIA had initially called the "New Eastern Route" because it was east of the main branch of the Annamitique Mountain chain before long had been retagged as the western supply corridor, referencing its placement inside the western border of South Vietnam. Reviewing NVA logistics developments in June 1973, the weekly CIA Memorandum summarized:

The heaviest activity is on the north-south routes constituting the Communists' western supply corridor. Recent evidence that North Vietnam has sent some 20,000 civilians into South Vietnam's Military Region 1 suggests that part of the logistic[al] activity in this region is to support the new civilian population. This may be particularly true along the western border, where road construction laborers reportedly were told recently that their work along the Route 608/616/548 complex will benefit "numerous settlements" North Vietnam plans to build in the A Shau Valley and Khe Sanh areas. . . . At least two and possibly three storage complexes were relocated from Route 922 in southern Laos to Route 548, north of A Shau in South Vietnam. In addition, Binh Tram 42, formerly operating in the Laotian Panhandle, was relocated as of 20 June along Route 548 in the same area.[26]

While the Communists continued a feverish pace of road construction down the western half of South Vietnam, their days of being satisfied with "limited all-weather" roads were also passing. "Since early

[1973], large sections of this part of the system between the Route 9 junction at Khe Sanh and the A Shau Valley have been extensively prepared, realigned, and upgraded by the application of a bitumen all-weather surface." In western Thua Thien, where steep mountains, adjoining hills, and deep ravines presented serious obstacles to road building, the 45-kilometer-long A Shau Valley was a natural super-highway corridor. South of A Shau, "the Communists intend the MR-1 portion to become a major resupply corridor serving a number of new or upgraded motorable feeder roads leading east into the coastal lowlands."[27]

And so the circle began to close, as the Binh Tram HQs forced out of the Punch Bowl in 1968 by the 1st Cavalry Division returned five years later to reclaim its domain, while the North Vietnamese endeavored to replace the local Montagnard population, long since departed, with a Communist-controlled citizenry. The volume of traffic on Route 548 continued to grow rapidly, with 183 trucks reported on 28 June alone (108 northbound, 34 southbound, and 41 parked).[28]

By mid-1973, the volume of vehicular traffic required enormous quantities of fuel, and so an enemy pipeline also wormed its way into the valley.

A further extension of the petroleum pipeline (PPL) originating in North Vietnam and crossing the DMZ into South Vietnam. This PPL has been extended into the A Shau Valley of South Vietnam. . . . It is visible . . . southward along route 548 to the recently constructed A Luoi petroleum storage area.[29]

Not all of the cargo, by any means, was just passing through. As the North Vietnamese looked to the final battles of the war, stockpiling of food and ordnance was a priority, "and as of 11 July one of the major NVA storage facilities on Route 548 north of A Shau had about 8,000 tons of cargo on hand, 75% of it rice."[30]

1974: Just a Matter of Time

In the fall of 1973, the enemy began to step up pressure against the ARVN 1st Infantry Division, which was for all intents and purposes occupying a slowly contracting enclave centered on Hue and including the important Phu Bai airfield complex to the south. For the most part, the enemy did not bring pressure up the well-defended Route 547 corridor, but rather initiated periodic small-scale offensives against the ARVN outpost line to the north and south.[31]

All of these actions were no doubt supplied through the A Shau complex from which the roads toward Hue had been much expanded and improved.

> Newly constructed or considerably improved arteries include Route 547 and Gorman's Road leading from storage depots in the A Shau Valley to units of the NVA 324B Division just west of Hue, Route 534 running to the Que Son Valley and the NVA 711th Division, and Communist-designated Route 105 (GVN Route 1404) extending from Kham Duc on the western supply route southeastward through Quang Tin and Quang Ngai provinces. As a result of new Communist construction since October 1973, this last route now extends more than 100 miles and is motorable into Binh Dinh Province in coastal MR 2.[32]

The importance of the A Shau Valley to the North Vietnamese logistical system — not only in Thua Thien Province and not only in Military Region 1 (formerly I Corps), but also to the enemy armies throughout South Vietnam — was finally made manifest when "the western highlands supply corridor, a network of roads stretching from north of the DMZ to Loc Ninh in MR 3, 500 miles to the south, was opened for limited traffic in late February 1974"[33](see Map 17). The construction of that corridor began with the upgrading of Route 548 in the A Shau Valley seven years earlier.

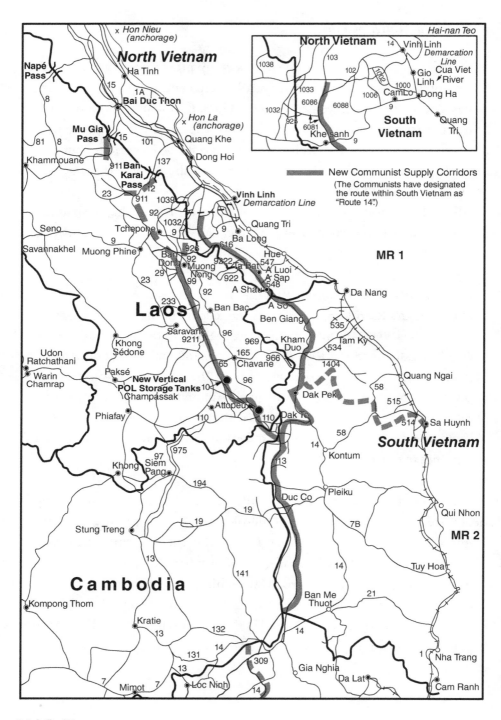

MAP 17

Source: *CIA MEMORANDUM: South Vietnam: A Net Military Assessment. 2 April 1974.*

1975: The End

The final enemy attack on Hue began as part of a country-wide NVA offensive on 8 March 1975. The ARVN fought hard, but even the best company in the Army, the Hac Bao (Black Panthers) on Hill 50, one of the few ARVN units to operate in the A Shau Valley on multiple occasions, could not withstand the enemy tanks and heavy artillery. There was, of course, no U.S. air support. For two weeks, the battle raged all around the Imperial City until, on 24 March, General Truong ordered its evacuation. By 30 March, the NVA controlled all of MR 1.[34] One month later, it was all over.

The tracks on the Russian-made bulldozers found by the 1st Cavalry in the A Shau Valley in April 1968 were the precursors of the tracks on the T-54 tanks that drove into Saigon seven years later. The North Vietnamese engineers of Binh Tram 7 (later 42) constructed a vast network into, through, and beyond the A Shau Valley, all the while assailed by the deadliest technology and ordnance the allies could devise and throw at them. They made an essential if little recognized contribution to the Communist victory in Vietnam. That victory was foretold in the A Shau Valley in 1961 when the Ho Chi Minh Trail reached South Vietnam, and came to fruition, in no small part, because the Peoples' Army of Vietnam prevailed in the A Shau Valley over the long run.

We weren't winning any more.

CONCLUSION

Throughout the years while the United States was committed to the Vietnam War, the North Vietnamese Army occupied the A Shau Valley. This was true even when the allies were also present. The role of the valley in that conflict began in 1961 with the arrival of the first troops of PAVN Transportation Group 559, and it continued until the end of the war in 1975. During the brief interlude when allied ground troops were (intermittently) present in force, from April 1968 through September 1969, the enemy possession of the A Shau was at best contested. The allied troops temporarily occupied their artillery firebases atop hills and mountains overlooking the valley, and launched "reconnaissance in force" and "search and destroy" patrols into the valley and the jungle nearby, while for a year and a half almost completely interdicting Route 548. Even then, however, the Communists continued to expand the logistics megalopolis in western Thua Thien Province.

Whether the Vietnam War was, at any point in time, "winnable" is beyond the scope of this work. What is clear is that by its failure to retain or regain occupancy of the A Shau Valley, the United States enabled the enemy to develop and use it as the key logistical base in the northern half of the country, if not all of South Vietnam. This wasn't so much a lost opportunity as one that was forgone. For whatever reasons, and for whatever purposes, troops that might have altered the course of the war by permanently occupying the A Shau were used elsewhere, for pacification or for pursuing the enemy's "big battalions" in the hinterlands. James McCoy reflected: "As long as the NVA held the initiative and could control its logistic[al] network without fear of serious ground disruption, it could maintain the momentum. However, if NVA supply depots had been continually hit by allied ground forces, the [NVA war] machine could have become paralyzed."[1] And reiterating the necessity for ground troops, the "NVA supply system was certainly vulnerable, but not from the air. It was vulnerable from the ground. Attacks upon the enemy supply lines would have won the war in Indochina."[2] There was no more critical locus for NVA supply in RVN than A Shau.

Were there signs of the eventual defeat that might have been ascertainable to an impartial observer, even during the period of apparent allied ascendancy in the A Shau Valley? The omnipresence of the enemy was, in retrospect, the key. PAVN perseverance prevailed, and although many allied commanders claimed to have achieved "control" of the valley, the best that could be accomplished was to contest and temporarily disrupt its use as a logistical artery and to capture and destroy enemy supplies. Except in those very short periods when allied infantry was physically present along Route 548, the enemy was able, at night and in poor flying weather, to move up and down the center of the valley almost at will, and during the day, to build its vast logistical complex in the jungles around the valley. We were never there long enough to take control.

If perseverance was the theme of the Communist commitment, I suspect that hubris was the corresponding dominant trait in the participation of the United States. Perhaps symbolic of our arrogance was a joke that made the rounds many years ago. It was about "Nguyen," the NVA soldier. Nguyen finished his training and was loaded up with a hundred pounds or so of gear, plus a couple of months' rice ration and two 82mm mortar rounds, and was sent down the Ho Chi Minh Trail. Three months later, after a bout with malaria and having dodged bombs, rockets, and miniguns countless times, he reached the Central Highlands just as a Communist offensive began. As he hiked up to his new unit, the mortar platoon leader called him over and said, "Give me those mortar bombs NOW!" THUNK . . . THUNK . . . was the sound as they left the tube. "Now go back and get me two more." So, who's the joke on? Perseverance prevailed.

The failure to comprehend that our "victorious" operations and battles, routinely measured in "body count" and "casualty ratios," were not taking us to victory in the conflict is the great tragedy of the Vietnam War. Because we thought we were winning, because our military commanders proclaimed every operation successful*, we didn't need to address questions such as whether we were using an appropriate strategy, what it would take *to* win, and whether we were willing to pay the price of increased troop commitments and casualties and/or expansion of the war zone into Laos, Cambodia, or North Vietnam, or whether to leave the South Vietnamese to battle on alone. Ultimately, two battles in Thua Thien – Hue in 1968 and Hamburger Hill in 1969 – one facilitated by the enemy base in the A Shau Valley and the other taking place inside the valley, forced America to face reality. When finally we saw what the price of victory might be, we realized Vietnam wasn't worth the cost. In the interim, we had increased exponentially the human cost of the war to the Vietnamese, North and South.

Bruce Jones, who served as a young intelligence officer, ruminated years later on his letters home from the war zone.

> I saw that constant, irresistible pattern of putting a good face on everything I sent home to try to relieve my parents' concern. Each of my mailings from APO 96307, San Francisco, was its own little exercise in smoke and mirrors. I think I even came to believe my own letters as time went by. I should have understood then that MACV would be no different in its public proclamations. But they would affect a lot more than just home-front morale. In time the incessant, insane optimism would lose the war.[3]

When I went to Vietnam, inspired by JFK's famous "Ask what you can do for your country" rhetoric, I believed that with a little American "know-how," we would soon prevail, for the good of the Vietnamese people. Instead, with a lot of American "know-it-all," we blundered to defeat.

In researching this book, I hoped to answer a few questions that had puzzled me since 1967. One concerned the allegations that recon teams

---

* An observation by NSC officer Michael Forrestal in a memo to McGeorge Bundy on 8 January 1964 is remarkably prescient on this point: "It is the old problem of having people who are responsible for operations also responsible for evaluating the results." Quoted in Gregory R. Clark, *Quotations on the Vietnam War* (Jefferson, NC: McFarland, 2001), p. 36.

simply vanished in the A Shau, never to be seen or heard from again. Another was "What was it really like for U.S. troops in the valley?" Was the enemy really everywhere? Was there an ambush around every bend in the trail? Did the NVA swat down helicopters like flies and overrun landing zones and firebases in human waves? Was it really *that* bad?

The rumors of lost recon teams that fueled the mystique of "A Shau" turned out to be, with one exception, just that. The 101st Airborne Division long-range patrol company, F/58th Infantry (LRP) lost Team 2-3 in the Ruong Ruong Valley on 20 November 1968, and the reflagged L/75th Infantry (Ranger) lost RT Kansas near Khe Sanh on 11 May 1970. Yet while there were many casualties, no teams from the Rangers were lost or disappeared in the A Shau Valley. MACVSOG lost 10 complete teams MIA and another 10 complete teams overrun or with their helicopters shot down over the course of the war, most in Laos and Cambodia. Two teams were lost near the A Shau, both in May 1971. On 10 May, Team Asp was overrun at abandoned FSB THOR (Hill 1084 at YC665835), between the Rao Nai and Song Huu Trach Valleys. On 18 May, Team Alaska was ambushed and overrun in the Da Krong Valley (YD036214). A number of teams were lost in Base Area 611 / Target Oscar Eight, west of the A Shau in Laos. *Only one team vanished in the A Shau.* On 20 May 1968, RT Idaho was inserted south of Tiger Mountain, directly on the Laotian border. After an initial radio check, they were never heard from again, as recounted in chapter 13.[4]

The reality of the fighting in the A Shau Valley was that it was much like the rest of the Vietnam War. It all depended on when you were there and just where "there" was. There was the ceaseless labor of "humping the boonies," just as Colin Powell had done, in the oppressive jungle and up and down countless hills. There was the constant toil of constructing and then tearing down fire support bases, many built and then razed repeatedly on the same hilltop. There were countless patrols that never sighted an enemy soldier or drew fire. There were sudden, violent, brief ambushes or meeting engagements with sometimes high casualties. There were similar firefights that evolved into days-long battles, such as occurred near Tiger Mountain, on Dong So Ridge, and at Hamburger Hill. There were also numerous attacks on fire bases, commonly with mortars or rockets, occasionally with artillery fire, and some, more terrible, by the dreaded and deadly NVA sappers.

From Colin Powell's adventure with snipers and punji stick booby traps to the Spectre AC-130A gunship shot down with a hand-held ground-to-air missile nine years later, the technology certainly evolved. Ironically, these two symbolic bookends of enemy technology belie the fact that it was the Americans who introduced dozens of technological innovations, most aiming to enhance the aerial interdiction of infiltration and the line of supply.

There were many conspicuous acts of bravery by U.S. soldiers during the years of conflict in the valley. Only a few of those, recognized by our nation's highest medals for valor, are related in the preceding narrative. There is no telling how many Silver and Bronze Stars, Distinguished Flying Crosses, and Air and Commendation Medals for Valor were awarded. But even if we knew those numbers and could relate all of the stories behind them, we would still fall far short of being able to take the full measure of the courage of the American servicemen who fought in the A Shau Valley. Above all, the courage of the helicopter crews of the air cavalry – the First of the Ninth Cavalry of the 1st Air Cavalry Division, which also accompanied the 101st Airborne Division during its 1968 operation there, and the Second of the Seventeenth Cavalry, which spent over two and a half years (March 1969 – late 1971) patrolling the hills and ravines of western Thua Thien, is immeasurable. The crews of the Hueys – slicks, gunships, Charlie-Charlies (command and control ships), the Cobras, including the Aerial Rocket Artillery – and most courageous of all, the men who flew in the little "Loach" scout helicopters. Day after day, with no Sundays off, the little egg-shaped choppers had only their maneuverability, the skills of their pilots and observers, and the threat of the punch of their Pink-team partners, with which to play a deadly game of hide-and-seek with the heavy machine guns, RPGs, and automatic rifles of the enemy hidden in the jungle below. I call that habitual heroism, for which no medals are awarded. It's not "above and beyond the call of duty"; it's just duty. But it still takes a hero. This book is dedicated to those helicopter crews.

A Shau Valley

CHRONOLOGY OF MAJOR EVENTS AND OPERATIONS

Operations and Major Battles

3/9–3/10/66	Fall of A Shau Special Forces Camp
4/18–6/13/67	Operation PIROUS
8/1–8/3/67	Operation CUMBERLAND II / CLOUD
4/19–5/17/68	Operation DELAWARE
8/4–8/20/68	Operation SOMERSET PLAIN
2/11–3/18/69	Operation DEWEY CANYON, Phase III
3/1–8/14/69	Operation KENTUCKY JUMPER (includes next three OPNs)
3/1–5/8/69	Operation MASSACHUSETTS STRIKER
5/10–6/7/69	Operation APACHE SNOW
5/11–5/20/69	Hamburger Hill
6/9–8/14/69	Operation MONTGOMERY RENDEZVOUS
6/14/69	FSB BERCHTESGADEN Sapper Attack

6/16/69	FSB CURRAHEE Sapper Attack
8/15–9/28/69	Operation LOUISIANA LEE
8/24/69	FSB BERCHTESGADEN Sapper Attack
12/7/69–3/31/70	Operation RANDOLPH GLEN
4/1–9/5/70	Operation TEXAS STAR
7/1–7/23/70	FSB RIPCORD siege
7/25–8/12/70	Operation CHICAGO PEAK
9/5/70–10/8/71	Operation JEFFSON GLEN
1/30–2/7/71	Operation DEWEY CANYON II
4/30–8/31/71	Operation LAM SON 720

Medals of Honor

3/10/66	MAJ Bernard F. Fisher, USAF 1st Air Commando Sqn
3/9–3/12/66	SFC Bennie G. Adkins Det A-102, 5th SF Group
4/25–4/26/68	1LT James "Mike" Sprayberry D/5/7th Cavalry
5/4/68	1LT Douglas B. Fournet (posthumous) B/1/7th Cavalry
7/11/69	SP4 Gordon Roberts B/1/506th Infantry
7/15–7/19/69	SSG John Gertsch (posthumous) E/1/327th Infantry

APPENDIX

Not By the Book: Mid–1960s United States Army Tactical Doctrine and Hamburger Hill

While Jay Sharbutt would not have been versed in U.S. Army tactical doctrine, GEN Zais, COL Conmy, and LTC Honeycutt were highly trained professional officers and certainly were well acquainted with the precepts applicable to attacks on fortified enemy positions. General Zais's after action memorandum (mis)focused the debate about the attacks by the 3/187th Infantry by implying that there were only two options: (1) attack Hill 937 as the Rakkasans did, or (2) pull back at least two miles and hit the enemy with B-52 bombers. Most historians and military writers in the subsequent half-century have concurred that as woeful as the American casualties were, the correct choice was made and the attacks by LTC Honeycutt were necessary.

One analysis that, as far as I am aware, has not been made is to evaluate whether the tactics employed were in accordance with contemporary U.S. Army doctrine. First and foremost, the Army taught its officers to follow the Principles of War. The following are excerpts from the field manual for battalion commanders:[1]

Principles of War

In each offensive action, the battalion commander must apply certain of the principles of war in the accomplishment of his mission. These principles, which are outlined below, are not inflexible rules, but are applied as required by each situation.
Objective. *The maximum available maneuver strength and fire power of the battalion is directed toward the seizure of the final objective. Each task assigned to a subordinate unit must contribute to the seizure of this objective.*

Mass. *When information is lacking or the enemy's weaknesses cannot be deter-mined, the battalion commander employs the smallest practicable force initially. He withholds reserves from the action so that he can employ them at a decisive time and place. The commander must also control his fire support so that he can shift and mass it rapidly at a critical point. When the opportunity for decisive action presents itself, the commander unhesitatingly commits his total resources to accomplish his mission.*

Offensive. *After the attack is launched, every effort is made to maintain its momentum until the objective is seized. Retention of the initiative is vital to successful offensive operations.*

Maneuver. *In the conduct of the attack, the commander maneuvers his forces so as to place the enemy at a relative disadvantage. He does this by avoiding enemy strength, engaging the enemy on terrain of the commander's choice, and exploiting located enemy weakness.*

Surprise. *The commander may achieve surprise by attacking the enemy at a time, place, and in a manner which the enemy has not anticipated. Attacking habitually at a certain hour or in a certain pattern should be avoided.*

Security. *In each offensive action, measures must be taken to prevent sur-prise, preserve freedom of action, and deny the enemy information of friendly forces.*

Simplicity. *Uncomplicated plans and clear, concise orders promote common understanding and intelligent execution. If other factors are equal, the simplest plan is preferred, since it can be more readily altered to exploit opportunities and meet unforeseen conditions.*

Unity of Command. *Although the commander allows maximum freedom of action in carrying out his orders, he ensures that there is a unity of effort and an effective integration of all elements of combat power.*

Economy of Force. *The commander allocates only those forces necessary for supporting attacks in order to ensure maximum combat power at the point of decision.*

The manual clearly states that "these principles . . . are not inflexi-ble rules." They are, however, the most fundamental guidance taught to infantry commanders and ought not to be lightly disregarded. Yet at the battalion and brigade levels during the battle for Hamburger Hill, they were more violated than adhered to:

Objective. While LTC Honeycutt did focus his effort on the objective, the same cannot be said for the brigade or division commanders. If destruction of the entrenched enemy on Hill 937 had indeed become the primary objective of operations in the valley, then the delay in bringing all subordinate units (battalions) into the battle until it had been raging for over a week is questionable. Notwithstanding LTC Honeycutt's belief that his battalion could accomplish the mission without assistance other than the slow-moving 1/506th Infantry, it was the responsibility of the brigade and division commanders to provide the forces needed to complete the envelopment of Dong Ap Bia, and their failure to do this contributed significantly to the enemy's ability to concentrate its firepower on the western approaches to the peak. One possible option would have been to bring the division reserve (General Zais did have a division reserve, didn't he?) to the FSB AIRBORNE area, and instead of pulling it back to AIRBORNE, bringing the 2/501st Infantry in on the 3/187th left flank. Prior to its move to AIRBORNE, the 2/501st was less than 10 kilometers northwest of Dong Ap Bia. This would also be in accordance with the principles of **Mass, Offensive, Maneuver,** and possibly **Surprise.** *As it was, the 2/501st Infantry spent five days (14–18 May) fruitlessly beating the bushes around AIRBORNE.*

Mass. By allowing Delta Company to muck around to the north of the hill until 16 May, LTC Honeycutt effectively split his battalion in the face of a stronger enemy force. He also had no reserve available to employ "at a decisive time and place," when the attacks on 14–15 May achieved some success, but the forward elements had to be withdrawn at the end of each day.

Offensive. Due in part to the lack of a reserve (**Mass**) and also to friendly fire, the 3/187th Infantry was unable to maintain momentum, when briefly achieved, and could not retain the initiative until the full attack force was assembled on 20 May.

Maneuver. With the exception of the abortive attempt of Delta Company to find the enemy right flank, until the battalion combat assaults on 19 May, this principle was nowhere in evidence on Dong Ap Bia. LTC

Honeycutt, for his part, did not "avoid enemy strength" or "engage the enemy on terrain of the commander's choice," and only the ARVN battalion finally "exploit[ed] located enemy weakness."

Surprise. If the enemy was ever surprised, it was only by the 2/3rd ARVN Infantry on 20 May. Indeed, the repeated frontal assaults up the same ridge fingers by the Rakkasans were "a certain pattern [that] should be avoided."

Security. LTC Honeycutt did secure his rear area(s), although he was still surprised by repeated enemy attacks from the flank and rear. Unfortunately, providing this security deprived his attacking force of a reserve.

Simplicity. The repeated assaults in the same places were certainly simple. Perhaps in this instance, it was too much of a good thing.

Unity of Command. COL Conmy does not appear to have had "an effective integration of all elements of combat power," as the 1/506th Infantry was slow to join the battle from the south, while LTC Honeycutt was permitted to continue attacks without the benefit of support from his right flank. Both battalion commanders had so much "freedom of action" that their combat power was not integrated.

Economy of Force. It is unclear how the division or brigade commanders might have employed this principle to gather more units, sooner, to support the 3/187th Infantry. But . . .

The field manual also discusses the appropriate "Forms of Maneuver" for use by the battalion in attack:[2]

Ground envelopment "requires that the enemy have an assailable flank";

Aerial envelopment "requires suppression of enemy air defense fires";

Penetration "is favored under the following conditions: when the enemy is overextended; weak spot(s) are detected in the enemy position; and strong fire support . . . is available."

The 29th NVA Regiment on Dong Ap Bia was not overextended, at least not when only assailed from one side, nor were any weak spots identified until 20 May. Thus, either aerial envelopment (as eventually happened) or a delay until ground envelopment could be implemented were more appropriate offensive tactics.

FM 31-50, COMBAT IN FORTIFIED AND BUILT-UP AREAS (March 1964, as modified 22 April 1970) provides more detailed instruction for assaults on positions such as Hill 937.[3] While emphasizing extensive and detailed planning, something that was not generally an option against an enemy as shadowy as the NVA/VC, this manual also emphasizes the importance of the reserve and the role of aerial envelopment: "Reserves follow closely behind the assault echelon to exploit the penetration, maintain the continuity of the attack, or defend critical areas against counterattack." "Airmobile and joint airborne forces may be used in conjunction with other attacks of the fortified area principally to block the movement of large enemy reserves, isolate and/or assault strongpoints, and attack fortifications from the rear."[4]*

As the battle went on, errors at brigade and division contributed materially to the cost in Rakkasan lives. Perhaps no oversight by higher headquarters had a more direct impact on LTC Honeycutt's failure to "penetrate" the enemy positon than failing to shut down the NVA line of communication. The field manual for division commanders, reiterating material in FM 31-50, addresses this explicitly: "The area selected for penetration must be isolated."[5] The 29th NVA Regiment was able to evacuate the wounded and reinforce and resupply the defensive position nightly, making the work of the 3/187th Infantry on Hamburger Hill a literal Sisyphean struggle.

While all the commanders involved endorsed the opinion that the enemy had to be quickly attacked lest it scamper to sanctuary in Laos, the NVA in fact had the almost unrestricted ability to flee or reinforce every night during the battle. The enemy troops remained on Hamburger Hill of their own volition (or at least the volition of their commanders). It was

---

* A more recent USMC manual on tactics has some very cogent comments under "Use of the Reserve in Combat": "The reserve is an important tool for exploiting success. The reserve is a part of the commander's combat power initially withheld from action in order to influence future action. The reason to create and maintain a reserve is to provide flexibility to deal with the uncertainty, chance, and disorder of war. The reserve is thus a valuable tool for maintaining adaptability. In general, the more uncertain the situation, the larger should be the reserve. Napoleon once said that 'war is composed of nothing but accidents, and . . . a general should never lose sight of everything to enable him to profit from [those] accidents.' These accidents take the form of opportunities and crises. The reserve is a key tactical tool for dealing with both." *USMC MCDP 1-3: Tactics* (Washington, DC: Headquarters, United States Marine Corps, 30 July 1997), 106.

not necessary for LTC Honeycutt to either attack or retreat two miles. A third option, remaining in close proximity to the enemy while refraining from assaults and instead ambushing the enemy line of communication until additional forces could close the circle, was not considered. This should have been ordered by COL Conmy or MG Zais.

ENDNOTES

Chapter 1

1 CIA Intelligence Memorandum: The Growth and Current Development of the Laotian-Based 559th Transportation Group. ER IM 71-25 February 1971.

2 Colin L. Powell, with Joseph E. Persico, *My American Journey* (New York: Random House, 1995), 80.

3 Ibid, p. 81.

4 Ray A. Bows, *Vietnam Military Lore: Legends, Shadows, and Heroes* (Hanover, MA: Bows & Sons, 1997), p. 1070.

5 Powell, *My American Journey*, 81.

6 Ibid. (see pp. 80–99 re: Powell's A Shau odyssey).

7 Shelby L. Stanton, *Green Berets at War: U.S. Army Special forces in Southeast Asia, 1958–1975* (Novato, CA: Presidio Press, 1985), 35.

8 Chalmers Archer, Jr., *Green Berets in the Vanguard, Inside Special Forces, 1953–1963* (Annapolis, MD: Naval Institute Press, 2001), 73–78.

9 John Prados, *Blood Road: The Ho Chi Minh Trail and the Vietnam War* (New York: John Wiley & Sons, 1999), 48.

10 Francis J. Kelly, *U.S, Army Special Forces, 1961–1971* (Washington, DC: Department of the Army, 1973), 35.

11 Stanton, *Green Berets at War*, 32–38.

12 Prados, *Blood Road*, 51.

13 Most of the information in this and the following paragraphs of this chapter comes from monthly summary reports for the A Shau SF camps from 20 May 1963 to 31 December 1965. 1st Special Forces Group. *Monthly Operational Summaries.* 20 May 1963–31 Dec 1965. Provided by Steve Sherman.

14 Stanton, *Green Berets at War*, 38.

15 *Area Handbook for South Vietnam.* DA Pam No. 550-55, April 1967. (Washington, D.C.: USGPO), 79.

16 Thomas R. Yarborough, *A Shau Valor: American Combat Operations in the Valley of Death, 1963–1971* (Havertown, PA: Casemate Publishers, 2016), 115–116.

17 MACV/CSD Working Paper No. 8, 18 December 1963: Border Surveillance Units.

18 William C. Westmoreland, *A Soldier Reports* (Garden City, NY: Doubleday & Co, 1976), 61.

19 Phone conversation with Steve Sherman, 9 May 2019.

20 Australian Army Training Team Vietnam (AATTV) Narrative, Annexes 1–31 May 1964.

21 1SFG, *Operational Summaries*, 20 May 64 to 20 Jun 64.

22 Bows, *Vietnam Military Lore*, 1067.

23 Andrew Wiest, *Vietnam's Forgotten Army: Heroism and Betrayal in the ARVN* (New York: New York University Press, 2008), 45–47.

24 *Defense Technical Information Agency (DTIC) Technical Report, revised edition, U.S. DefenseLogistics Agency − A Study of Strategic Lessons Learned in Vietnam*. Volume VI: Conduct of the War, Book 1: Operational Analysis − Part 2 of 4: Chapter 3, 9 May 1980. Earl Tilford Collection, doc 2850112001. Volume I, The Enemy, 3–14.

25 Ibid., 3–16.

26 Bennie G. Adkins and Katie L. Jackson, *A Tiger Among Us* (New York: Da Capo Press, 2018), 40–41.

27 Kelly, *U.S. Army Special Forces*, 1961–1971, Appendix C.

28 1SFG, *Operational Summaries*, 1–31 October 1965.

29 Samuel Zaffiri, *Hamburger Hill* (Novato, CA: Presidio Press, 1988), 19.

30 Kelly, *U.S. Army Special Forces*, 1961–1971, 92.

31 5th Special Forces Group (Airborne) *Quarterly Command Report* for the Period Ending 31 December 1965.

32 Charles M. Simpson, III, *Inside the Green Berets: The First Thirty Years* (Novato, CA: Presidio Press, 1983), 106.

Chapter 2

1 Combined Intelligence Center Vietnam, Untitled Report − Re: A Shau Valley Terrain, undated. Sam Johnson Collection, doc F031100090021, p. 2.

2 Combined Intelligence Center Vietnam, *Special Study: Avenues of Approach − A Shau Valley*, 14 Jul 1969, Sam Johnson Collection, doc F031100090003.

3 *Project CHECO Southeast Asia Report #91 − Operation Delaware, 19 April − 17 May 1968, 2 Sep 1968*. Sam Johnson Collection, doc F031100041363, p. 9.

4 A Shau Valley Terrain, 16.

5 Ibid., 25.

Chapter 3

1 Stanton, Green Berets at War, 66.

2 Ibid., 118.

3 Adkins, A Tiger Among Us, 44.

4 Ibid., 37.

5 John D. Blair, IV. Operations of Special Forces Detachment A-102, 5th Special Forces Group (Airborne), 1st Special Forces, in the Defense of the Special Forces Camp at A Shau, Republic of Vietnam, 9-12 March, 1966. United States Army Infantry School, Fort Benning, GA. 2 January 1968.

6 Ibid., 7.

7 Ibid., 8.

8 5th Special Forces Group (Airborne), Operational Report − Lessons Learned for the Period Ending 30 April 1966.

9 Blair, Operations of Special Forces Detachment A-102, 8.

10 5 SFG, OR–LL 30 Apr 66.

11 Adkins, A Tiger Among Us, 52.

12 Project CHECO Southeast Asia Report #1, The Fall of A Shau, issued 18 April 1966 by HQ PACAF, p. 1.

13 Blair, Operations of Special Forces Detachment A-102; Adkins, A Tiger Among Us; Yarborough, A Shau Valor; 220th Aviation Company, Battle and Fall of A Shau, March 1966 at https://www.catkillers.org/history-1966-Battle-of-A-Shau.pdf accessed 7/6/2019

14 5 SFG, OR–LL 30 Apr 66.

15 Ibid. At 0230 hours, according to Captain Blair.

16 Blair, Operations of Special Forces Detachment A-102, 59.

17 220th Aviation Company, Battle and Fall of A Shau, 26.

18 Stanton, Green Berets at War, 118.

19 Blair, Operations of Special Forces Detachment A-102, 12–13.

20 Ibid.

21 Stanton, Green Berets at War, 118.

22 Samuel Zaffiri, Westmoreland (New York: William Morrow, 1994), 168.

Chapter 4

1 Prados, *Blood Road*, 83.

2 B-52 (Project Delta) *After Action Report – Operation 9-6*, p. 65-09-2.

3 Stephen A. Carpenter, *Boots on the Ground: The History of Project Delta* (Stephen A. Carpenter, 2010), 126.

4 B-52 (Project Delta) *After Action Report - Operation 4-66*.

5 Ibid., 12–13.

6 B-52 (PROJECT DELTA) *After Action Report - Operation PIROUS*, provided by Steve Sherman, p. 170.

7 Adkins, *A Tiger Among Us*, 147–151.

8 Jack Shulimson, *U.S. Marines in Vietnam: An Expanding War, 1966* (Washingon, DC: History and Museums Division, HQ, USMC, 1982), 149–150.

9 George L MacGariggle, *Taking the Offensive: October 1966 – October 1967* (Washington, DC: Center of Military History, U.S. Army, 1998), 28–29.

10 I Corps Mike Force Facebook page. "Special Forces Mike Force Lineage in I Corps, 1964-1970." Accessed 12/22/19.

11 The following is based largely on the "Handwritten Report: Provisional Det A-100 Members, Interviewees," *Operation Longstreet*, 28 Nov 1966. Dale W. Andrade Collection, doc 24990313006.

12 R.C. Morris, *The Ether Zone: U.S. Army Special Forces Detachment B-52, Project Delta.* (Ashland, OR: Hellgate Press, 2009), 166.

13 Aaron Gritzmaker, *Personal Narrative.* Aaron Gritzmaker Collection, doc 4450101001.

14 5th Special Forces Group (Airborne) *Operational Report -- Lessons Learned* for the period ending 30 April 1967.

15 Company C, 5th SFG (Abn), 1st SF, *AAR to OPORD 3-67 OCONEE (Blackjack 12)*, 24 Mar 1967. Dale W. Andrade Collection, doc 24990313005.

16   I Corps Mike Force Facebook page. "Blackjack 13/Task Force 13 Operational Summaries." Accessed 12/22/19.

Chapter 5

1   Jack Anderson, "Rainmakers Proving Success Over Ho Chi Minh Trail." *New York Times*, 20 March 1971.

2   *The Straits Times*, 24 June 1972. "US denies 'rain-making' in Vietnam."

3   *Hearings before the Subcommittee on Oceans and International Environment of the Committee on Foreign Relations, United States Senate, re: Weather Modification*, 20 Mar 1974. Douglas Pike Collection, doc 2390601002, p. 109.

4   Seymour Hersh in the *New York Times* 7 July 1972 attached to CIA Memorandum for the Record. OLC72-0768. 5 July 1972.

5   CIA Chronology of Buddhist Crisis in South Vietnam in 1963. 25 October 1963; Current Intelligence Memorandum OCI No. 1561/63 3 June 1963.

6   *Foreign Relations of the United States, 1964-1968, Volume XXVIII*, LAOS, Para 274.

7   Ibid.

8   David Reade. "The VMFA-115 Rainmakers of SEA: Now you can know the rest of the story." P-3publications.com/PDF/VMFA-115Rainmakers2016.pdf. Accessed 21 May 2019.

9   *Hearings before the Subcommittee on Oceans and International Environment of the Committee on Foreign Relations, United States Senate. re: Weather Modification*, p. 89.

10   Ibid., 111.

11   Ibid., Chart 6.

12   Ibid., 102.

13   "388th Operation POPEYE 'Motorpool'" at www.vva388.com/uploads/2/3/9/8/23987277/2-2016.pdf. Accessed 19 May 2019.

14   Rostow, W. W., memo to President Lyndon B. Johnson re: "Landslides in A Shau Valley," 10 Jul 1967. Veteran Members of the 109th Quartermaster Company (Air Delivery) Collection, doc 0010124006.

15   Hearings before the Subcommittee on Oceans and International Environment of the Committee on Foreign Relations, United States Senate. re: Weather Modification, p. 118.

16   Ibid., 109.

17   Ibid., 111.

18   https://en.wikipedia.org/wiki/Environmental_Modification_Convention

19   Westmoreland, *A Soldier Reports*, 281.

20   Jacob Van Staaveren, *Interdiction in Southern Laos 1960-1968* (Washington, DC: Center for Air Force History, 1993), 238.

21   *Foreign Relations of the United States, 1964-1968, Volume XXVIII*, LAOS, Para 289.

22   *MACV Command History 1967, Vol III*, 16 Sep 1968. Bud Harton Collection, doc 16830001072, pp. 1093–1094.

23   Van Staaveren, *Interdiction in Southern Laos 1960-1968*, 239.

24   Robert E. Cushman, Jr., *Oral History Transcript*, 1984, USMC History Division Oral History Collection, doc 10970111001, p. 21.

25 William A. Buckingham, *Operation Ranch Hand: The Air Force and Herbicides in Southeast Asia 1961-1971* (Washington, DC: Office of Air Force History, 1982), 33.

26 Ibid., 63–68.

27 Agent Orange Data Warehouse at http://www.workerveteranhealth.org/milherbs/new/index.php

28 Ibid. See Psywar and Civil Affairs Annex to Plan 2t-20t, HQ 11th DTZ, MANG CA. January 1964.

29 Ibid. See MACJ315 Memorandum for the Record 10 April 1964. Re: Reconnaissance of Proposed Aerial Defoliation Targets in I Corps.

30 Ibid. See Republic of Vietnam, Ministry of Interior, Thua Thien Province Confirmation Slip of Highways To Be Defoliated, 11 April 1964. Re: Defoliation Target 20-24.

31 Ibid. See MAGTN HU Execution of 20P&T Plan 30 April 1964. Re: Defoliation Target 20-24.

32 Ibid. See MACJ315 Memorandum for the Record, 5 May 1964. Re: Defoliation Target 20-24.

33 "U.S. Military and the Herbicide Program in Vietnam." https://www.ncbi.nlm.nih.gov/books/NBK236347/

34 Agent Orange Data Warehouse. See Mr. Nes to Melvin L. Manfull Proposed Defoliation in Thua Thien 15 May 1964. Re: Defoliation Target 20-24.

35 Ibid. See MAGTN-LC Report on Herbicide Operations, 22 July 1964.

36 Ibid. See MACJ311 Memorandum for Mr. P C. Habib, Chief, Political Section, US Embassy, re: Re-initiation of Defoliation Mission, Thua Thien Province, early Sep 1965. Re: Defoliation Target 2-24.

37 "U.S. Military and the Herbicide Program in Vietnam"

38 Project CHECO Southeast Asia Report #250 – The War in Vietnam, 1966, 23 Oct 1967, 120. Sam Johnson Collection, doc F031100440437.

39 "U.S. Military and the Herbicide Program in Vietnam"

40 William E. Thomas, National Academy of Sciences–National Research Council, *Effects of Herbicides in South Vietnam, Part B: Working Papers*, February 1974. Distributed by NTIS. Defense Technical Information Center.

41 "Herbicide Operation Evaluation," 26 Aug 1968. Paul Cecil Collection, doc 2520308006.

Chapter 6

1 Michael A. Hennessy, Strategy in Vietnam: The Marines and Revolutionary Warfare in I Corps, 1965-1972 (Westport, CT: Praeger, 1997), 74.

2 Harry G. Summers, Jr. On Strategy: The Vietnam War in Context (Carlisle Barracks, PA: Strategic Studies Institute, US Army War College, 1981), 56.

3 Shulimson, U.S. Marines in Vietnam: An Expanding War, 1966, 13.

4 Ibid., re: USMC operations during 1966; Hennessy, Strategy in Vietnam, 82.

5 Westmoreland, A Soldier Reports, 165.

6 Ibid., 166.

7 Ibid., 165.

8   Ibid.
9   William R. Fails, Marines and Helicopters, 1962-1973 (Washington, DC: History and Museums Division, Headquarters, USMC, 1978), 54–55.
10 Shulimson, An Expanding War, 1966, 319.
11 Summers, On Strategy, 73.
12 B-52 (Project DELTA), After Action Report – Operation PIROUS, 26.
13 Morris, The Ether Zone, 203.
14 B-52 (Project DELTA), After Action Report – Operation PIROUS, 29.
15 Yarborough, A Shau Valor, 70.
16 B-52 (Project DELTA), After Action Report – Operation PIROUS, 63.
17 Ibid., 114–115.
18 Yarborough, A Shau Valor, 73.
19 B-52 (Project DELTA), After Action Report – Operation PIROUS, 37.
20 Ibid., 209.
21 Sources for the RR 107/1 Rgr Co action include Ibid.; Yarborough, A Shau Valor, 80–81; Carpenter, Boots on the Ground, 211–212; and 281st Assault Helicopter Company at https://www.281st.com/281history/281history_html/ICTZ_opns281.htm.
22 References for the footnote are Hugh Shelton, Without Hesitation: The Odyssey of an American Warrior (New York: St. Martin's Press, 2010), 56–61 and B-52 (Project DELTA), After Action Report - Operation PIROUS, 69–70, 232–234.
23 Ibid., 54–56.

Chapter 7

1   MacGariggle, *Taking the Offensive*, 205–206.
2   Gary L. Telfer, Lane Rogers, and G.K. Fleming, Jr., *U.S. Marines in Vietnam: Fighting the North Vietnamese, 1967* (Washington, DC: History and Museums Division, HQ, USMC, 1984), 75.
3   MacGarrigle, *Taking the Offensive*, 205.
4   LTGEN Cushman, message to GEN Westmoreland, 31 July 1967. Courtesy of Bill McBride.
5   James W. Hammond, Jr., letter to James W. Hammond, III, 1 Feb 1999. Courtesy of Bill McBride.
6   2d Bn, 4th Marines, 010600h Aug 67 OPERATION ORDER 11-67 (OPERATION CLOUD).
7   James W. Hammond, Jr., message to Bill McBride, 17 Sep 1998. Courtesy of Bill McBride.
8   B-52 (PROJECT DELTA) *After Action Report - OPERATION PIROUS*, Para 9a2.
9   Lawrence C. Vetter, Jr., *Never Without Heroes: Marine 3rd Reconnaissance Battalion in Vietnam, 1965-1970* (New York: Ivy Books, 1996), 169.
10 Ibid., 171.
11 Ibid., 170.
12 Ibid., 173.

13 Ibid., 175. Patrol Reports for Party Line One and Mono type II are included in 3rd Reconnaissance Battalion – Command Chronology for the period 1 August 1967 to 31 August 1967, see re: OPN ORD 311-67 and 314-67. Courtesy of Bill McBride.

14 Robert Bliss, email to Bill McBride, 28 Sep 1998. Courtesy of Bill McBride.

15 Michael Gavlick, email to Bill McBride, 9 Sep 1998. Courtesy of Bill McBride.

16 LTGEN Cushman, message to LTG Krulak, 3 August 1967. Courtesy of Bill McBride.

17 *Operation Cumberland OP [Operational Procedures] File,* 17 Jul–16 Sep 1967. USMC History Division Collection, doc 1201064063.

Chapter 8

1   National Intelligence Estimate (NIE) 14-3-66 7 July 1966.

2   Ibid., Map 1 June 1966.

3   CIA memorandum, "The Situation in Vietnam 28 February 1967."

4   NPIC memorandum, 24 March 1967.

5   NPIC MSG 062335Z APR 67.

6   NPIC MSG 150041Z APR 67.

7   Intelligence memorandum, "Road Construction in the Laotian Panhandle and Adjacent Areas of South Vietnam 1967-1968." May 1968.

8   Intelligence information cable, 17 April 1967.

9   CIA memorandum, "The Situation in Vietnam 24 April 1967."

10 Intelligence memorandum, "Recent Trends in Infiltration into South Vietnam 22 June 1967."

11 CIA memorandum, "The Situation in Vietnam 20 November 1967."

12 Lt Gen John A Chaisson, USMC (Ret), *Oral History Transcripts*, 1 Aug 1967 & 2 Jan 1968. USMC History Division Oral History Collection, doc 10970103001, 60–61.

13 Ibid., 127–128.

14 CIA, Summary of November 21 Meeting with Saigon Advisors.

15 *MACV COMMAND HISTORY 1968, VOL I*, 30 Apr 1969. Bud Harton Collection, doc 168300010746, 23–24.

16 Jack Shulimson, Leonard A. Blaisol, Charles R. Smith, and David A. Dawson, *U.S. Marines in Vietnam: The Defining Year, 1968* (Washington, DC: History and Museums Division, HQs, USMC, 1997), 107.

17 Ibid., 108

18 Westmoreland, *A Soldier Reports*, 347.

19 Ibid., 328–329.

20 *Combined Campaign Plan 1968, AB 143*. Glenn Helm Collection, doc 1071816001.

21 NPIC / R-37/68, February 1968.

22 CIA intelligence memorandum, 6 February 1968.

23 CIA memorandum, "The Situation in Vietnam 9 February 1968."

24 See e.g., CIA memorandum, "The Situation in Vietnam 28 November 1967"; CIA memorandum, "The Situation in Vietnam 30 November 1967"; CIA memorandum, "The Situation in Vietnam 14 February 1968"; and CIA memorandum, "The Situation in Vietnam 21 February 1968."

25 III MAF, *Periodic Intelligence Report* (PerIntRep) 2-68, 14 January 1968, p. 5.

26 CIA memorandum, "The Situation in Vietnam 3 March 1968."

27 Intelligence memorandum, "Road Construction in the Laotian Panhandle and Adjacent Areas of South Vietnam 1967-1968." May 1968, 9, see Figure 1 re: Road Construction 1967-8 Dry Season.

28 NPIC photographic interpretation memorandum, "Telecommunications Network Construction in the Laotian Panhandle." NPIC / R-46/68 March 1968.

Chapter 9

1 Rostow, W.W., memoes to President Lyndon B. Johnson, re: *Reports #1-19 on the Situation in the Khe Sanh Area,* 5 Feb–23 Feb 1968. Veteran Members of the 109th Quartermaster Company (Air Delivery) Collection, doc 0010113001, Report #18.

2 *Plans for Northern I Corps Operations - re: priority to Thua Thien and Quang Tri,* 15 Mar 1968. Veteran Members of the 109th Quartermaster Company (Air Delivery) Collection, doc 0010120004.

3 1st Air Cavalry Division, *AAR: Operation DELAWARE,* 11 Jul 1968. Richard Detra Collection, doc 0690401002, interview with MG Tolson, 5/27/68.

4 Eric Villard, *Staying the Course: October 1967 to September 1968* (Washington, DC: Center of Military History, United States Army, 2017), 520.

5 Operation Delaware AAR, interview with MGEN Tolson, 5/27/68.

6 Villard, *Staying the Course,* 519.

7 COMUSMACV (CDEC) Department of Defense Intelligence Information Report re: *Organization and Activities of the 559th Transportation Group,* 13 Jun 1968. Glenn Helm Collection, doc 1071618023.

8 James F. Dunnigan and Albert A. Nofi, *Dirty Little Secrets of the Vietnam War* (New York, NY: St. Martin's Press, 1999), 205–208.

9 *Organization and Activities of the 559th Transportation Group* Report.

10 Shulimson et al., *U.S. Marines in Vietnam: The Defining Year, 1968,* 107.

11 Robert M. Gillespie, *Black Ops Vietnam: The Operational History of MACVSOG* (Annapolis, MD: Naval Institute Press, 2011), 146.

12 W.W. Rostow, memo to President Lyndon B. Johnson, re: *Report #46 on the Situation in the Khe Sanh/DMZ/A Shau Valley Area,* 21 Mar 1968. Veteran Members of the 109th Quartermaster Company (Air Delivery) Collection, doc 0010313004.

13 W.W. Rostow, memo to President Lyndon B. Johnson, re: *Report #47 on the Situation in the Khe Sanh/DMZ/A Shau Valley Area,* 22 Mar 1968. Veteran Members of the 109th Quartermaster Company (Air Delivery) Collection, doc 0010313001.

14 W.W. Rostow, memo to President Lyndon B. Johnson, re: *Report #51 on the Situation in the Khe Sanh/DMZ/A Shau Valley Area,* 26 Mar 1968. Veteran Members of the 109th Quartermaster Company (Air Delivery) Collection, doc 0010319007.

15 W.W. Rostow, memo to President Lyndon B. Johnson, re: *Report #53 on the Situation in the Khe Sanh/DMZ/A Shau Valley Area,* 28 Mar 1968. Veteran Members of the 109th Quartermaster Company (Air Delivery) Collection, doc 0010315010.

16 See Distinguished Service Cross citation.

17 W.W. Rostow, memo to President Lyndon B. Johnson, re: *Report #54 on the Situation in the Khe Sanh/DMZ/A Shau Valley Area*, 29 Mar 1968. Veteran Members of the 109th Quartermaster Company (Air Delivery) Collection, doc 0010315007.

18 W.W. Rostow, memo to President Lyndon B. Johnson, re: *Report #66 on the Situation in the Khe Sanh/DMZ/A Shau Valley Area*, 10 Apr 1968. Veteran Members of the 109th Quartermaster Company (Air Delivery) Collection, doc 00103121005.

19 See Carpenter, *Boots on the Ground*, 258–271, and Yarborough, *A Shau Valor*, 107–114, for a description of some of the patrols and ground combat actions of Operation SAMURAI IV.

20 B-52 (Project Delta) After Action Report – *Operation 2-68 (Samurai IV)*

21 *Project CHECO Southeast Asia Report #91*, 57.

22 John J. Tolson, *Airmobility, 1961-1971* (Washington, DC: Department of the Army, 1973), 184.

23 Gaylor, Capt Wayne, ALO, 1st Squadron, 9th Cavalry (Airmobile), *Interview Transcript re: Operation Delaware*, 24 May 1968. Sam Johnson Collection, doc F031100041620.

24 Ibid.

25 SGT Raymond Mills, Jr., US Army, Intelligence NCO, 1st Squadron, 9th Cavalry (Airmobile), *Interview Transcript re: Operation Delaware*, 24 May 1968. Sam Johnson Collection, doc F031100041630.

26 *Project CHECO Southeast Asia Report #91*, Appendix IV.

27 Mills *Interview Transcript* and Shelby L. Stanton, *Anatomy of a Division: The 1st Cav in Vietnam* (Novato, CA: Presidio Press, 1987), 145.

28 1st Air Cavalry Division, *AAR: Operation DELAWARE*, interview with MG Tolson, 5/27/68.

29 Mills, *Interview Transcript*.

30 Stanton, *Anatomy of a Division*, 144.

31 *Project CHECO Southeast Asia Report #91*, 24–25.

32 Ibid., 21, 27.

33 Robert C. Ankony, "No Peace in the Valley," *Vietnam Magazine*, October 2008, 29.

34 227 Assault Helicopter Battalion AAR at http://a227ahb.org/DelawareLamSon216.html. Last update 3/16/2009. accessed 8/7/2019.

35 Ibid.

36 Ankony, "No Peace in the Valley," 30.

37 *Operation Delaware/Lamson 216 OP [Operational Procedures] File*, 22 Apr–20 May 1968. USMC History Division Collection, doc 1201064124, SITREP #1, 120.

38 1st Air Cavalry Division, *AAR: Operation DELAWARE*, 9.

39 Vietnam Helicopter Pilots Association at https://www.vhpa.org/KIA; cited hereafter as VHPA website with listings by helicopter serial numbers. 66-15052.

40 A/227 Assault Helicopter Battalion *Valley of Sorrow - A Shau 25 April 1968* at http://stories.a227ahb.org/Ashau.html accessed 5/8/2019; see also VHPA 66-15052 & 66-19080.

41 *The Air Cavalry Division*, September 1968. J.D. Coleman Collection, doc 2770Serial610702.

42 Villard, *Staying the Course*, 520.

Chapter 10

1   227 Assault Helicopter Battalion AAR.
2   *Project CHECO Southeast Asia Report #91,* Figure 7.
3   Charles Baker, *Gray Horse Troop: Forever Soldiers,* 2nd revision (Powder River Publications, 2014), 283).
4   VHPA, 66-16799; 227 Assault Helicopter Battalion AAR.
5   Baker, *Gray Horse Troop,* 283.
6   Per phone conversation with Mike Sprayberry 30 May 2020, the NVA used Lang Ka Kou as the cantonment for anti-aircraft weapons during the rainy season, accounting for the enemy concentration.
7   Baker, *Gray Horse Troop,* 284–289.
8   Ibid., 285.
9   Mike Sprayberry, phone conversation, 30 May 2020.
10 Ibid.
11 Ibid.
12 Villard, *Staying the Course,* 522.
13 VHPA, 66-19063.
14 Ibid., 66-14205.
15 Baker, *Gray Horse Troop,* 296–297.
16 VHPA, 64-13124.
17 *ARMY 1966-72 CH 47 Strike Aircraft (Chinook) Destroyed,* 31 Aug 1972. Bud Harton Collection, doc 168300010067.
18 Baker, *Gray Horse Troop,* 299.
19 Sprayberry conversation, 30 May 2020.
20 Baker, *Gray Horse Troop,* 298–299.
21 1st Air Cavalry Division, *AAR: Operation DELAWARE,* Tab H.
22 *Operation Delaware Valley/Lamson 216 OP [Operational Procedures] File,* See FMFPAC CC Msg 26 Apr 68; Baker, *Gray Horse Troop,* 299.
23 227th Assault Helicopter Battalion AAR. FMFPAC SITREP #1120.
24 *Operation Delaware Valley/Lamson 216 OP [Operational Procedures] File,* SITREP #1120.
25 Baker, *Gray Horse Troop,* 302.
26 Operation Delaware Valley/Lamson 216 OP [Operational Procedures] File, SITREP #1124.
27 Baker, *Gray Horse Troop,* 305.
28 Ibid., 306
29 Sprayberry conversation, 30 May 2020.
30 Ibid., 310.
31 Ibid.
32 Baker, *Gray Horse Troop,* 314.
33 *Operation Delaware Valley/Lamson 216 OP [Operational Procedures] File,* SITREP #1136.
34 Sprayberry conversation, 11 Jun 2020.
35 Baker, *Gray Horse Troop,* 315–318.
36 Ibid., 319.
37 Ibid.
38 Ibid., 320.

39 Ibid., 321–323

40 Sprayberry conversation, 30 May 2020.

41 Baker, *Gray Horse Troop*, 325; Author's database.

42 1st Air Cavalry Division, *AAR: Operation DELAWARE*, Tab C.

43 Baker, *Gray Horse Troop*, 327–328.

44 Ibid., 329.

45 guarding a base that would not be penetrated until almost a year later, Ibid., 332; 1st Air Cavalry Division, *AAR: Operation DELAWARE*, 4; *Operation Delaware Valley/ Lamson 216 OP [Operational Procedures] File*, SITREP #1136; 1st Cavalry Division, *Operational Report - Lessons Learned [OR-LL]*, Period Ending 31 July 1968. Dated 26 December 1968.

46 Baker, *Gray Horse Troop*, 332.

47 Ibid., 334.

48 Sprayberry e-mail to author, 9 June 2020.

49 227 Assault Helicopter Battalion AAR.

50 VHPA, 66-16567.

51 227 Assault Helicopter Battalion AAR.

52 Time according to Sprayberry conversation 30 May 2020. This is a significantly earlier cessation of air activities in the afternoon than given in the "official" record.

53 Yarborough, *A Shau Valor*, 123.

54 Baker, *Gray Horse Troop*, 290.

55 *Operation Delaware Valley/Lamson 216 OP [Operational Procedures] File*, SITREP #1120.

56 Stanton, *Anatomy of a Division*, 146.

57 Tolson, *Airmobility, 1961-1971*, 187.

58 1st Air Cavalry Division, *AAR: Operation DELAWARE*, Tab C; *Operation Delaware Valley/Lamson 216 OP [Operational Procedures] File*, FMPAC CC Msg 26 Apr 68; *The Air Cavalry Division*, September 1968, p.4.

59 1st Air Cavalry Division, *AAR: Operation DELAWARE*, 9.

60 Villard, *Staying the Course*, 523.

61 Author's database.

62 1st Air Cavalry Division, *AAR: Operation DELAWARE*, 9.

63 Ibid., 4.

64 Author's database.

65 *The Air Cavalry Division*, September 1968, p. 7.

66 Medal of Honor Citation.

67 1st Air Cavalry Division, *AAR: Operation DELAWARE*, 4.

68 Author's database.

69 *Operation Delaware Valley/Lamson 216 OP [Operational Procedures] File*, SITREP #1136.

70 Operation Delaware After Action Report, 4.

71 Ibid., 11.

72 227 Assault Helicopter Battalion AAR.

73 Ibid.

74 Villard, *Staying the Course*, 523.

75 1st Air Cavalry Division, *AAR: Operation DELAWARE*, Tab C.

76 *Operation Delaware Valley/Lamson 216 OP [Operational Procedures] File*, SITREP #1124.

77 Author's database.

78   Ibid.
79   1st Air Cavalry Division, *AAR: Operation DELAWARE*, p. 5 and Tab C.
80   1st Cavalry Division, *Operational Report - Lessons Learned [OR-LL]*, Period Ending 31 July 1968, p. 9
81   Author's database.
82   *Operation Delaware Valley/Lamson 216 OP [Operational Procedures] File*, SITREP #1136.
83   Ibid., FMFPAC SITREP #1137.
84   Author's database.
85   1st Air Cavalry Division, *AAR: Operation DELAWARE*, 11.

Chapter 11

1   *Operation Delaware Valley/Lamson 216 OP [Operational Procedures] File*, FMFPAC CCMsg 26 Apr 68.
2   A/227 Assault Helicopter Battalion *Valley of Sorrow*.
3   *Operation Delaware Valley/Lamson 216 OP [Operational Procedures] File*, FMFPAC CCMsg 26 Apr 68.
4   227 Assault Helicopter Battalion AAR.
5   VHPA, 66–19066.
6   1st Air Cavalry Division, *AAR: Operation DELAWARE*, 9.
7   B Company, 2nd Battalion, 8th Cavalry report on Khe Sanh and A Shau Valley Operations at http://www.eagerarms.com/khesanhashauvalleyoperations.html. Accessed 6/5/2019.
8   1st Air Cavalry Division, *AAR: Operation DELAWARE*, 9, Tab C.
9   B Company, 2nd Battalion, 8th Cavalry report on Khe Sanh and A Shau Valley Operations.
10   Ibid.
11   Author's database.
12   *The Air Cavalry Division*, September 1968, 7.
13   1st Air Cavalry Division, *AAR: Operation DELAWARE*, 5.
14   Author's database.
15   1st Air Cavalry Division, *AAR: Operation DELAWARE*, 10.
16   Ibid., 4.
17   *Operation Delaware Valley/Lamson 216 OP [Operational Procedures] File*, FMFPAC SITREP #1128.
18   John Ulfer's Diary (D/1/8 Cav) at http://www.jumpingmustangs.com/delta6571/deltaulferdiary.html. Accessed 8/15/2019.
19   1st Cavalry Division, *Operational Report - Lessons Learned [OR-LL]*, Period Ending 31 July 1968, p. 8.
20   1st Air Cavalry Division, *AAR: Operation DELAWARE*, 5; Author's database.
21   John Ulfer, diary.

22 B Company, 2nd Battalion, 8th Cavalry report on Khe Sanh and A Shau Valley Operations.

23 Author's database.

24 1st Cavalry Division, *Operational Report - Lessons Learned [OR-LL]*, Period Ending 31 July 1968, p. 9; *Operation Delaware Valley/Lamson 216 OP [Operational Procedures] File*, FMFPAC SITREP #1139.

25 *Operation Delaware Valley/Lamson 216 OP [Operational Procedures] File*, FMFPAC SITREP #1135.

26 Ibid., FMFPAC SITREP #1137.

27 Ibid., FMFPAC SITREP #1135; 1st Air Cavalry Division, *AAR: Operation DELAWARE*, 5.

28 Ibid., FMFPAC SITREP #1139.

29 Ibid.

30 1st Air Cavalry Division, *AAR: Operation DELAWARE*, 5.

31 John Ulfer's Diary.

32 1st Air Cavalry Division, *AAR: Operation DELAWARE*, Tab I.

33 *Operation Delaware Valley / Operation Lam Son 216* [OP FILE], FMFPAC CCMsg 26 Apr 68 & FMFPAC SITREP #1124.

34 A Company, 1st Battalion, 8th Cavalry website. http://vnwarstories.com/Odyssey/vn.ACompanyOdyssey25April68.html. Based on Daily Staff Journals of the 3 1/8 Cav at Texas Tech, Collection 369, Daily Staff Journals 1965-1969. Accessed 8/15/2019.

35 Ibid.

36 J.D. Coleman (ed.), *First Cavalry Division Vietnam, August 1965 to December 1969* (Paducah, KY: Turner Publishing, 1995), 288–289; *Memorandum for the Press re: Lieutenant General Cushman's Press Brief*, 2 May 1968. Sam Johnson Collection, doc F031100041480.

37 1st Air Cavalry Division, *AAR: Operation DELAWARE*, 4.

38 *The Air Cavalry Division*, September 1968, 7.

39 A Company, 1st Battalion, 8th Cavalry website.

40 227 Assault Helicopter Battalion AAR.

41 *Project CHECO #203 – Southeast Asia Report Short Rounds, June 1967 – June 1968, 25 Aug 1968*. Sam Johnson Collection, doc F031100180684, 15–21.

42 A Company, 1st Battalion, 8th Cavalry website.

43 Author's database.

44 Ibid; *Operation Delaware Valley/Lamson 216 OP [Operational Procedures] File*, FMFPAC SITREP #1135.

45 1st Air Cavalry Division, *AAR: Operation DELAWARE*, Tab C.

46 VHPA, 63-09097.

47 *Operation Delaware Valley/Lamson 216 OP [Operational Procedures] File*, FMFPAC SITREP #1141.

48 A Company, 1st Battalion, 8th Cavalry website.

49 III MAF PerIntRep 20-68, 19 May 68, 2.

50 C Company, 1st Battalion, 8th Cavalry website at http://www.c1-8cav68.org/ashauvalley.html. Accessed 8/15/2019.

Chapter 12

1   1st Air Cavalry Division, *AAR: Operation DELAWARE*, 1

2   Ibid., Tab C.

3   *Memorandum for the Press re: Lieutenant General Cushman's Press Brief*, 7 May 1968. Sam Johnson Collection, doc F031100041487.

4   Tolson, *Airmobility, 1961-1971*, 190.

5   *Memorandum for the Press re: Lieutenant General Cushman's Press Brief*, 7 May 1968.

6   1st Air Cavalry Division, *AAR: Operation DELAWARE*, 10–11.

7   Ngo Quang Truong, *Republic of Viet Nam Armed Forces and US Operational Cooperation and Coordination* (Indochina Monographs), 1979, CMH Pub 92-16. J.D. Coleman Collection, doc 27700602001, 104.

8   Ibid., 105.

9   *Project CHECO Southeast Asia Report #91.*

10   Ibid., 7.

11   Ibid., 31.

12   Ibid., 69.

13   Ibid., 32.

14   Ibid., 34.

15   Ibid., 31.

16   Chris Hobson, *Vietnam Air Losses,* (Hinckley, England: Midland Publishing, 2001), 146–147.

17   *Project CHECO Southeast Asia Report #91,* 39–40.

18   Donald L. Abbott, 1st Cavalry Division (Airmobile), *Interview Transcript re: Operation Delaware*, 24 May 1968. Sam Johnson Collection, doc F031100041606.

19   Hobson, *Vietnam Air Losses*, 146.

20   Donald L. Abbott, 1st Cavalry Division (Airmobile), *Interview Transcript re: Operation Delaware*.

21   *Project CHECO Southeast Asia Report #91,* 35.

22   Ibid., 36.

23   Ibid., 18.

24   1st Air Cavalry Division, *AAR: Operation DELAWARE*, Tab T.

25   McGowan, Sam, "A Luoi Airstrip in the A Shau Valley." At http://stories.a227ahb.org/Aluoi.html. Accessed 5/8/2019.

26   *Project CHECO Southeast Asia Report #91,* 60.

27   1st Air Cavalry Division, *AAR: Operation DELAWARE*, Tab T.

28   Ibid., 11.

29   *Memorandum for the Press re: Lieutenant General Cushman's Press Brief*, 2 May 1968.

30   Memorandum for the Press re: Lieutenant General Cushman's Press Brief, 17 May 1968. Sam Johnson Collection, doc F031100041500.

31   Stanton, *Anatomy of a Division*, 144.

32   Graham A. Cosmas, *The Joint Chiefs of Staff and The War in Vietnam 1960-1968, Part 3* (Washington, DC: Office of Joint History, Office of the Chairman of the Joint Chiefs of Staff, 2009), 205.

33   www.militaryfactory.com.

34   Ibid.

35 *Tactical Principles and Logistics for Cavalry,* Academic Division, The Cavalry School, Fort Riley, KS, 1937.

36 *Memorandum for the Press re: Lieutenant General Cushman's Press Brief,* 17 May 1968. Sam Johnson Collection, doc F031100041500.

37 1st Air Cavalry Division, *AAR: Operation DELAWARE,* 13.

38 Seventh Air Force, *Weekly Air Intelligence Summary re: Operation Delaware,* 1 Jun 1968. Sam Johnson Collection, doc F031100041654, 3.

39 Donald L. Abbott, 1st Cavalry Division (Airmobile), *Interview Transcript re: Operation Delaware,* 24 May 1968.

40 Intelligence Information Report, Captured Documents (CDEC): *re: Observation of Two 100mm AA Guns in Thua Thien Prov,* Rpt Nr 60-5868-68, 12 Dec 1968. doc F034604700729.

41 Stanton, *Anatomy of a Division,* 148.

42 *Memorandum for the Press re: Lieutenant General Cushman's Press Brief,* 11 May 1968. Sam Johnson Collection, doc F031100041494, 5.

43 Combined Intelligence Center, Vietnam, *Order of Battle Study ST 67-029: VC/NVA Employment of Snipers,* 6 Jan 1967. Sam Johnson Collection, doc F015900210639.

44 1st Air Cavalry Division, *AAR: Operation DELAWARE,* 6–7.

45 Ibid., 7.

46 227 Assault Helicopter Battalion AAR.

47 Villard, *Staying the Course,* 524.

48 Quoted in Ankony, "No Peace in the Valley," 31.

49 Villard, *Staying the Course,* 520.

50 1st Air Cavalry Division, *AAR: Operation DELAWARE,* Tab T.

51 Ibid., 7.

52 *MACV COMMAND HISTORY 1968, VOL I,* 161–163.

53 Tolson, *Airmobility, 1961-1971,* 191–192.

54 Willard Pearson, *The War in the Northern Provinces 1966-1968* (Washington, DC: Department of the Army, 1975), 92.

55 Villard, *Staying the Course,* 539.

56 III MAF, Periodic Intelligence Reports, PerIntRep 21-68 26 May 68, p. 2, and 22-68 2 Jun 68, p. 3.

Chapter 13

1 1st Air Cavalry Division, *AAR: Operation DELAWARE,* 11.

2 1st Cavalry Division, *Operational Report - Lessons Learned [OR-LL],* Period Ending 31 July 1968, p. 8.

3 Tolson, *Airmobility, 1961-1971,* 190.

4 *Project CHECO Southeast Asia Report #117 – Igloo White (Initial Phase), 31 Jul 1968.* Sam Johnson Collection, doc F031100410008, 1.

5 Ibid., 26.

6 *Project CHECO Southeast Asia Report #124 – Interdiction in Seasia, November 1966 – October 1968, 20 Jun 1969.* CHECO Reports, doc 0390213001, 95.

7 *Project CHECO Southeast Asia Report #91,* 5

8   *Project CHECO Southeast Asia Report #124*, 96.

9   *Project CHECO Southeast Asia Report #91*, 6.

10   *Project CHECO Southeast Asia Report #124*, 90–91.

11   *Project CHECO Southeast Asia Report #176 – Visual Reconnaissance in I Corps, 30 Sep 1968*. Sam Johnson Collection, doc F031100181082, 6.

12   *Project CHECO Southeast Asia Report #124,* 97–98.

13   *Project CHECO Southeast Asia Report #91*, 53.

14   *Project CHECO Southeast Asia Report #124*, 99–100.

15   Ibid., 101.

16   *Project CHECO Southeast Asia Report #107 – Impact of Geography on Air Operations in Sea, 11 Jun 1970*. Sam Johnson Collection, doc F031100101565, 51.

17   Committee on Armed Services, United States Senate, Ninety-Second Congress, First Session, *Investigation into Electronic Battlefield Program*, 1971. Douglas Pike Collection, doc 2390812004, 17.

18   *Project CHECO Southeast Asia Report #124*, 131.

19   McLaughlin, Brigadier General George W., USAF, Director, Tactical Control Center, *letter re: MACV Year End Review of Vietnam, 1968,* 4 Jan 1969. Sam Johnson Collection, doc F031100151073.

20   *Project CHECO Southeast Asia Report #124*, 115.

21   Task Force Omega at http://taskforceomegainc.org/; Virtual Wall.

22   Ibid.

23   *Project CHECO Southeast Asia Report #124*, 102.

24   MACV, *Significant Problem Areas, May 1968,* 6 Jun 1968. Sam Johnson Collection, doc F015800220378, 29.

25   MACV, *Significant Problem Areas Report, Nov 1969,* 26 Nov 1969. Sam Johnson Collection, doc F015800220558, 8.

26   *Interdiction Program in I Corps*, Sam Johnson Collection, doc F031100090196.

27   John L. Plaster, *SOG: The Secret Wars of America's Commandos in Vietnam* (New York: Simon & Schuster, 1997), 183.

28   Yarborough, *A Shau Valor*, 130.

29   Plaster, *Secret Wars*, 183–190.

30   DSC Citation (lists location as Vietnam, but that is incorrect. This was in Laos.)

31   TF Omega website; MACVSOG at http://www.macvsog.cc

32   Ibid.

33   Ibid.

34   Gillespie, *Black Ops Vietnam*, 147.

35   MACVSOG website; Virtual Wall.

36   Kenn Miller, *Six Silent Men: 101st LRP/Rangers, Book Two* (New York: Ivy Books, 1997), 119–136.

37   Ibid., 101–112; Gary A. Linderer, *The Eyes of the Eagle: F Company LRPs in Vietnam, 1968*. (New York: Ivy Books, 1991), 33–38.

38   Linderer, *Eyes of the Eagle*, 76–77.

Chapter 14

1   *III MAF Periodic Intelligence Report 24-68*, 16 Jun 68, 7.
2   *III MAF Periodic Intelligence Report 28-68*, 14 July 68, 3–4.
3   *III MAF Periodic Intelligence Report 31-68, 4 August 68, 3.*
4   101st Airborne Division, *CASE STUDY: OPERATION "SOMERSET PLAIN," 4-20 AUGUST 1968*, Prepared by Major Emmett Kelly. Sam Johnson Collection, doc F031100090062; see also Richard Detra Collection, doc 0690418002, p. 1.
5   *III MAF, COMMAND CHRONOLOGY, August 1968.* USMC History Division Collection, doc 1201004039.
6   *Operation Somerset Plain After Action Report* at http://www.fearfulodds.com/Ashau, p. 30.
7   101st Airborne Division, *CASE STUDY: OPERATION "SOMERSET PLAIN,"* 10.
8   VHPA, 66-15055.
9   1st Squadron, 9th Cavalry, 1st Air Cavalry Division, *Operational Report for Quarterly Period Ending October 31 1968, 9 Nov 1968.* Bud Harton Collection, doc 168300010267.
10 *SUMMARY OF ARC LIGHT ACTIVITY, 1-31 August 1968.* Sam Johnson Collection, doc F031100140241.
11 101st Airborne Division, *CASE STUDY: OPERATION "SOMERSET PLAIN,"* 5.
12 *Operation Somerset Plain After Action Report, 7.*
13 Ibid., 8.
14 Ibid., 23.
15 *Operation Somerset Plain Daily Staff Journals* at http://www.fearfulodds.com/Ashau
16 101st Airborne Division, *CASE STUDY: OPERATION "SOMERSET PLAIN,"* 11.
17 1/327 Infantry *Operation Somerset Plain After Action Report* at http://www.fearfulodds.com/Ashau, para 10.
18 VHPA, 66-00680.
19 *Operation Somerset Plain After Action Report,* 13.
20 *Operation Somerset Plain Daily Staff Journals.*
21 101st Airborne Division, *CASE STUDY: OPERATION "SOMERSET PLAIN,"* 17.
22 *Operation Somerset Plain After Action Report,* 18.
23 *III MAF Periodic Intelligence Report 32-68*, 11 August 68, 4.
24 *Operation Somerset Plain Daily Staff Journals.*
25 VHPA, 66-15227.
26 *III MAF Periodic Intelligence Report 33-68*, 18 August 68, B-2.
27 *Operation Somerset Plain After Action Report,* 14; *Operation Somerset Plain Daily Staff Journals.*
28 *Operation Somerset Plain After Action Report*, 15.
29 Ibid.
30 Virtual Wall.
31 101st Airborne Division, *CASE STUDY: OPERATION "SOMERSET PLAIN,"* 16.

32   *Project CHECO Southeast Asia Report #204 – Short Rounds, June 1968 – May 1969, 15 Aug 1969.* Sam Johnson Collection, doc F031100180731, 24.

33   Virtual Wall.

34   Linderer, *Eyes of the Eagle*, 90–91; Miller, *Six Silent Men*, 176.

35   *Operation Somerset Plain Daily Staff Journals.*

36   *Operation Somerset Plain After Action Report,* 15; *III MAF Periodic Intelligence Report 33-68,* 18 August 68, 4.

37   *Operation Somerset Plain Daily Staff Journals.*

38   *Operation Somerset Plain After Action Report,* 16; *III MAF Periodic Intelligence Report 33-68,* 18 August 68, 4.

39   *Operation Somerset Plain Daily Staff Journals;* Virtual Wall.

40   Charles W. Newhall III, *Fearful Odds: A Memoir of Vietnam and Its Aftermath,* (Owingsmills, MD: Bibliotheca Brightside, 2015), 61, 64.

41   Ibid., 62-63.

42   *Operation Somerset Plain After Action Report,* 16; *Operation Somerset Plain Daily Staff Journals.*

43   Newhall, *Fearful Odds,* 66.

44   Ibid., 67.

45   Ibid., 67–73.

46   Ibid., 79–80.

47   *Operation Somerset Plain After Action Report,* 16.

48   Newhall, *Fearful Odds,* 82.

49   *Operation Somerset Plain After Action Report,* 16; *Operation Somerset Plain Daily Staff Journals.*

50   *Operation Somerset Plain After Action Report,* 17.

51   1/327 Infantry *Operation Somerset Plain After Action Report.*

52   *Operation Somerset Plain After Action Report,* 17; *Operation Somerset Plain Daily Staff Journals.*

53   www.wikitree.com.

54   *Operation Somerset Plain After Action Report,* 17.

55   Ibid., 6.

56   Villard, *Staying the Course,* 608.

57   *III MAF Periodic Intelligence Report 34-68,* 25 August 1968, 2–3.

58   *Operation Somerset Plain After Action Report,* 18–19.

59   101st Airborne Division, *CASE STUDY: OPERATION "SOMERSET PLAIN,"* 22–23.

60   Villard, *Staying the Course,* 607–608.

61   Shulimson, *U.S. Marines in Vietnam: The Defining Year, 1968.*

Chapter 15

1   Plaster, *The Secret Wars of America's Commandos,* 193–194; Yarborough, *A Shau Valor,* 134–136; VHPA, 65-12782.

2   Task Force Omega website; MACVSOG website; Virtual Wall.

3 *Resume of the A Shau Valley Air Interdiction Campaign, 9 Dec 1968 to 28 Feb 1969.* Sam, Johnson Collection doc F031100090085.

4 COMUSMACV msg to JCS *re: Commander's Estimate of ARC LIGHT Effectiveness,* 17 Dec 1968. Sam Johnson Collection, doc F031100140278.

5 III MAF, *AIR CONTROL (SPECAT VOL 3),* 1 Dec 1968. USMC History Division Collection, doc 1201005003, CG III MAF 220608Z DEC68.

6 *Resume of the A Shau Valley Air Interdiction Campaign.*

7 III MAF, *AIR CONTROL (SPECAT VOL 3),* 1 Dec 1968, COMUSMACV 280029Z DEC68.

8 *Resume of the A Shau Valley Air Interdiction Campaign.*

9 *National Security Study Memorandum (NSSM) One – March 1969: The Situation in Vietnam.* Douglas Pike Collection, doc 2120107001, 10-3.

10 Ibid., 28a-2.

11 Ibid., 27-2–27-5.

12 *Project CHECO Southeast Asia Report #2 – A Shau Valley Campaign, December 1968 – May 1969, 15 Oct 1969.* Sam Johnson Collection, doc F031100081306, 14.

13 Hobson, *Vietnam Air Losses,* 172.

14 *Resume of the A Shau Valley Air Interdiction Campaign.*

15 Ibid.

16 Ibid.

17 Ibid.

Chapter 16

1 Charles R. Smith, *U.S. Marines in Vietnam: High Mobility and Standdown, 1969* (Washington, DC: History and Museums Division, Headquarters, USMC, 1988), 12.

2 Ibid., 10–11.

3 Ibid., 11.

4 Quoted in Ibid., 9.

5 Untitled CIA document "Memorandum for Chief [redacted] from Acting Chief [redacted]" dated 14 November 1967.

6 See Lewis Sorley, "Outgunned: The ARVN Under Westmoreland," in *VVA Veteran* online, March/April 2012, for a comprehensive discussion of GEN Westmoreland and the arming of the ARVN.

7 Lewis Sorley, *Vietnam Chronicles: The Abrams Tapes, 1968-1972* (Lubbock Texas: Texas Tech University Press, 2004), 26, 3 AUG 1968: WIEU.

8 Smith, *U.S. Marines in Vietnam: High Mobility and Standdown,* 1969, 12.

9 CIA MEMORANDUM FOR: The Honorable Henry A. Kissinger. Data on Communist Military and Terrorist Activity in Vietnam. 5 February 1969.

10 *Project CHECO Southeast Asia Report #105 – The Fourth Offensive,* 1 Oct 1969. Sam Johnson Collection, doc F031100070740, 17.

11 Ibid., 50.

12 Smith, *U.S. Marines in Vietnam: High Mobility and Standdown, 1969,* 28.

13 9th Marines, *Combat Operations AAR (DEWEY CANYON),* 8 Apr 1969. Sam Johnson Collection, doc F031100081398, 12.

14   Ibid., 3–4.

15   Ibid., 14.

16   Smith, *U.S. Marines in Vietnam: High Mobility and Standdown, 1969*, 33–34.

17   This battle occurred outside the focus area of this book. LCPL Noonan's citation may be seen at Smith, *U.S. Marines in Vietnam: High Mobility and Standdown, 1969*, 369, and the battle and subsequent move of G Company to safety on 6–8 February are described at Ibid., 35–36, and Yarborough, *A Shau Valor*, 145–148.

18   3rd Marine Division, *SPECIAL FILE: UNCLASSIFIED OFFICIAL RECORDS, "OPERATION DEWEY CANYON."* USMC History Division Collection, doc 120103300, Tab C, p. 3.

19   Smith, *U.S. Marines in Vietnam: High Mobility and Standdown, 1969*, 38–40.

20   9th Marines, *Combat Operations AAR (DEWEY CANYON)*, 9.

21   Ibid., 15–16.

22   Ibid., 16.

23   Ibid., 16–18; Smith, *U.S. Marines in Vietnam: High Mobility and Standdown, 1969*, 38–39.

24   9th Marines, *Combat Operations AAR (DEWEY CANYON)*, 18; Smith, *U.S. Marines in Vietnam: High Mobility and Standdown, 1969*, 39.

25   This battle occurred outside the focus area of this book. 1LT Fox's citation may be seen at Smith, *U.S. Marines in Vietnam: High Mobility and Standdown, 1969*, 365, and the battle is detailed in his biography, Wesley L. Fox, *Marine Rifleman: Forty-three Years in the Corps* (Washington, DC: Brassey's, 2002), 250–263, and at Smith, *U.S. Marines in Vietnam: High Mobility and Standdown, 1969*, 45–46.

26   9th Marines, *Combat Operations AAR (DEWEY CANYON)*, 21.

27   Lt Col. George C. Fox, USMC, CO, 2nd Btn/9th Mar, Interview by Lt Col Bert Aton, 3 Aug 1969. Sam Johnson Collection, doc F031100090148.

28   Fox, *Marine Rifleman*, 257.

29   3rd Marine Division, *Special File: Unclassified Official Records, "Operation Dewey Canyon,"* Tab C, 18.

30   9th Marines, *Combat Operations AAR (DEWEY CANYON)*, 12.

31   3rd Marine Division, *special file: unclassified official records, "operation dewey canyon,"* Tab C, 12–14.

32   *Project CHECO Southeast Asia Report #2*, 9.

33   Medal of Honor Citation is at Smith, *U.S. Marines in Vietnam: High Mobility and Standdown, 1969*, 373.

34   Smith, *U.S. Marines in Vietnam: High Mobility and Standdown, 1969*, 41.

35   Ibid.

36   Ibid., 43–44.

37   Ibid., 44–45.

38   Ibid., 47.

39   NPIC Attachment to M-5968. OPEN WIRE TELECOMMUNICATIONS NETWORK LAOTIAN PANHANDLE.

40   Smith, *U.S. Marines in Vietnam: High Mobility and Standdown, 1969*, 33–34.

41   James S. Rayburn, "Direct Support during Operation DEWEY CANYON (U)." In *The NAVSECGRU Bulletin*, Vol. XXIV, No. 11, Nov/Dec 1980 (a now declassified NSA publication), 17.

42   Ibid.

43   Ibid.

44   Smith, *U.S. Marines in Vietnam: High Mobility and Standdown, 1969*, 41.

45   G. C. Fox, Interview by Lt Col Bert Aton.

46   Rayburn, "Direct Support during Operation DEWEY CANYON," 17.

47   Ibid., 20.

48   Ibid., 19.

49   9th Marines, *Combat Operations AAR* (DEWEY CANYON), 37.

50   Medal of Honor Citation is at Smith, *U.S. Marines in Vietnam: High Mobility and Standdown, 1969*, 368. The citation states that the action took place in Quang Tri Province, RVN. That is false, as it occurred in Laos. All of the Marine casualties listed on the Vietnam Memorial from the incursion into Laos are similarly mislabeled.

51   G. C. Fox, Interview by Lt Col Bert Aton.

52   Smith, *U.S. Marines in Vietnam: High Mobility and Standdown, 1969*, 47.

53   Rod Burns, "DEWEY CANYON, 1969-PRAIRIE FIRE!" at http://www.macvsog. cc/dewey_canyon,_1969.htm Accessed 10/5/2019.

54   Ibid.

55   Alan Hoe, *The Quiet Professional: Major Richard J. Meadows of the U.S. Army Special Forces* (University Press of Kentucky, 2011), 99.

56   Ibid., 100–101.

57   Ibid., 103–104.

58   Operational Report of Headquarters, XXIV Corps for Period Ending 30 April 1969. dated 4 June 1969, p. 4.

59   *Project CHECO Southeast Asia Report #2*, 16.

60   Fox, *Marine Rifleman*, 263.

61   Ibid., 256.

62   Ibid., 259.

Chapter 17

1   2nd Brigade, 101st Airborne Division, *Combat Operations AAR, Operation MASSACHUSETTS STRIKER*, 25 May 1969. Sam Johnson Brigade, 101st Airborne Collection, doc F031100081690, 3–4.

2   101 ABN DIV OR-LL, 31 Jul 69, downloaded at www.3-5cavblackknights.org. Accessed 1/5/2020, 2.

3   Ibid.

4   2/17th Cavalry Squadron *Daily Staff Journals*. Unless otherwise noted, material in this section is documented in these journals.

5   *Combat Operations AAR, Operation MASSACHUSETTS STRIKER*, 5.

6   Ibid.

7   *III MAF Periodic Intelligence Report 14-69*, 6 April 1969, 5.

8   Project CHECO Southeast Asia Report #2, 18–19.

9   Capt Albert W. Estes, USAF, FAC, 3rd Bde, 101st Abn, Summary of interview by Lt Col Bert Aton, 8 Aug 69. Sam Johnson Collection, doc F031100090132, p. 4; *The Screaming Eagle,* 12 May 1969, 8, gives the spelling "Friedrich," which is likely correct.

10   *III MAF Periodic Intelligence Report 18-69,* 4 May 1969, 10.

11   Estes, Summary of interview, 4.

12   VHPA, 67-17697; Virtual Wall.

13   VHPA, 65-09428.
14   Estes, Summary of interview, 5.
15   *III MAF Periodic Intelligence Report 18-69,* 4 May 1969, 10–11.
16   Information in this paragraph is from Estes, Summary of interview, 5; VHPA 67-17640 & 67-17647; and the Virtual Wall.
17   VHPA, 67-16385.
18   *The Screaming Eagle,* 12 May 1969, 8.
19   *III MAF Periodic Intelligence Report 18-69,* 4 May 1969, 5.
20   Silver Star citation at homeofheroes.com.
21   *Project CHECO Southeast Asia Report #2,* 19.
22   *Combat Operations AAR, Operation MASSACHUSETTS STRIKER,* 17.
23   Ibid.
24   Report, Department of Defense – re: Organization of activities of the 559th Transportation Group, 13 June 1968, 9.
25   Lt. Frank Hair, "Massachusetts Striker," *Rendezvous with Destiny* (101st Airborne Division magazine), Summer 1969. Barry Wain Collection, doc 28810616035., p. 6.

Chapter 18

1   Zaffiri, *Hamburger Hill.*
2   1st Battalion, 506th Airborne Infantry, *Combat Operations AAR, Operation APACHE SNOW,* 18 Jun 1969. Sam Johnson Collection, doc F031100081508, Encl 11, p. 3.
3   XXIV Corps, *Combat Operations AAR, Operation Apache Snow, 10 May–7 June 1969,* 27 Aug 1969. Bud Harton Collection, doc 168300010547, 9; Smith, *U.S. Marines in Vietnam: High Mobility and Standdown, 1969,* 67–68.
4   3rd Brigade, 101st Airborne Division (Airmobile), *Combat Operations AAR – Summary APACHE SNOW, 10 May–7 June 1969,* 25 Jun 69. Bud Harton Collection, doc 168300010449, 3.
5   1st Battalion, 506th Airborne Infantry, *Combat Operations AAR, Operation APACHE SNOW,* 4–7.
6   3rd Battalion, 187th Airborne Infantry, *Combat Operations AAR, Operation APACHE SNOW 9 May–21 May 1969,* 20 Jun 1969. Sam Johnson Collection, doc F031100081508, Encl 10, Appendix 1.
7   *III MAF Periodic Intelligence Report 19-69,* 11 May 1969, 4.
8   3rd Brigade, 101st Airborne Division (Airmobile), *Combat Operations AAR – Summary APACHE SNOW,* 4.
9   *III MAF Periodic Intelligence Report 20-69,* 18 May 1969, 11.
10   22nd Military History Detachment, *Narrative of Sapper Attack on FSB Airborne,* 13 May 1969. Sam Johnson Collection, doc F031100090122, 1–2.
11   Ibid., 7.
12   Zaffiri, *Hamburger Hill,* 106; 22nd Military History Detachment, *Narrative of Sapper Attack on FSB Airborne,* 5–7.
13   Virtual Wall.
14   Zaffiri, *Hamburger Hill,* 112.

15 22nd Military History Detachment, *Narrative of Sapper Attack on FSB Airborne*, 7–8.

16 Zaffiri, *Hamburger Hill*, 112.

17 XXIV Corps, *Combat Operations AAR, Operation Apache Snow*, 8.

18 Ibid., 9; 22d Military History Detachment, *Narrative – Operation "Apache Snow"*, 101st Airborne – 7 June 1969. Bud Harton Collection, doc 168300010493.

19 *III MAF Periodic Intelligence Reports 20-69*, 18 May 1969, p. 14, *21-69*, 25 May 1969, pp. 3, 7, *23-69*, 7 June 1969, pp. 8–9; XXIV Corps, *Operational Report - Lessons Learned, for Period Ending 31 July 1969*, 3–4.

20 22d Military History Detachment, *Narrative – Operation "Apache Snow."*

21 *III MAF Periodic Intelligence Report 23-69*, 7 June 1969, p. 3.

22 2/17 Cavalry Squadron, *Daily Staff Journal.*

23 Chapters 18 and 19 draw heavily from the following sources: Zaffiri, *Hamburger Hill*; Frank Boccia, *The Crouching Beast* (Jefferson, NC: McFarland & Company, 2013); 22d Military History Detachment, *Narrative – Operation "Apache Snow"*; 3rd Battalion, 187th Airborne Infantry, *Combat Operations AAR, Operation APACHE SNOW*; 1st Battalion, 506th Airborne Infantry, *Combat Operations AAR, Operation APACHE SNOW*.

24 Zaffiri, *Hamburger Hill*; James H. Willbanks, "Hamburger Hill," *Vietnam Magazine*, June 2009, 22–31.

25 Boccia, *Crouching Beast.*

26 Citation quoted at the Virtual Wall.

27 3rd Battalion, 187th Airborne Infantry, *Combat Operations AAR, Operation APACHE SNOW*, Appendix 1 to ANNEX C.

28 Zaffiri, *Hamburger Hill.*

29 22d Military History Detachment, *Narrative – Operation "Apache Snow,"* 7.

30 Zaffiri, *Hamburger Hill*, 95–96; 3rd Battalion, 187th Airborne Infantry, *Combat Operations AAR, Operation APACHE SNOW*. Appendix 3 to ANNEX C.

31 3rd Battalion, 187th Airborne Infantry, *Combat Operations AAR, Operation APACHE SNOW*, Appendix 4 to ANNEX C.

32 Boccia, *Crouching Beast*, 356–360.

33 1st Battalion, 506th Airborne Infantry, *Combat Operations AAR, Operation APACHE SNOW*, 6.

34 Ibid., 7

35 Zaffiri, *Hamburger Hill*, 131.

36 22d Military History Detachment, *Narrative – Operation "Apache Snow,"* 7.

37 Zaffiri, *Hamburger Hill*, 141.

38 3rd Battalion, 187th Airborne Infantry, *Combat Operations AAR, Operation APACHE SNOW*, Para 12a.

39 22d Military History Detachment, *Narrative – Operation "Apache Snow,"* 9–10.

40 Zaffiri, *Hamburger Hill*, 146.

41 Ibid., 150.

42 3rd Battalion, 187th Airborne Infantry, *Combat Operations AAR, Operation APACHE SNOW*, Appendix 5 to ANNEX C.

43 Zaffiri, *Hamburger Hill*, 151.

44 Ibid., 154.

45 1st Battalion, 506th Airborne Infantry, *Combat Operations AAR, Operation APACHE SNOW*, 7–8.

46   3rd Battalion, 187th Airborne Infantry, *Combat Operations AAR, Operation APACHE SNOW,* Appendix 5 to ANNEX C.

47   Zaffiri, *Hamburger Hill*, 156.

48   Ibid., 155–156.

49   Ibid., 158.

50   Boccia, *Crouching Beast*, 402.

51   3rd Battalion, 187th Airborne Infantry, *Combat Operations AAR, Operation APACHE SNOW,* Appendix 6 to ANNEX C.

52   1st Battalion, 506th Airborne Infantry, *Combat Operations AAR, Operation APACHE SNOW*, 8.

53   Ibid., 9.

54   Zaffiri, *Hamburger Hill*, 176.

55   Ibid., 177.

56   William M. Hammond, *Public Affairs: The Military and the Media, 1968–1973* (Washington, DC: Center of Military History, U.S. Army, 1996), 86.

57   William Head, "Battle of 'Hamburger Hill,' May 10–20, 1969: The Beginning of the End of America's Commitment to the Republic of Vietnam," *Virginia Review of Asian Studies* 17 (2015): 106.

58   Willard J. Webb, *The Joint Chiefs of Staff and The War in Vietnam 1969-1970* (Washington, DC: Office of Joint History, Office of the Chairman of the Joint Chiefs of Staff, 2002), 44.

59   Hammond, *Public Affairs: The Military and the Media*, 86–87.

60   Harry G. Summers, "30th Anniversary: Hamburger Hill Revisited," *Vietnam Magazine*, June 1999, 38.

61   Ibid., 42.

62   Joseph B. Conmy, Jr., "Crouching Beast Cornered," *Vietnam Magazine*, August 1990, 33.

63   Quoted in Summers, "30th Anniversary: Hamburger Hill Revisited," 41.

64   James W. McCoy, *Secrets of the Viet Cong* (New York: Hippocrene Books, 1992), 301–313.

65   Ibid., 307.

66   Ibid., 308.

Chapter 19

1   Zaffiri, *Hamburger Hill*, 184.

2   3rd Battalion, 187th Airborne Infantry, *Combat Operations AAR, Operation APACHE SNOW,* Appendix 9 to ANNEX C.

3   Distinguished Service Cross citation.

4   Zaffiri, *Hamburger Hill*, 204.

5   Ibid., 201.

6   22d Military History Detachment, *NARRATIVE – OPERATION "APACHE SNOW,"* 20–21.

7   3rd Battalion, 187th Airborne Infantry, *Combat Operations AAR, Operation APACHE SNOW*, Appendix 11 to ANNEX C.

8   Wiest, *Vietnam's Forgotten Army*, 169–170.

9   Ibid., 171.

10 3rd Battalion, 187th Airborne Infantry, *Combat Operations AAR, Operation APACHE SNOW*, Appendix 11 to ANNEX C.

11 2nd Battalion, 501st Airborne Infantry, *Combat Operations AAR, Operation Apache Snow*, 22 Jun 1969. Sam Johnson Collection, doc F031100081508, Encl 13, p. 3.

12 22d Military History Detachment, *NARRATIVE – OPERATION "APACHE SNOW."*

13 2/17 Cavalry Squadron, *Daily Staff Journal*; Virtual Wall.

14 1st Battalion, 506th Airborne Infantry, *Combat Operations AAR, Operation APACHE SNOW.*

15 Virtual Wall.

16 *III MAF Periodic Intelligence Report 21-69*, 25 May 1969, 3.

17 2nd Battalion, 506th Airborne Infantry, *Combat Operations AAR, OPORD 3-69 (APACHE SNOW)*, 20 Jun 1969. Sam Johnson Collection, doc F031100081508, Encl 12.

18 XXIV Corps, *Combat Operations AAR, Operation Apache Snow*, 3, 10.

19 Ibid., 9–10.

20 22d Military History Detachment, *NARRATIVE – OPERATION "APACHE SNOW,"* 27–28.

21 *III MAF, INCOMING MESSAGES*, 18 Jun 1969, Box __, Folder 006, USMC History Division Collection, doc 1201006026, Msg, SSO XXIV CORPS TO CO III MAF, 180336Z.

22 Joseph H. Conmy, Jr., "I Led a Brigade at Hamburger Hill," *The Washington Post*, 27 May 1989.

23 Sorley, *Vietnam Chronicles*, p. 187.

24 Ibid., 15, 112, 135.

25 DEPCOMUSMACV (Abrams) msg to CJCS (Wheeler, Chairman, Joint Chiefs of Staff), *re: protect Dak To from by NVA 1st Division, Kontum battlefield*, 22 Nov 1967. Veteran Members of the 109th Quartermaster Company (Air Delivery) Collection, doc 0010105003.

26 Edward F. Murphy, *Dak To* (Novato, CA: Presidio Press, 1993), 301–304.

27 DEPCOMUSMACV (Abrams) msg to CJCS (Wheeler, Chairman, Joint Chiefs of Staff), *re: protect Dak To from by NVA 1st Division*.

28 Sorley, *Vietnam Chronicles*, 199.

29 Ibid., 394–395.

30 See Combined Intelligence Center, Vietnam, *Vol I, ORDER OF BATTLE SUMMARY 31 MAY 1969*. Sam Johnson Collection, doc F0159000607147, p. IV-35; *1-31 JUNE 1969*. Sam Johnson Collection, doc F015900060420, p. IV-35; *1-31 JULY 1969*. Sam Johnson Collection, doc F015900060702, pp. IV-35–36.

31 Conmy, "I Led a Brigade at Hamburger Hill."

32 Willbanks, "Hamburger Hill," 30.

33 Lewis Sorley, *Thunderbolt* (New York: Simon & Schuster, 1992), 261.

34 Summers, "30th Anniversary: Hamburger Hill Revisited," 38.

35 Willbanks, "Hamburger Hill," 31.

Chapter 20

1   101 ABN DIV OR–LL, 31 Jul 69, 7.

2   Engineer Troops Vietnam, *The KYSU,* Vol. 1 No. 3, Fall 1969. James Bussey Collection, doc 10720101012, 6–9.

3   3/5 CAV Memo 9 Jun 69 at www.3-5cavblackknights.org. Accessed 1/5/2020.

4   2/17 Cavalry Squadron, *Daily Staff Journal.*

5   Silver Star citation.

6   *III MAF COMMAND CHRONOLOGY, June 1969.* USMC History Division Collection, doc 1201006028, 15; *MACV Monthly Summary, June 1969.* John M. Shaw Collection, doc 7390115001, 37.

7   2/17 Cavalry Squadron, *Daily Staff Journals.*

8   Ibid., Virtual Wall.

9   101 ABN DIV OR–LL, 31 Jul 69, 40–41.

10 2/17 Cavalry Squadron, *Daily Staff Journal.*

11 http://www.charlie1-506.com/html/pages%20from%20main%20nav/Create
Space%20Charlie%20Co.htm Accessed 12/22/2019.

12 101 ABN DIV OR–LL, 31 Jul 69, 9.

13 *MACV Monthly Summary, June 1969*, 48.

14 Virtual Wall.

15 Distinguished Service Cross citation.

16 Headquarters, 3d Brigade, 101st Airborne Division: Sapper Attack – FB Berchtesgaden, 4 August 1969.

17 101 ABN DIV OR–LL, 31 Jul 69, 9–10.

18 *III MAF Periodic Intelligence Report 25-69*, 21 Jun 1969, 3, 12.

19 *Monthly Summary, June 1969*, 57; *III MAF Periodic Intelligence Report 25-69*, 21 Jun 1969, 3.

20 http://www.charlie1-506.com

21 Engineer Troops Vietnam, *The KYSU,* Fall 1969, 9.

22 www.3-5blackknights.org

23 http://www.charlie1-506.com; 2/17 CAV DSJ.

24 2/17 Cavalry Squadron, *Daily Staff Journals.*

25 Ibid.

26 1st Battalion, 506th Airborne Infantry, *Combat Operations AAR, Operation APACHE SNOW*, 4.

27 http://old *MACV*.506infantry.org/hisvietnam/his1stbnvnarticle32.html

28 Silver Star citation.

29 http://old.506infantry.org/hisvietnam/his1stbnvnarticle32.html

30 Medal of Honor citation.

31 *Roberts, Gordon, Medal of Honor Award Case File,* 11 Jul 1969. Tim Frank Collection, doc 18700263001.

32 *MACV Monthly Summary, July 1969.* John M. Shaw Collection, doc 7390201001, 25.

33 101 ABN DIV OR–LL, 31 Jul 69, 4.

34 *III MAF Periodic Intelligence Report 29-69*, 22 Jul 1969, 7–8.

35 *III MAF COMMAND CHRONOLOGY, July1969.* USMC History Division Collection, doc 1201006048, 10.

36 *MACV Monthly Summary, July 1969*, 36.

37 101 ABN DIV OR-LL, 31 Jul 69, 7.

38 Medal of Honor citation.

39 *MACV Monthly Summary, July 1969*, 38.

40 Ibid., 40.

41 Silver Star citation.

42 *MACV Monthly Summary, July 1969*, 49.

43 *III MAF Periodic Intelligence Report 29-69*, 22 Jul 1969, 12.

44 *III MAF Periodic Intelligence Report 30-69*, 29 Jul 1969, 3.

45 101 ABN DIV OR-LL, 31 Oct 69, downloaded at www.3-5cavblackknights.org. Accessed 1/5/2020, p. 14.

46 2/17 Cavalry Squadron, *Daily Staff Journals*.

47 *MACV Monthly Summary, August 1969*. Bud Harton Collection, doc 168300010776, 12; 101 ABN DIV OR-LL, 31 Oct 69, 3.

48 *MACV Monthly Summary, August 1969*, 12.

49 *III MAF Periodic Intelligence Report 32-69*, 12 Aug 1969, 3.

50 http://www.charlie1-506.com/

51 *MACV Monthly Summary, August 1969*, 19; *III MAF, COMMAND CHRONOLOGY, August 1969*. USMC History Division Collection, doc 1201007008.

52 Virtual Wall.

53 *MACV Monthly Summary, August 1969*, 27; *III MAF, COMMAND CHRONOLOGY, August 1969*.

54 *MACV Monthly Summary, August 1969*, 236–237; *III MAF, COMMAND CHRONOLOGY, August 1969*.

55 http://www.charlie1-506.com/

56 *III MAF Periodic Intelligence Report 33-69*, 9 Aug 1969, 3.

57 Combat Operations After Action Report re: Operation Massachusetts Striker, 3.

58 22d Military History Detachment, *NARRATIVE – OPERATION "APACHE SNOW"*.

59 Prados, *Blood Road*, 347.

60 *Project CHECO Southeast Asia Report #2*, 6–7.

61 XXIV Corps, *Operational Report - Lessons Learned, for Period Ending 31 July 1969*, 4–5.

62 *Project CHECO Southeast Asia Report #2*, 11.

Chapter 21

1 101 ABN DIV OR-LL, 31 Oct 69, 7.

2 Ibid., 14–16.

3 Virtual Wall.

4 *MACV Monthly Summary, August 1969*, 68; http://www.charlie1-506.com/html/pages%20from%20main%20nav/CreateSpace%20Charlie%20Co.htm, 53; Virtual Wall.

5 101 ABN DIV OR-LL, 31 Oct 69, 7, 10; Virtual Wall.

6 *III MAF Periodic Intelligence Report 35-69,* 2 Sep 1969, 2.

7 http://www.charlie1-506.com/html/pages%20from%20main%20nav/CreateSpace%20Charlie%20Co.htm, 54–55.

8   2/17 Cavalry Squadron, Daily Staff Journals.
9   3/5 Cav AAR, OPN Louisiana Lee at www.3-5cavblackknights.org.Accessed 1/5/2020.
10 Virtual Wall
11 *III MAF Periodic Intelligence Report 34-69,* 26 Aug 1969, 3.
12 VHPA, 67-16254.
13 3/5 Cav AAR, OPN Louisiana Lee.
14 Ibid.; Virtual Wall.
15 3/5 Cav AAR, OPN Louisiana Lee.
16 Ibid.
17 2/17 Cavalry Squadron, Daily Staff Journals.
18 Ibid.
19 Ibid.
20 101 ABN DIV OR-LL, 31 Oct 69, 7.
21 Ibid., 47.
22 Ibid., 26.
23 *III MAF Periodic Intelligence Report 38-69*, 23 Sep 1969, 3.
24 101 ABN DIV OR-LL, 31 Oct 69, p. 47.
25 XXIV Corps, *Operational Report for Period Ending 31 October 1969*, 22.
26 *Project CHECO Southeast Asia Report #2*, 8.
27 Ibid., 39
28 Ibid., 22.
29 Smith, *U.S. Marines in Vietnam, High Mobility and Standdown, 1969*, 133.
30 Ibid., 135.
31 Sorley, *Vietnam Chronicles*, 223–224.
32 MG John M. Wright, Jr., CG 101st Airborne Division (Airmobile), Period 25 May 1969 to 25 May 1970. Senior Officer Debriefing Report. Distributed by NTIS. Defense Technical Information Center, 8.
33 Ibid., 17.

Chapter 22

1   2/17 Cavalry Squadron, *Daily Staff Journals.*
2   XXIV Corps, *Operational Report for Period Ending 31 Oct 1969*, 4.
3   XXIV Corps, *Operational Report for Period Ending 31 January 1970*, 17 Feb 1970, 3.
4   2/17 Cavalry Squadron, *Daily Staff Journals.*
5   Michael L. Lanning and Ray W. Stubbe, *Inside Force Recon* (New York: Ivy Books, 1989), 178–179.
6   Ibid., 180–181.
7   Ibid.
8   Bruce H. Norton, *Force Recon Diary, 1969* (New York: Ballantine Books, 1991), 182.
9   2/17 Cavalry Squadron, *Daily Staff Journals.*
10 3rd FORCE RECONNAISSANCE COMPANY, III MAF, *Command Chronology, December 1969*. USMC History Division Collection, doc 1201025002.

11 Graham A. Cosmas and Terrence P. Murray, *U.S. Marines in Vietnam: Vietnamization and Redeployment, 1970-1971* (Washington, DC: History and Museums Division, Headquarters, U.S. Marine Corps, 1986), 307.

12 3rd FORCE RECONNAISSANCE COMPANY, III MAF, *Command Chronology, December 1969.*

13 Ibid.

14 Ibid.

15 Ibid.

16 HQS, 101 Airborne Division (Airmobile) Combat Operations After Action Report, RANDOLPH GLEN, 12.

17 3rd FORCE RECONNAISSANCE COMPANY, III MAF, *Command Chronology, December 1969;* Alex Lee, *Force Recon Command* (New York, NY: Ivy Books, 1995), 124–139.

18 XXIV Corps, *Operational Report for Period Ending 31 January 1970.* 17 Feb 1970, 3.

19 3rd FORCE RECONNAISSANCE COMPANY, III MAF, *Command Chronology, January 1970.* USMC History Division Collection, doc 1201025003.

20 Ibid.

21 VHPA, 67-17349; Virtual Wall.

22 3rd FORCE RECONNAISSANCE COMPANY, III MAF, *Command Chronology, January 1970.*

23 Ibid.

24 Ibid.

25 Ibid.

26 Ibid.

27 Ibid.

28 Ibid.

29 XXIV Corps, *Operational Report for Period Ending 31 January 1970,* 4.

30 Navy Cross Citation

31 Virtual Wall.

32 2/17 Cavalry Squadron, *Daily Staff Journals.*

33 Ibid.; Norton, *Force Recon Diary, 1969,* 271–281.

34 2/17 Cavalry Squadron, *Daily Staff Journals.*

35 See Lee, *Force Recon Command,* for a discussion of this event, which took place in late February, probably to the north of the A Shau Valley.

36 2/17 Cavalry Squadron, *Daily Staff Journals.*

37 Lanning, *Inside Force Recon,* 183; North, *Force Recon Diary,* 1969, 289.

38 Lee, *Force Recon Command,* 232.

39 2/17 Cavalry Squadron, *Daily Staff Journals;* 101 Airborne Division (Airmobile) Combat Operations After Action Report, RANDOLPH GLEN, 15; VHPA, 66-17772.

40 *"The OV-1 Mohawk Remembered Firsthand: Piloting the Mohawk in Vietnam,"* at https://www/defensemedianetwork.com/stories. Accessed 2/18/2020.

41 2/17 Cavalry Squadron, *Daily Staff Journals.*

42 Jack S. Ballard, *Development and Employment of Fixed-Wing Gunships, 1962-1972* (Washington, DC: Office of Air Force History, United States Air Force, 1982).

43   2/17 Cavalry Squadron, *Daily Staff Journals.*
44   101 Airborne Division (Airmobile) *Combat Operations After Action Report, RAN-DOLPH GLEN,* 19.
45   Frank Johnson, *Diary of an Airborne Ranger* (New York: Ballantine Books, 2001), 111.
46   2/17 CAV DSJ.
47   101 Airborne Division (Airmobile) Combat Operations After Action Report, RAN-DOLPH GLEN HQS, 30.
48   Ibid., 83.

Chapter 23

1   U.S.G. Sharp and W.C. Westmoreland, *Report on the War in Vietnam (As of 30 June 1968)* (Washington, DC: U.S. Government Printing Office, 1968), 178.
2   Benjamin L. Harrison, *Hell on a Hill Top,* (Lincoln, NE: iUniverse, 2004), 21.
3   Cosmas, *U.S. Marines in Vietnam: Vietnamization and Redeployment,* 45.
4   XXIV Corps, *Operational Report for Period Ending 31 July 1970.* 1 Dec 1970, 12.
5   VHPA, 67-16229; www.armyaircrews.com/cayuse.html.
6   http://www.vietnamproject.ttu.edu/banshee/History/TexStar70.html. Accessed 2/8/2020.
7   2/17 Cavalry Squadron, *Daily Staff Journals.*
8   Hobson, *Vietnam Air Losses,* 202.
9   2/17 Cavalry Squadron, *Daily Staff Journals.*
10   Ibid.
11   Ibid.
12   Ibid.;   http://www.vietnamproject.ttu.edu/banshee/History/TexStar70.html. Accessed 2/8/2020.
13   XXIV Corps, *Operational Report for Period Ending 31 July 1970.*
14   Ibid., 22, 26.
15   Ibid., 3.
16   *The 324B Division,* Hanoi, Vietnam: People's Army Publishing House, 1992. Quoted in Harrison, *Hell on a Hilltop,* 67–68.
17   Harrison, *Hell on a Hilltop,* 72, 171–172.
18   John G. Roberts, *Operation Texas Star: The Last Battles of the American War* (John G. Roberts, 2017), 24.
19   Ibid., 97.
20   Ibid., 105.
21   Ibid., 400.
22   David B. Sigler, *Vietnam Battle Chronology* (Jefferson, NC: McFarland & Company, 1992), 114–115.
23   Harrison, *Hell on a Hilltop,* 65.
24   XXIV Corps, *Operational Report for Period Ending 31 July 1970,* 4.
25   Keith W. Nolan, *Ripcord,* (Novato, CA: Presidio Press, 2000).
26   Roberts, *Operation Texas Star.*
27   Harrison, *Hell on a Hill Top.*
28   Roberts, *Operation Texas Star,* 31–41.
29   Ibid., 387.
30   Ibid., 233.

31 Ibid., 23.
32 Nolan, *Ripcord*, 14.
33 Roberts, *Operation Texas Star*, 393.
34 Nolan, *Ripcord*, 3.
35 Hobson, *Vietnam Air Losses*, 207.
36 2/17 Cavalry Squadron, *Daily Staff Journals*.
37 XXIV Corps, *Operational Report for Period Ending 31 July 1970,* 9–10.
38 Ibid., 3.
39 Lewis Sorley, *A Better War* (New York: Harcourt Brace, 1999), 217.

Chapter 24

1   XXIV Corps, *Operational Report – Lessons Learned, Period Ending 30 April 1971.* 17 May 1971, 13.
2   *MACV Significant R&D Problem Areas Report (RCS: CICPAC 3960, 2)*, 1 Oct 1969. Sam Johnson Collection, doc F015800220516, Incl. 7.
3   2nd Squadron, 17th Cavalry Regiment, *Unit History for 1970*, 15 Jan 1971. James Matthews Collection, doc 5440123001.
4   2/17 Cavalry Squadron, *Daily Staff Journals*.
5   Ibid.
6   Ibid.
7   Ibid.
8   2nd Squadron, 17th Cavalry Regiment, *Unit History for 1970.*
9   101 ABN DIV OR-LL, 30 Apr 71, downloaded at www.3-5cavblackknights.org. Accessed 1/5/2020, pp. 62, 94.
10 2nd Squadron (Ambl), 17th Cavalry, *Combat Operations AAR, Operation JEFFERSON GLEN*, 13 Oct 1971. James Matthews Collection, doc 5440114001; C Troop, 2/17th Cavalry, *Summary of Activity 3 August 1971*, 04 Aug 1971. Michael Sloniker Collection, doc 8850601003.
11 2nd Squadron (Ambl), 17th Cavalry, *Combat Operations AAR, Operation JEFFERSON GLEN.*
12 VHPA, 67-16161.
13 Keith W. Nolan, *Into Laos* (Novato, CA: Presidio Press, 1986), 12–15.
14 Ibid., 33, 103, 271.
15 Robert D. Sander, *Invasion of Laos, 1971* (Norman, OK: University of Oklahoma Press, 2014), 105.
16 Ibid.
17 101 ABN DIV OR-LL, 30 Apr 71, 15.
18 Ibid., 74, 76, 101.
19 Sander, *Invasion of Laos*, 180.
20 101 ABN DIV OR-LL, 30 Apr 71, 18.
21 VHPA, 66-17765.
22 2nd Squadron (Ambl), 17th Cavalry, *Combat Operations AAR, Operation JEFFERSON GLEN.*
23 Gillespie, *Black Ops Vietnam*, 228.
24 Plaster, *SOG: The Secret Wars of America's Commandos in Vietnam*, 320.
25 VHPA, 68-15255.

26   Hobson, *Vietnam Air Losses*, 212.

27   Plaster, *SOG: The Secret Wars of America's Commandos in Vietnam*, 320–323; Gillespie, *Black Ops Vietnam*, 228–229; Yarborough, *A Shau Valor*, 243–259.

28   2/17 Cavalry Squadron, *Daily Staff Journals*.

29   https://www.75thrra.org/history/175_hx.html, Accessed 3/6/2020.

30   Virtual Wall.

31   Ibid.; Gary A. Linderer, *Six Silent Men: 101st LRP/Rangers, Book Three* (New York, NY: Ivy Books, 1997), 275–280, but the insertion date and location given in this account are incorrect.

32   James H. Willbanks, *A Raid Too Far* (College Station, TX: Texas A&M University Press, 2014), 132.

33   Sorley, *Vietnam Chronicles: The Abrams Tapes*, 577.

34   Linderer, *Six Silent Men: 101st LRP/Rangers, Book Three*, 281–284; 2/17 Cavalry Squadron, *Daily Staff Journals*; Virtual Wall.

35   VHPA, 68-15753; Distinguished Service Cross Citation.

36   Linderer, *Six Silent Men: 101st LRP/Rangers, Book Three*, 287–293; C Troop, 2/17th Cavalry, *Summary of Activity 3 August 1971*; Virtual Wall.

37   Linderer, *Six Silent Men: 101st LRP/Rangers, Book Three*, 295; C Troop, 2/17th Cavalry, *Summary of Activity 3 August 1971*; 2/17 Cavalry Squadron, *Daily Staff Journals*.

38   101 ABN DIV OR-LL, 30 Apr 71, 111.

39   Ibid., 112.

40   *LRP/Rangers, Book Three,* pp. 304-315; C Troop, 2/17th Cavalry, *Summary of Activity 3 August 1971*; 2/17 Cavalry Squadron, *Daily Staff Journals*; VHPA 66-16588 & 68-16199; Virtual Wall.

41   C Troop, 2/17th Cavalry, *Summary of Activity 3 August 1971*.

42   101 ABN DIV OR-LL, 30 Apr 71, 112.

43   Task Force Omega at http://taskforceomegainc.org/

44   2/17 Cavalry Squadron, *Daily Staff Journals*.

45   Linderer, *Six Silent Men: 101st LRP/Rangers, Book Three*, 295–301.

46   Virtual Wall.

47   LTC Lloyd N. Cosby, Co, 2/502 Infantry (Airborne), in *The Screaming Eagle*, "Vietnam Eagles: Medal of Honor," January–February 1999. Gary Jestes Collection, doc 1160101022; VHPA, 68-16303; Virtual Wall.

48   LTC Lloyd N. Cosby, Co, 2/502 Infantry (Airborne), in *The Screaming Eagle*, "Vietnam Eagles: Medal of Honor."

49   2nd Squadron (Ambl), 17th Cavalry, *Combat Operations AAR, Operation JEFFERSON GLEN*.

50   2/17 Cavalry Squadron, *Daily Staff Journals*.

51   2nd Squadron, (Ambl), 17th Cavalry, *Combat Operations AAR, Operation JEFFERSON GLEN*; Virtual Wall. The official records report only one KIA, but two are listed on the Wall.

52   2nd Squadron (Ambl), 17th Cavalry, *Combat Operations AAR, Operation JEFFERSON GLEN*.

53   VHPA, 67-16670.

54   *Pacific Stars & Stripes*, Viet Marines Kill 200 Reds, 15 May 1971. Douglas Pike Collection, doc 2131804084.

55   https://www.75thrra.org/history/175_hx.html; 2/17th Cavalry Squadron, *Commander's Situation Reports. Michael Sloniker Collection.*

56   Yarborough, *A Shau Valor*, 262;VHPA.

57   Hobson, *Vietnam Air Losses*, 214.

58   Wiest, *Vietnam's Forgotten Army*, 234–235.

59   2/17th Cavalry Squadron, *Commander's Situation Reports.*

60   2nd Squadron (Ambl), 17th Cavalry, *Combat Operations AAR, Operation JEFFERSON GLEN.*

61   Ibid., 2/17th Cavalry Squadron, *Commander's Situation Reports*

62   2nd Squadron (Ambl), 17th Cavalry, *Combat Operations AAR, Operation JEFFERSON GLEN*; 2/17th Cavalry Squadron, *Commander's Situation Reports.*

63   VHPA;Virtual Wall.

64   *History of the 2nd Squadron 17th Cavalry Regiment from 1916-1976*, No Date. James Matthews Collection, doc 544011900, 25.

Chapter 25

1   NPIC MSG 032257Z APR 70

2   CIA Msg 14 May 1970, 4.

3   Intelligence Memorandum, The Growth and Current Deployment of the Laotian-Based 559th Transportation Group. ER IM 71-25 February 1971,Table 1.

4   Intelligence Memorandum, Recent Construction on the Ho Chi Minh Trail. ERIM 71-185 September 1971, 6.

5   Ibid., 6–8.

6   COL Hillman Dickinson, Sr. Advisor, 1st ARVN Infantry Division, *Memorandum. re: Analysis of TFA Sensor Data*, 26 Mar 1972. Dale W. Andrade Collection, doc 24992209005.rd

7   *MACDI Study 73-01, The Nguyen Hue Offensive,* 12 Jan 1973. Dale W. Andrade Collection, doc 24991801005, see map on p. 10.

8   First Regional Assistance Command, 1st Infantry Division (ARVN), *Significant Activities Report*, 15 Apr 1972. Dale W. Andrade Collection, doc 24992207008, 2–3.

9   David Biggs, *Footprints of War* (Seattle, WA: University of Washington Press, 2018), 179.

10   See, e.g., Dale Andrade, *Trial by Fire* (New York: Hippocrene Boks, 1995), 57; First Regional Assistance Command, *Command History*, 1972–1973, 28 Mar 1973. Dale W. Andrade Collection, doc 24991612002, 5, 11.

11   MACV *Command History - Annex L: Quang Tri and Hue*, 14 Oct 1975. Dale W. Andrade Collection, doc 24992506004, p. L-5.

12   Kroesen, MG, Acting CG XXIV Corps, msg to GEN ABRAMS, re: *Commanders Daily Evaluation*, 19 Mar 1972. Dale W. Andrade Collection, doc 24991703009.

13   First Regional Assistance Command, 1st Infantry Division (ARVN), *Significant Activities Report*, 15 Apr 1972, 6.

14   Kroesen, MG, Acting CG XXIV Corps, msg to GEN ABRAMS, re: *Commanders Daily Evaluation*, 19 Mar 1972.

15   Kroesen, MG, Acting CG XXIV Corps, msg to GEN ABRAMS, re: *Commanders Daily Evaluation*, 22 Mar 1972. Dale W. Andrade Collection, doc 24991703010.

16 MACV *Command History – Annex L: Quang Tri and Hue*, L-5.
17 Truong, Lt. Gen. Ngo Quang, *The Easter Offensive of 1972*, Indochina Monographs, U.S. Army Center of Military History. Garnett Bell Collection, doc 1127171100, 48.
18 Ibid., 49–50.
19 First Regional Assistance Command, *Command History*, 1972–1973, 28, 55.
20 MACV *Command History – Annex L: Quang Tri and Hue*, L-31.
21 1st ARVN Division, *Daily Staff Journal, 20 Apr 1972. TFA Sensor Analysis for the period 14–20 April 1972*. Dale W. Andrade Collection, doc 24992207004.
22 Millener, COL George A., Jr., Senior Advisor, 1st Infantry Division (ARVN), period 20 May 1972 to 29 January 1973, Senior Officer Debriefing Report. Distributed by NTIS. Defense Technical Information Center, 6.
23 Intelligence Memorandum: Recent Communist Logistical and Manpower Developments in Indochina. ER IM 73-19-12 1 May 1973, 2.
24 TCS No. 3263/73: North Vietnamese Capabilities and Intentions Through the Rainy Season and Beyond, B-2.
25 Intelligence Memorandum: Recent Communist Logistical and Manpower Developments in Indochina. ER IM 73-19-13 8 May 1973, 2–3.
26 Ibid.
27 TCS No. 3263/73: North Vietnamese Capabilities and Intentions Through the Rainy Season and Beyond, B-3.
28 Twentieth Report: COMMUNIST VIOLATIONS OF THE VIETNAM AND LAOS SETTLEMENT AGREEMENTS. 6 July 1973.
29 NPIC 15498 202054Z SEP 73.
30 CIA document (unnamed) dated 16 July 1973
31 William E. Le Gro, *Vietnam from Cease-Fire to Capitulation* (Washington, DC: U.S. Army Center of Military History, Second Printing, 1985), 125–131.
32 CIA MEMORANDUM: South Vietnam: A Net Military Assessment. 2 April 1974, 3.
33 Ibid.
34 Le Gro, *Vietnam from Cease-Fire to Capitulation*, 155–161.

Conclusion

1 McCoy, *Secrets of the Viet Cong*, 52.
2 Ibid., 54.
3 Bruce E. Jones, *War Without Windows* (New York: Vanguard Press, 1987), 48.
4 John L. Plaster, *SOG: A Photo History of the Secret War* (Boulder, CO: Paladin Press, 2000), 270–271.

Appendix

1 *FM 7-20 Department of the Army Field Manual: Infantry, Airborne Infantry, and Mechanized Infantry Battalions* (Washington, DC: Headquarters, Department of the Army, January 1962).
2 Ibid., 98–99.

3 *FM 31-50 Department of the Army Field Manual: Combat in Fortified and Built-up Areas* (Washington, DC: Headquarters, Department of the Army, March 1964, as modified 22 April 1970).

4 Ibid., 12.

5 *FM 61-100 Department of the Army Field Manual: The Division* (Washington, DC: Department of the Army, June 1965), 141.

BIBLIOGRAPHY

BOOKS

The 324B Division. Hanoi, Vietnam: People's Army Publishing House, 1992.

Coleman, J.D. editor-in-chief. *First Cavalry Division Vietnam, August 1965 to December 1969*. Paducah, KY: Turner Publishing, 1995.

Adkins, Bennie G., and Katie L. Jackson. *A Tiger Among Us*. New York: Da Capo Press, 2018.

Andrade, Dale. *Trial by Fire*. New York: Hippocrene Boks, 1995.

Archer, Chalmers, Jr. *Green Berets in the Vanguard, Inside Special Forces, 1953-1963*. Annapolis, MD: Naval Institute Press, 2001.

Baker, Charles. *Gray Horse Troop: Forever Soldiers*, 2nd rev. Powder River Publications, 2014.

Biggs, David. *Footprints of War*. Seattle, WA: University of Washington Press, 2018.

Boccia, Frank. *The Crouching Beast*. Jefferson, NC: McFarland & Company, 2013.

Bows, Ray A. *Vietnam Military Lore: Legends, Shadows, and Heroes*. Hanover, MA: Bows & Sons, 1997.

Carpenter, Stephen A. *Boots on the Ground: The History of Project Delta*. Stephen A. Carpenter, 2010.

Chambers, Larry. *Death in the A Shau Valley: L Company LRRPs in Vietnam, 1969-1970*. New York: Ivy Books, 1998.

Clark, Gregory R., editor. *Quotations on the Vietnam War*. Jefferson, NC: McFarland, 2001.

Colby, William. *Lost Victory*. Chicago, IL: Contemporary Books, 1989.

Davidson, Phillip B. *Vietnam at War*. Novato, CA: Presidio Press, 1988.

Dunnigan, James F., and Albert A. Nofi. *Dirty Little Secrets of the Vietnam War*. New York: St. Martin's Press, 1999.

Fox, Wesley L. *Marine Rifleman: Forty-three Years in the Corps*. Washington, DC: Brassey's, 2002.

Gillespie, Robert M. *Black Ops Vietnam: The Operational History of MACVSOG*. Annapolis, MD: Naval Institute Press, 2011.

Hammel, Eric. *Fire in the Streets: The Battle for Hue, Tet 1968*. Chicago: Contemporary Books, 1991.

Harrison, Benjamin L. *Hell on a Hill Top*. Lincoln, NE: iUniverse, 2004.

Hastings, Max. *Vietnam: An Epic Tragedy, 1945-1975*. New York: HarperCollins, 2018.

Hennessy, Michael A. *Strategy in Vietnam: The Marines and Revolutionary Warfare in I Corps, 1965-1972.* Westport, CT: Praeger, 1997.

Hobson, Chris. *Vietnam Air Losses.* Hinckley, England: Midland Publishing, 2001.

Hoe, Alan. *The Quiet Professional: Major Richard J. Meadows of the U.S. Army Special Forces.* University Press of Kentucky, 2011.

Johnson, Frank. *Diary of an Airborne Ranger.* New York: Ballantine Books, 2001.

Jones, Bruce E. *War Without Windows.* New York: Vanguard Press, 1987.

Karnow, Stanley. *Vietnam: A History.* New York: Viking Press, 1983.

Kelley, Michael P. *Where We Were in Vietnam.* Central Point, OR: Hellgate Press, 2002.

Lee, Alex. *Force Recon Command.* New York: Ivy Books, 1995.

Lanning, Michael L., and Ray W. Stubbe. *Inside Force Recon.* New York: Ivy Books, 1989.

Linderer, Gary A. *Six Silent Men: 101st LRP/Rangers, Book Three.* New York: Ivy Books, 1997.

———. *The Eyes of the Eagle: F Company LRPs in Vietnam, 1968.* New York: Ivy Books, 1991.

McCoy, James W. *Secrets of the Viet Cong.* New York: Hippocrene Books, 1992.

Miller, Kenn. *Six Silent Men: 101st LRP/Rangers, Book Two.* New York: Ivy Books, 1997.

Morris, R.C. *The Ether Zone: U.S. Army Special Forces Detachment B-52, Project Delta.* Ashland, OR: Hellgate Press, 2009.

Murphy, Edward F. *Dak To.* Novato, CA: Presidio Press, 1993.

Nesser, John A. *The Ghosts of Thua Thien.* Jefferson, NC: McFarland, 2008.

Newhall, Charles W., III. *Fearful Odds: A Memoir of Vietnam and Its Aftermath.* Owingsmills, MD: Bibliotheca Brightside, 2015.

Nolan, Keith W. *Battle of Hue: Tet 1968.* Novato, CA: Presidio Press, 1983.

———. *Into Laos.* Novato, CA: Presidio Press, 1986.

———. *Ripcord.* Novato, CA: Presidio Press, 2000.

Norton, Bruce H. *Force Recon Diary, 1969.* New York: Ballantine Books, 1991.

Oberdorfer, Don. *Tet!* Garden City: Doubleday & Company, 1971.

Plaster, John L. *SOG: The Secret Wars of America's Commandos in Vietnam.* New York: Simon & Schuster, 1997.

———. *SOG: A Photo History of the Secret War.* Boulder, CO: Paladin Press, 2000.

Powell, Colin L., with Joseph E. Persico. *My American Journey.* New York: Random House, 1995.

Prados, John. *Blood Road: The Ho Chi Minh Trail and the Vietnam War.* New York: John Wiley & Sons, 1999.

———. *Vietnam: The History of an Unwinnable War.* Lawrence, KS: University Press of Kansas, 2009.

Roberts, John G. *Operation Texas Star: The Last Battles of the American War.* John G. Roberts, 2017.

Sander, Robert D. *Invasion of Laos, 1971.* Norman, OK: University of Oklahoma Press, 2014.

Schandler, Herbert Y. *America in Vietnam.* Lanham, MD: Rowman & Littlefield, 2009.

Shelton, Hugh. *Without Hesitation: The Odyssey of an American Warrior.* New York: St. Martin's Press, 2010

Sigler, David B. *Vietnam Battle Chronology.* Jefferson, NC: McFarland & Company, 1992.

Simpson, Charles M., III. *Inside the Green Berets: The First Thirty Years.* Novato, CA: Presidio Press, 1983.

Sorley, Lewis. *Thunderbolt: General Creighton Abrams and the Army of His Times.* New York: Simon & Schuster, 1992.

Sorley, Lewis. *A Better War.* New York: Harcourt Brace, 1999.

———. *Thunderbolt. From the Battle of the Bulge to Vietnam and Beyond: General Creighton Abrams and the Army of His Times.* New York: Simon & Schuster, 1992.

———. *Vietnam Chronicles: The Abrams Tapes, 1968-1972.* Lubbock Texas: Texas Tech University Press, 2004.

Spector, Ronald H. *After Tet: The Bloodiest Year in Vietnam.* New York: The Free Press, 1993.

Stanton, Shelby L. *Green Berets at War: U.S. Army Special forces in Southeast Asia, 1958-1975.* Novato, CA: Presidio Press, 1985.

———. *Anatomy of a Division: The 1st Cav in Vietnam.* Novato, CA: Presidio Press, 1987.

Vetter, Lawrence C., Jr. *Never Without Heroes: Marine 3rd Reconnaissance Battalion in Vietnam, 1965-1970.* New York: Ivy Books, 1996.

Westmoreland, William C. *A Soldier Reports.* Garden City: Doubleday & Co, 1976.

Wiest, Andrew. *Vietnam's Forgotten Army: Heroism and Betrayal in the ARVN.* New York: New York University Press, 2008.

Willbanks, James H. *A Raid Too Far.* College Station, TX: Texas A&M University Press, 2014.

Yarborough, Thomas R. *A Shau Valor: American Combat Operations in the Valley of Death, 1963-1971.* Havertown, PA: Casemate Publishers, 2016.

Zaffiri, Samuel. *Hamburger Hill.* Novato, CA: Presidio Press, 1988.

———. *Westmoreland.* New York: William Morrow, 1994.

MILITARY AND GOVERNMENT PUBLICATIONS

Area Handbook for South Vietnam. DA Pam No. 550-55, April 1967. Washington, DC: U.S. Government Printing Office.

Buckingham, William A. *Operation Ranch Hand: The Air Force and Herbicides in Southeast Asia 1961-1971*. Washington, DC: Office of Air Force History, United States Air Force, 1982.

Fails, William R. *Marines and Helicopters, 1962-1973*. Washington, DC: History and Museums Division, Headquarters, USMC, 1978.

FM 7-20 Department of the Army Field Manual: Infantry, Airborne Infantry, and Mechanized Infantry Battalions. Washington, DC: Headquarters, Department of the Army, January 1962.

FM 31-50 Department of the Army Filed Manual: Combat in Fortified and Built-up Areas, Washington, DC: Headquarters, Department of the Army, March 1964 (as modified 22 April 1970).

FM 61-100 Department of the Army Field Manual: The Division, Washington, DC: Department of the Army, June 1965.

Foreign Relations of the United States, 1964-1968, Volume XXVIII, LAOS.

Hammond, William M. *Public Affairs: The Military and the Media, 1968-1973*, Washington, DC: Center of Military History, U.S. Army, 1996.

Kelly, Francis J. *U.S, Army Special Forces, 1961-1971*. Washington, DC: Department of the Army, 1973.

Le Gro, William E., *Vietnam from Cease-Fire to Capitulation*. Washington, DC: U.S Army Center of Military History, 1985.

MacGariggle, George L. *Taking the Offensive: October 1966–October 1967*. Washington, DC: Center of Military History, U.S. Army, 1998.

Neel, Spurgeon. *Medical Support, 1965-1970*. Washington, DC: Department of the Army, 1970.

Pearson, Willard, *The War in the Northern Provinces 1966-1968*. Washington, DC: Department of the Army, 1975.

Sharp, U.S.G., and W.C. Westmoreland. *Report on the War in Vietnam (as of 30 June 1968)*. Washington, DC: Superintendent of Documents, U.S. Government Printing Office, 1968.

Shulimson, Jack. *U.S. Marines in Vietnam: An Expanding War, 1966*. Washingon, DC: History and Museums Division, HQ, USMC, 1982.

Shulimson, Jack, Leonard A. Blasiol, Charles R. Smith, and, David A Dawson. *U.S. Marines in Vietnam: The Defining Year, 1968*. Washington, DC: History and Museums Division, HQs, USMC, 1997.

Smith, Charles R. *U.S. Marines in Vietnam: High Mobility and Standdown, 1969*. Washington, DC: History and Museums Division, Headquarters, USMC, 1988.

Summers, Harry G., Jr. *On Strategy: The Vietnam War in Context*. Carlisle Barracks, PA: Strategic Studies Institute, US Army War College, 1981.

Tactical Principles and Logistics for Cavalry. Fort Riley, KS: Academic Division, The Cavalry School, 1937.

Telfer, Gary L., Lane Rogers, and G.K. Fleming, Jr. *U.S. Marines in Vietnam: Fighting the North Vietnamese, 1967*. Washington, DC: History and Museums Division, HQ, USMC, 1984.

Tolson, John J. *Airmobility, 1961-1971*. Washington, DC: Department of the Army, 1973.

Van Staaveren, Jacob. *Interdiction in Southern Laos 1960-1968.* Washington, DC. Center for Air Force History, 1993.

Villard, Eric, *Staying the Course: October 1967 to September 1968.* Washington, DC: Center of Military History, United States Army, 2017.

Webb, Willard J. *The Joint Chiefs of Staff and The War in Vietnam 1969-1970*. Washington, DC: Office of Joint History, Office of the Chairman of the Joint Chiefs of Staff, 2002.

ARTICLES

Anderson, Jack. "Rainmakers Proving Success Over Ho Chi Minh Trail." *New York Times*, 20 March 1971.

Ankony, Robert C. "No Peace in the Valley." *Vietnam Magazine*, October 2008, 26–31.

Boccia, Frank. "Deadly Mistake on Hamburger Hill." *Vietnam Magazine*, October, 2014, 46–53.

Burns, Rod "DEWEY CANYON, 1969–PRAIRIE FIRE!" at http://www. macvsog.cc/dewey_canyon,_1969.htm. Accessed 10/5/2019.

Conmy, Joseph B., Jr. "Crouching Beast Cornered." *Vietnam Magazine*, August 1990, 28–36.

———. "I Led a Brigade at Hamburger Hill." *The Washington Post,* 27 May 1989.

Head, William. "Battle of 'Hamburger Hill,' May 10-20, 1969: The Beginning of the End of America's Commitment to the Republic of Vietnam." in *Virginia Review of Asian Studies* 17 (2015): 93–112.

McGowan, Sam. "A Luoi Airstrip in the A Shau Valley." http://stories.a227ahb. org/Aluoi.html. Accessed 5/8/2019.

"The OV-1 Mohawk Remembered Firsthand: Piloting the Mohawk in Vietnam." https://www/defensemedianetwork.com/stories. Accessed 2/18/2020.

Rayburn, James S. "Direct Support during Operation DEWEY CANYON (U)." *The NAVSECGRU Bulletin* XXIV, no. 11 (Nov/Dec 1980). (A now-declassified NSA publication)

Reade, David. "The VMFA-115 Rainmakers of SEA: Now you can know the rest of the story." P-3publications.com/PDF/VMFA-115Rainmakers2016. pdf. Accessed 5/21/19.

Sharbutt, Jay. "Mountain Battle Tough, Bloody for GIs." *New York Times,* 18 May 1969.

————. Americans Stained with Blood, Sweat, and Mud – 10th Assault on Hill Fails." *Washington Star*, 19 May 1969.

————. "U.S. Assault on Mountain Continues, Despite Heavy Toll." *New York Times,* 19 May 1969.

————. "Allied Troops Capture Mountain on Eleventh Try in Ten Days." *New York Times,* 20 May 1969.

Sorley, Lewis. "Outgunned: The ARVN Under Westmoreland." *VVA Veteran* (March/April 2012).

Summers, Harry G. "30th Anniversary: Hamburger Hill Revisited." *Vietnam Magazine* (June 1999), 38–44.

Willbanks, James H. "Hamburger Hill." *Vietnam Magazine* (June 2009), 22–31.

MILITARY DOCUMENTS – VARIOUS SOURCES

1st Special Forces Group. *Monthly Operational Summaries*, 20 May 1963–31 Dec 1965. Provided by Steve Sherman.

"U.S. Military and the Herbicide Program in Vietnam." https://www.ncbi.nlm.nih.gov/books/NBK236347/

101 ABN DIV OR-LL, 30 Apr 71, downloaded at www.3-5cavblackknights.org. Accessed 1/5/2020.

101 ABN DIV OR-LL, 31 Jul 69, downloaded at www.3-5cavblackknights.org. Accessed 1/5/2020.

101 ABN DIV OR-LL, 31 Oct 69, downloaded at www.3-5cavblackknights.org. Accessed 1/5/2020.

1st Cavalry Division. *Operational Report – Lessons Learned [OR-LL]*, period ending 31 July 1968. Dated 26 December 1968.

2d Bn, 4th Marines 010600h Aug 67 OPERATION ORDER 11-67 (OPERATION CLOUD).

3/5 Cav AAR, OPN Louisiana Lee at www.3-5cavblackknights.org. Accessed 1/5/2020.

3/5 CAV Memo 9 Jun 69 at www.3-5cavblackknights.org. Accessed 1/5/2020.

3rd Reconnaissance Battalion – Command Chronology for the period 1 August 1967 to 31 August 1967. Provided by Bill McBride.

5th Special Forces Group (Airborne). *Operational Report – Lessons Learned*, for the period ending 30 April 1966.

5th Special Forces Group (Airborne). *Operational Report – Lessons Learned,* for the period ending 30 April 1967.

5th Special Forces Group (Airborne). *Quarterly Command Report*, for the period ending 31 December 1965.

Australian Army Training Team Vietnam (AATTV). *Narrative, Annexes*, 1–31 May 1964.

B-52 (Project Delta). *After Action Report – Operation 2-68 (Samurai IV)*.

B-52 (Project Delta). *After Action Report – Operation 4-66*.

B-52 (Project Delta). *After Action Report – Operation 9-65*.

B-52 (PROJECT DELTA) *AFTER ACTION REPORT – OPERATION PIROUS*. Provided by Steve Sherman.

Blair, John D., IV. *Operations of Special Forces Detachment A-102, 5th Special Forces Group (Airborne), 1st Special Forces, in the Defense of the Special Forces Camp at A Shau, Republic of Vietnam, 9-12 March 1966*. United States Army Infantry School, Fort Benning, GA, 2 January 1968.

Headquarters, 3d Brigade, 101st Airborne Division: Sapper Attack – FB Berchtesgaden, dtd 4 August 1969.

101 Airborne Division (Airmobile). *Combat Operations After Action Report, RANDOLPH GLEN*.

Millener, COL George A., Jr., Senior Advisor, 1st Infantry Division (ARVN), period 20 May 1972 to 29 January 1973, Senior Officer Debriefing Report. Distributed by NTIS. Defense Technical Information Center.

MSG, LTG Cushman to LTG Krulak, 3 August 1967. Provided by Bill McBride.

MSG, LTG Cushman to GEN Westmoreland, 31 July 1967. Provided by Bill McBride.

Project CHECO Southeast Asia Report #1, *The Fall of A Shau*, issued 18 April 1966 by HQ PACAF.

Thomas, William E. National Academy of Sciences – National Research Council. *Effects of Herbicides in South Vietnam, Part B: Working Papers*, February 1974. Distributed by NTIS. Defense Technical Information Center.

Wright, MG John M., Jr., CG 101st Airborne Division (Airmobile), Period 25 May 1969 to 25 May 1970, Senior Officer Debriefing Report. Distributed by NTIS. Defense Technical Information Center.

USMC MCDP 1-3: Tactics. Washington, DC: Headquarters, United States Marine Corps, 30 July 1997.

XXIV Corps. *Operational Report for Period Ending 30 April 1969*. 4 Jun 1969.

XXIV Corps. *Operational Report - Lessons Learned for Period Ending 31 July 1969*.

XXIV Corps. *Operational Report for Period Ending 31 Oct 1969*.

XXIV Corps. *Operational Report for Period Ending 31 January 1970*. 17 Feb 1970.

XXIV Corps. *Operational Report for Period Ending 31 July 1970*. 1 Dec 1970.

XXIV Corps. *Operational Report for Period Ending 31 October 1970*. 12 Nov 1970.

XXIV Corps. *Operational Report - Lessons Learned, Period Ending 30 April 1971*. 17 May 1971.

CENTRAL INTELLIGENCE AND NPIC DOCUMENTS

CIA. CHRONOLOGY OF BUDDHIST CRISIS IN SOUTH VIETNAM IN 1963. 25 October 1963.

CIA INTELLIGENCE MEMORANDUM 6 February 1968.

CIA MEMORANDUM FOR THE RECORD. OLC72-0768. 5 July 1972.

CIA MEMORANDUM FOR: The Honorable Henry A. Kissinger. Data on Communist Military and Terrorist Activity in Vietnam. 5 February 1969.

CIA MEMORANDUM: South Vietnam: A Net Military Assessment. 2 April 1974.

CIA MEMORANDUM: The Situation in Vietnam 14 February 1968.

CIA MEMORANDUM: The Situation in Vietnam 20 November 1967.

CIA MEMORANDUM: The Situation in Vietnam 21 February 1968.

CIA MEMORANDUM: The Situation in Vietnam 24 April 1967.

CIA MEMORANDUM: The Situation in Vietnam 28 February 1967.

CIA MEMORANDUM: The Situation in Vietnam 28 November 1967.

CIA MEMORANDUM: The Situation in Vietnam 3 March 1968.

CIA MEMORANDUM: The Situation in Vietnam 30 November 1967.

CIA MEMORANDUM: The Situation in Vietnam 9 February 1968.

CIA Msg 14 May 70.

CURRENT INTELLIGENCE MEMORANDUM OCI No. 1561/63 3 June 1963.

Intelligence Information Cable 17 April 1967.

Intelligence Memorandum: Recent Communist Logistical and Manpower Developments in Indochina. ER IM 73-19-12 1 May 1973.

Intelligence Memorandum: Recent Communist Logistical and Manpower Developments in Indochina. ER IM 73-19-13 8 May 1973.

Intelligence Memorandum: Recent Communist Logistical and Manpower Developments in Indochina. ER IM 73-19-20 26 June 1973.

Intelligence Memorandum: Recent Construction on the Ho Chi Minh Trail. ER IM 71-185 September 1971.

Intelligence Memorandum: Recent Trends in Infiltration into South Vietnam. 22 June 1967.

Intelligence Memorandum: Road Construction in the Laotian Panhandle and Adjacent Areas of South Vietnam 1967-1968. May 1968.

Intelligence Memorandum: The Growth and Current Deployment of the Laotian-Based 559th Transportation Group. ER IM 71-25 February 1971.

MACV/CSD Working Paper No. 8, 18 December 1963: BORDER SURVEILLANCE UNITS.

National Intelligence Estimate (NIE) 14-3-66 7 July 1966.

November 21 Meeting with Saigon Advisors (Summary by CIA).

NPIC / R-37/68 February 1968.

NPIC 15498 202054Z SEP 73.
NPIC Attachment to M-5968. OPEN WIRE TELECOMMUNICATIONS
NETWORK LAOTIAN PANHANDLE.
NPIC Memorandum 24 March 1967.
NPIC MSG 032257Z APR 70.
NPIC MSG 062335Z APR 67.
NPIC MSG 150041Z APR 67.
NPIC Photographic Interpretation Memorandum: TELECOMMUNICATIONS
NETWORK CONSTRUCTION IN THE LAOTIAN PANHANDLE.
NPIC / R-46/68 March 1968.
NPIC Photographic Interpretation Report: CONTINUING ROAD
CONSTRUCTION SOUTHERN LAOTIAN PANHANDLE NPIC /
R-28/70 MAY 1970.
NSA review completed 16 July 1973.
TCS No. 3263/73: North Vietnamese Capabilities and Intentions Through the
Rainy Season and Beyond.
Twentieth Report: COMMUNIST VIOLATIONS OF THE VIETNAM AND
LAOS SETTLEMENT AGREEMENTS. 6 July 1973.
Untitled CIA document MEMORANDUM FOR Chief [redacted] FROM
Acting Chief [redacted] dated 14 November 1967.

LETTERS AND E-MAILS

Bliss, Robert. E-mail to Bill McBride, 28 Sep 1998. Courtesy of Bill McBride.
Gavlick, COL Michael, RET. E-mail to Bill McBride, 9 Sep 1998. Courtesy of
Bill McBride.
Hammond, COL James W., Jr. Letter to James W. Hammond, III, 1 Feb 1999.
Courtesy of Bill McBride.
Hammond, COL James W., Jr. Letter to Bill McBride, 17 Sep 1998. Courtesy of
Bill McBride.

VIETNAM CENTER AND ARCHIVE AT TEXAS TECH UNIVERSITY[]

1st Air Cavalry Division. *AAR: Operation DELAWARE*, 11 Jul 1968. Richard
Detra Collection, doc 0690401002.
1st ARVN Division. *Daily Staff Journal, 20 Apr 1972. TFA Sensor Analysis for the
Period 14-20 April 1972.* Dale W. Andrade Collection, doc 24992207004.
1st Battalion 9th Marines. *Command Chronology for period 1 February to
28 February 1969*, 21 Mar 1969. Jim Ginther Collection, doc 10220103001.

1st Battalion, 506th Airborne Infantry. *Combat Operations AAR, Operation APACHE SNOW,* 18 Jun 1969. Sam Johnson Collection, doc F031100081508, Encl 11.

1st Squadron, 9th Cavalry, 1st Air Cavalry Division. *Operational Report for Quarterly Period Ending October 31 1968,* 9 Nov 1968. Bud Harton Collection, doc 168300010267.

2nd Battalion, 501st Airborne Infantry. *Combat Operations AAR, Operation Apache Snow,* 22 Jun 1969. Sam Johnson Collection, doc F031100081508, Encl 13.

2nd Battalion, 506th Airborne Infantry. *Combat Operations AAR, OPORD 3-69 (APACHE SNOW),* 20 Jun 1969. Sam Johnson Collection, doc F031100081508, Encl 12.

2d Brigade, 101st Airborne Division. *Combat Operations AAR, Operation MASSACHUSETTS STRIKER,* 25 May 1969. Sam Johnson Collection, doc F031100081690.

2/17th Cavalry Squadron. *Commander's Situation Reports.* Michael Sloniker Collection.

2/17th Cavalry Squadron. *Daily Staff Journals.* Michael Sloniker Collection.

2nd Squadron (Ambl), 17th Cavalry. *Combat Operations AAR, Operation JEFFERSON GLEN,* 13 Oct 1971. James Matthews Collection, doc 5440114001.

2nd Squadron, 17th Cavalry Regiment. *Unit History for 1970,* 15 Jan 1971. James Matthews Collection, doc 5440123001.

3rd Battalion, 187th Airborne Infantry. *Combat Operations AAR, Operation APACHE SNOW 9 May–21 May 1969,* 20 Jun 1969. Sam Johnson Collection, doc F031100081508, Encl 10.

3rd Brigade, 101st Airborne Division (Airmobile). *Combat Operations AAR – Summary APACHE SNOW, 10 May–7 June 1969,* 25 Jun 69. Bud Harton Collection, doc 168300010449.

3rd FORCE RECONNAISSANCE COMPANY, III MAF. *Command Chronology, December 1969.* USMC History Division Collection, doc 1201025002.

3rd FORCE RECONNAISSANCE COMPANY, III MAF. *Command Chronology, January 1970.* USMC History Division Collection, doc 1201025003.

3rd Marine Division. *SPECIAL FILE: UNCLASSIFIED OFFICIAL RECORDS, "OPERATION DEWEY CANYON."* USMC History Division Collection, doc 1201033005.

9th Marines. *Combat Operations AAR (DEWEY CANYON),* 8 Apr 1969. Sam Johnson Collection, doc F031100081398.

22d Military History Detachment. *NARRATIVE – OPERATION "APACHE SNOW," 101ST AIRBORNE DIVISION, 10 MAY–7 JUNE 1969.* Bud Harton Collection, doc 168300010493.

22nd Military History Detachment. *NARRATIVE OF SAPPER ATTACK ON FSB AIRBORNE,* 13 May 1969. Sam Johnson Collection, doc F031100090122.

101st Airborne Division. *BATTLE OF DONG AP BIA FACT SHEET,* 24 May 1969. Sam Johnson Collection, doc F031100081508, Encl 9.

101st Airborne Division. *CASE STUDY: OPERATION "SOMERSET PLAIN," 4-20 AUGUST 1968.* Prepared by Major Emmett Kelly. Sam Johnson Collection, doc F031100090062; see also Richard Detra Collection, doc 0690418002.

Abbott, Capt Donald L., FAC, 1st Cavalry Division (Airmobile). *Interview Transcript re: Operation Delaware,* 24 May 1968. Sam Johnson Collection, doc F031100041606.

The Air Cavalry Division, September 1968. J.D. Coleman Collection, doc 2770Serial610702.

Annual Military Casualties – 1961-1972, Table. Douglas Pike Collection, doc 2234403039.

ARMY 1966-72 CH 47 STRIKE AIRCRAFT (CHINOOK) DESTROYED, 31 Aug 1972. Bud Harton Collection, doc 168300010067.

C Troop, 2/17th Cavalry. *Summary of Activity 3 August 1971,* 04 Aug 1971. Michael Sloniker Collection, doc 8850601003.

Chaisson, Lt Gen John A., USMC (Ret). *Oral History Transcripts,* 1 Aug 1967 & 2 Jan 1968. USMC History Division Oral History Collection, doc 10970103001.

Combined Campaign Plan 1968, AB 143. Glenn Helm Collection, doc 1071816001.

Combined Intelligence Center, Vietnam. *559th TRANSPORTATION GROUP, UNIT OB [Order of Battle] SUMMARY,* 5 May 1969. Sam Johnson Collection, doc F031100090246.

Combined Intelligence Center, Vietnam. *Order of Battle Study ST 67-029: VC/ NVA Employment of Snipers,* 6 Jan 1967. Sam Johnson Collection, doc F015900210639.

Combined Intelligence Center, Vietnam. *Vol I, ORDER OF BATTLE SUMMARY 1-31 MAY 1969.* Sam Johnson Collection, doc F0159000607147.

Combined Intelligence Center, Vietnam. *Vol I, ORDER OF BATTLE SUMMARY 1-31 JUNE 1969.* Sam Johnson Collection, doc F015900060420.

Combined Intelligence Center, Vietnam. *Vol I, ORDER OF BATTLE SUMMARY 1-31 JULY 1969.* Sam Johnson Collection, doc F015900060702.

Combined Intelligence Center, Vietnam. *Special Study: Avenues of Approach – A Shau Valley,* 14 Jul 1969, Sam Johnson Collection, doc F031100090003.

Combined Intelligence Center, Vietnam. *Untitled report – Re: A Shau Valley Terrain,* undated. Sam Johnson Collection, doc F031100090021.

Committee on Armed Services, United States Senate, Ninety-Second Congress, First Session. *Investigation into Electronic Battlefield Program,* 1971. Douglas Pike Collection, doc 2390812004.

Company C, 5th SFG (Abn), 1st SF. *AAR to OPORD 3-67 OCONEE (Blackjack 12)*, 24 Mar 1967. Dale W. Andrade Collection, doc 24990313005.

COMUSMACV. Msg to JCS *re: Commander's Estimate of ARC LIGHT Effectiveness*, 17 Dec 1968. Sam Johnson Collection, doc F031100140278.

COMUSMACV (CDEC). Department of Defense Intelligence Information Report re: *Organization and Activities of the 559th Transportation Group*, 13 Jun 1968. Glenn Helm Collection, doc 1071618023.

Cosby, Lloyd N., LTC, Co, 2/502 Infantry (Airborne), in *The Screaming Eagle*, "Vietnam Eagles: Medal of Honor," January-February 1999. Gary Jestes Collection, doc 1160101022.

Cushman, General Robert E., Jr., USMC (Ret). *Oral History Transcript*, 1984. USMC History Division Oral History Collection, doc 10970111001.

Defense Technical Information Agency (DTIC) Technical Report, revised edition, U.S. Defense

Logistics Agency – A Study of Strategic Lessons Learned in Vietnam. Volume VI: Conduct of the War, Book 1: Operational Analysis – Part 2 of 4: Chapter 3, 9 May 1980. Earl Tilford Collection, doc 2850112001.

DEPCOMUSMACV (Abrams). Msg to CJCS (Wheeler, Chairman, Joint Chiefs of Staff), *re: protect Dak To from by NVA 1st Division, Kontum battlefield*, 22 Nov 1967. Veteran Members of the 109th Quartermaster Company (Air Delivery) Collection, doc 0010105003.

Dickinson, COL Hillman, Sr. Advisor, 1st ARVN Infantry Division. *MEMORANDUM. re: Analysis of TFA Sensor Data*, 26 Mar 1972. Dale W. Andrade Collection, doc 24992209005.

Engineer Troops Vietnam. *The KYSU* 1, no. 3 (Fall 1969). James Bussey Collection, doc 10720101012.

Estes, Capt Albert W., USAF, FAC, 3rd Bde, 101st Abn. *Summary of interview* by Lt Col Bert Aton, 8 Aug 69. Sam Johnson Collection, doc F031100090132.

First Regional Assistance Command. *Command History, 1972-1973*, 28 Mar 1973. Dale W. Andrade Collection, doc 24991612002.

First Regional Assistance Command, 1st Infantry Division (ARVN). *Significant Activities Report*, 15 Apr 1972. Dale W. Andrade Collection, doc 24992207008.

Fox, Lt Col George C., USMC, CO, 2nd Btn/9th Mar. *Interview with by Lt Col Bert Aton*, 3 Aug 1969. Sam Johnson Collection, doc F031100090148.

Gaylor, Capt Wayne, ALO, 1st Squadron, 9th Cavalry (Airmobile). *Interview Transcript re: Operation Delaware*, 24 May 1968. Sam Johnson Collection, doc F031100041620.

Gritzmaker, Aaro. *Personal Narrative*. Aaron Gritzmaker Collection, doc 4450101001.

Guam conference transcript: Speech and questions by President Johnson with responses from General Westmoreland and Secretary Rusk, 25 Mar 1967. Larry Berman Collection, doc 0240613003.

Hair, Lt. Frank. "Massachusetts Striker," *Rendezvous with Destiny* (101st Airborne Division magazine, Summer 1969). Barry Wain Collection, doc 28810616035.

Hearings before the Subcommittee on Oceans and International Environment of the Committee on Foreign Relations, United States Senate, re: Weather Modification, 20 Mar 1974. Douglas Pike Collection, doc 2390601002.

Helms, Richard, CIA Director. *Memo for President Lyndon B. Johnson, re: The Validity and Significance of Viet Cong Loss Data*, 22 Nov 1967. Larry Berman Collection, doc 0240816010.

HERBICIDE OPERATION EVALUATION, 26 Aug 1968. Paul Cecil Collection, doc 2520308006.

History of the 2nd Squadron 17th Cavalry Regiment from 1916-1976, no date. James Matthews Collection, doc 5440119001.

Intelligence Information Report, Captured Documents (CDEC): *re: Observation of Two 100mm AA Guns in Thua Thien Prov*, Rpt Nr 60-5868-68, 12 Dec 1968. doc F034604700729.

III MAF. *AIR CONTROL (SPECAT VOL 3)*, 1 Dec 1968. USMC History Division Collection, doc 1201005003.

III MAF. *COMMAND CHRONOLOGY, August 1968*. USMC History Division Collection, doc 1201004039.

III MAF. *COMMAND CHRONOLOGY, June 1969*. USMC History Division Collection, doc 1201006028.

III MAF. *COMMAND CHRONOLOGY, July 1969*. USMC History Division Collection, doc 1201006048.

III MAF. *COMMAND CHRONOLOGY, August 1969*. USMC History Division Collection, doc 1201007008.

III MAF. *INCOMING MESSAGES*, 18 Jun 1969, Box ___, Folder 006, USMC History Division Collection, doc 1201006026.

III MAF. *Periodic Intelligence Reports*. USMC Military History Collection. There are many of these documents on file. Use website search function to locate files.

Interdiction Program in I Corps, Sam Johnson Collection, doc F031100090196.

Kroesen, MG, acting CG XXIV Corps. Msg to GEN Abrams, re: *Commanders Daily Evaluation*, 19 Mar 1972. Dale W. Andrade Collection, doc 24991703009.

Kroesen, MG, acting CG XXIV Corps. Msg to GEN Abrams, re: *Commanders Daily Evaluation*, 22 Mar 1972. Dale W. Andrade Collection, doc 24991703010.

MACDI Study 73-01, The Nguyen Hue Offensive, 12 Jan 1973. Dale W. Andrade Collection, doc 24991801005.

MACV Command History - Annex L: Quang Tri and Hue, 14 Oct 1975. Dale W. Andrade Collection, doc 24992506004.

MACV COMMAND HISTORY 1967, VOL III, 16 Sep 1968. Bud Harton
    Collection, doc 168300010725.
MACV COMMAND HISTORY 1968, VOL I, 30 Apr 1969. Bud Harton
    Collection, doc 168300010746.
MACV Monthly Summary, June 1969. John M. Shaw Collection, doc
    7390115001.
MACV Monthly Summary, July 1969. John M. Shaw Collection, doc
    7390201001.
MACV Monthly Summary, August 1969. Bud Harton Collection, doc
    168300010776.
MACV. *Significant Problem Areas Report, Nov 1969*, 26 Nov 1969. Sam Johnson
    Collection, doc F015800220558.
MACV. *Significant Problem Areas, May 1968,* 6 Jun 1968. Sam Johnson Collection,
    doc F015800220378.
MACV Significant R&D Problem Areas Report (RCS: CICPAC 3960, 2), 1 Oct
    1969. Sam Johnson Collection, doc F015800220516.
McLaughlin, Brigadier General George W., USAF, Director, Tactical Control
    Center. *Letter re: MACV Year End Review of Vietnam, 1968,* 4 Jan 1969. Sam
    Johnson Collection, doc F031100151073.
Memorandum for the Press re: Lieutenant General Cushman's Press Brief, 2 May 1968.
    Sam Johnson Collection, doc F031100041480.
Memorandum for the Press re: Lieutenant General Cushman's Press Brief, 7 May 1968.
    Sam Johnson Collection, doc F031100041487.
Memorandum for the Press re: Lieutenant General Cushman's Press Brief, 11 May 1968.
    Sam Johnson Collection, doc F031100041494.
Memorandum for the Press re: Lieutenant General Cushman's Press Brief, 17 May 1968.
    Sam Johnson Collection, doc F031100041500.
Mills, Sergeant Raymond, Jr., US Army, Intelligence NCO, 1st Squadron, 9th
    Cavalry (Airmobile). *Interview Transcript re: Operation Delaware*, 24 May 1968.
    Sam Johnson Collection, doc F031100041630.
*NATIONAL SECURITY STUDY MEMORANDUM (NSSM) ONE - March
    1969: The Situation in Vietnam.* Douglas Pike Collection, doc 2120107001.
Operation Cumberland OP [Operational Procedures] File, 8 Jun–11 Jul 1967. USMC
    History Division Collection, doc 1201064052.
Operation Cumberland OP [Operational Procedures] File, 17 Jul–16 Sep 1967.
    USMC History Division Collection, doc 1201064063.
Operation Delaware Valley/Lamson 216 OP [Operational Procedures] File,
    22 Apr–20 May 1968. USMC History Division Collection, doc 1201064124.
Pacific Stars & Stripes, Viet Marines Kill 200 Reds, 15 May 1971. Douglas Pike
    Collection, doc 2131804084.
Percentage of Allied Wounded Returned to Duty in Vietnam – CIA Research Reports
    (Supplement), 07 Jan 1967. Sam Johnson Collection, doc F029200040019.

Plans for Northern I Corps Operations – re: priority to Thua Thien and Quang Tri, 15 Mar 1968. Veteran Members of the 109th Quartermaster Company (Air Delivery) Collection, doc 0010120004.

Project CHECO Southeast Asia Report #2 – A Shau Valley Campaign, December 1968–May 1969, 15 Oct 1969. Sam Johnson Collection, doc F031100081306.

Project CHECO Southeast Asia Report #87 - Direct Air Support in I Corps, July 1965– January 1969, 31 Aug 1969. Sam Johnson Collection, doc F031100201112.

Project CHECO Southeast Asia Report #91 – Operation Delaware, 19 April–17 May 1968, 2 Sep 1968. Sam Johnson Collection, doc F031100041363.

Project CHECO Southeast Asia Report #105 – The Fourth Offensive, 1 Oct 1969. Sam Johnson Collection, doc F031100070740.

Project CHECO Southeast Asia Report #107 – Impact of Geography on Air Operations in Sea, 11 Jun 1970. Sam Johnson Collection, doc F031100101565.

Project CHECO Southeast Asia Report #117 – Igloo White (Initial Phase), 31 Jul 1968. Sam Johnson Collection, doc F031100410008.

Project CHECO Southeast Asia Report #118 – Igloo White, July 1968–December 1969, 10 Jan 1970. Sam Johnson Collection, doc F031100160005.

Project CHECO Southeast Asia Report #124 – Interdiction in Seasia, November 1966–October 1968, 20 Jun 1969. CHECO Reports, doc 0390213001.

Project CHECO Southeast Asia Report #176 – Visual Reconnaissance in I Corps, 30 Sep 1968. Sam Johnson Collection, doc F031100181082.

Project CHECO Southeast Asia Report #203 – Short Rounds, June 1967–June 1968, 25 Aug 1968. Sam Johnson Collection, doc F031100180684.

Project CHECO Southeast Asia Report #204 – Short Rounds, June 1968–May 1969, 15 Aug 1969. Sam Johnson Collection, doc F031100180731.

Project CHECO Southeast Asia Report #250 – The War in Vietnam, 1966, 23 Oct 1967. Sam Johnson Collection, doc F031100440437.

Provisional Det A-100 members, interviewees, *Operation Longstreet*, 28 Nov 1966. Dale W. Andrade Collection, doc 24990313006.

RESUME OF THE A SHAU VALLEY AIR INTERDICTION CAMPAIGN, 9 Dec 1968 to 28 Feb 1969. Sam Johnson Collection, doc F031100090085.

Roberts, Gordon, *Medal of Honor Award Case File*, 11 Jul 1969. Tim Frank Collection, doc 18700263001.

Rostow, W. W. Memo to President Lyndon B. Johnson re: *Landslides in A Shau Valley*, 10 Jul 1967. Veteran Members of the 109th Quartermaster Company (Air Delivery) Collection, doc 0010124006.

Rostow, W. W. Memoes to President Lyndon B. Johnson, re: *Reports #1-19 on the Situation in the Khe Sanh Area*, 5 Feb–23 Feb 1968. Veteran Members of the 109th Quartermaster Company (Air Delivery) Collection, doc 0010113001.

Rostow, W. W. Memo to President Lyndon B. Johnson, re: *Report #46 on the Situation in the Khe Sanh/DMZ/A Shau Valley Area*, 21 Mar 1968. Veteran

Members of the 109th Quartermaster Company (Air Delivery) Collection, doc 0010313004.

Rostow, W.W. Memo to President Lyndon B. Johnson, re: *Report #47 on the Situation in the Khe Sanh/DMZ/A Shau Valley Area*, 22 Mar 1968. Veteran Members of the 109th Quartermaster Company (Air Delivery) Collection, doc 0010313001.

Rostow, W.W. Memo to President Lyndon B. Johnson, re: *Report #51 on the Situation in the Khe Sanh/DMZ/A Shau Valley Area*, 26 Mar 1968. Veteran Members of the 109th Quartermaster Company (Air Delivery) Collection, doc 0010319007.

Rostow, W.W. Memo to President Lyndon B. Johnson, re: *Report #53 on the Situation in the Khe Sanh/DMZ/A Shau Valley Area*, 28 Mar 1968. Veteran Members of the 109th Quartermaster Company (Air Delivery) Collection, doc 0010315010.

Rostow, W.W. Memo to President Lyndon B. Johnson, re: *Report #54 on the Situation in the Khe Sanh/DMZ/A Shau Valley Area*, 29 Mar 1968. Veteran Members of the 109th Quartermaster Company (Air Delivery) Collection, doc 0010315007.

Rostow, W.W. Memo to President Lyndon B. Johnson, re: *Report #66 on the Situation in the Khe Sanh/DMZ/A Shau Valley Area*, 10 Apr 1968. Veteran Members of the 109th Quartermaster Company (Air Delivery) Collection, doc 00103121005.

The Screaming Eagle, Vol. IV, No. 20 (Excerpt), 27 September 1971. Gary Jestes Collection, doc 1160101006.

Seventh Air Force. *Weekly Air Intelligence Summary re: Operation Delaware,* 1 Jun 1968. Sam Johnson Collection, doc F031100041654.

SUMMARY OF ARC LIGHT ACTIVITY, 1-31 August 1968. Sam Johnson Collection, doc F031100140241.

Truong, Lt. Gen. Ngo Quang. *The Easter Offensive of 1972,* Indochina Monographs, U.S. Army Center of Military History. Garnett Bell Collection, doc 11271711001.

Truong, Lt Gen. Ngo Quang. *Republic of Viet Nam Armed Forces and US Operational Cooperation and Coordination* (Indochina Monographs), 1979, CMH Pub 92-16. J.D. Coleman Collection, doc 27700602001.

XXIV Corps. *Combat Operations AAR, Operation Apache Snow, 10 May–7 June 1969,* 27 Aug 1969. Bud Harton Collection, doc 168300010547.

INTERNET

1/327 Infantry Operation Somerset Plain After Action Report. http://www.fearfulodds.com/Ashau

220th Aviation Company. *Battle and Fall of A Shau, March 1966.* https://www.catkillers.org/history-1966-Battle-of-A-Shau.pdf. Accessed 7/6/2019.

227 Assault Helicopter Battalion AAR. http://a227ahb.org/DelawareLamSon216.html. Last update 3/16/2009. Accessed 8/7/2019.

281st Assault Helicopter Company. https://www.281st.com/281history/281history_html/ICTZ_opns281.htm

388TH OPERATION POPEYE "MOTORPOOL." www.vva388.com/uploads/2/3/9/8/23987277/2-2016.pdf. Accessed 5/19/19.

A Company, 1st Battalion, 8th Cavalry Website. http://vnwarstories.com/Odyssey/vn.ACompanyOdyssey25April68.html. Based on Daily Staff Journals of the3 1/8 Cav at Texas Tech, Collection 369, Daily Staff Journals 1965-1969. Accessed 8/15/2019.

A/227 Assault Helicopter Battalion. Valley of Sorrow – A Shau 25 April 1968. http://stories.a227ahb.org/Ashau.html. Accessed 5/8/2019.

Agent Orange Data Warehouse. http://www.workerveteranhealth.org/milherbs/new/index.php.

Author's database. I have compiled a database referencing names on the Vietnam Memorial in Washington, DC to published books and Vietnam Magazine articles. It contains almost 80,000 entries and close to 50,000 names. Information (where known) includes unit, operation and/or battle, position, and other fields.

B Company, 2nd Battalion, 8th Cavalry. Report on Khe Sanh and A Shau Valley Operations. http://www.eagerarms.com/khesanhashauvalleyoperations.html. Accessed 6/5/2019

C Company, 1st Battalion, 8th Cavalry. Website at http://www.c1-8cav68.org/ashauvalley.html. Accessed 8/15/2019.

http://old.506infantry.org/hisvietnam/his1stbnvnarticle32.html

http://www.charlie1-506.com/html/pages%20from%20main%20nav/CreateSpace%20Charlie%20Co.htm Accessed 12/22/2019.

http://www.vietnamproject.ttu.edu/banshee/History/TexStar70.html. Accessed 2/8/2020.

https://en.wikipedia.org/wiki/Environmental_Modification_Convention

https://sofrep.com/specialoperations/recon-team-python-ruled-valley-death/. Accessed 3/7/2020.

https://www.75thrra.org/history/175_hx.html, Accessed 3/6/2020.

I Corps Mike Force facebook page. *Special Forces Mike Force Lineage in I Corps, 1964-1970.* Accessed 12/22/19.

I Corps Mike Force facebook page. *Blackjack 13/Task Force 13 Operational Summaries.* Accessed 12/22/19.

John Ulfer's Diary (D/1/8 Cav). http://www.jumpingmustangs.com/delta6571/deltaulferdiary.html. Accessed 8/15/2019.

MACVSOG. http://www.macvsog.cc

Operation Somerset Plain After Action Report. http://www.fearfulodds.com/
     Ashau
Operation Somerset Plain Daily Staff Journals. http://www.fearfulodds.com/
     Ashau
The Screaming Eagle, 12 May 1969.
Task Force Omega. http://taskforceomegainc.org/
"US denies 'rain-making' in Vietnam." *The Straits Times*, 24 June 1972.
Vietnam Helicopter Pilots Association. https://www.vhpa.org/KIA; cited as
     VHPA website with listings by helicopter serial numbers.
www.armyaircrews.com/cayuse.html

INDEX

CPSIA information can be obtained
at www.ICGtesting.com
Printed in the USA
LVHW091415200521
688012LV00005B/12/J